Essential Microbiology

Essential Microbiology

Second Edition

Stuart Hogg

University of South Wales, UK

⊛WILEY-BLACKWELL

A John Wiley & Sons, Ltd., Publication

This edition first published 2013 © 2013 by John Wiley & Sons, Ltd

Wiley-Blackwell is an imprint of John Wiley & Sons, formed by the merger of Wiley's global Scientific, Technical and Medical business with Blackwell Publishing.

Registered office: John Wiley & Sons, Ltd, The Atrium, Southern Gate, Chichester, West Sussex, PO19 8SQ, UK

Editorial offices: 9600 Garsington Road, Oxford, OX4 2DQ, UK
 The Atrium, Southern Gate, Chichester, West Sussex, PO19 8SQ, UK
 111 River Street, Hoboken, NJ 07030-5774, USA

For details of our global editorial offices, for customer services and for information about how to apply for permission to reuse the copyright material in this book please see our website at www.wiley.com/wiley-blackwell.

The right of the author to be identified as the author of this work has been asserted in accordance with the UK Copyright, Designs and Patents Act 1988.

All rights reserved. No part of this publication may be reproduced, stored in a retrieval system, or transmitted, in any form or by any means, electronic, mechanical, photocopying, recording or otherwise, except as permitted by the UK Copyright, Designs and Patents Act 1988, without the prior permission of the publisher.

Designations used by companies to distinguish their products are often claimed as trademarks. All brand names and product names used in this book are trade names, service marks, trademarks or registered trademarks of their respective owners. The publisher is not associated with any product or vendor mentioned in this book.

Limit of Liability/Disclaimer of Warranty: While the publisher and author(s) have used their best efforts in preparing this book, they make no representations or warranties with respect to the accuracy or completeness of the contents of this book and specifically disclaim any implied warranties of merchantability or fitness for a particular purpose. It is sold on the understanding that the publisher is not engaged in rendering professional services and neither the publisher nor the author shall be liable for damages arising herefrom. If professional advice or other expert assistance is required, the services of a competent professional should be sought.

Library of Congress Cataloging-in-Publication Data

Hogg, Stuart (Stuart I.)
  Essential microbiology / Stuart Hogg. – 2nd ed.
    p. ; cm.
  Includes index.
  ISBN 978-1-119-97891-6 (cloth) – ISBN 978-1-119-97890-9 (pbk.)
  I. Title.
  [DNLM: 1. Microbiological Phenomena. 2. Microbiological Techniques. 3. Microbiology. QW 4]
    579–dc23   2012051595

A catalogue record for this book is available from the British Library.

Wiley also publishes its books in a variety of electronic formats. Some content that appears in print may not be available in electronic books.

Cover image: Gary Thompson
Cover design by Stain 2.png supplied by publisher

Set in 10.5/12.5pt Times Ten by Aptara Inc., New Delhi, India

Printed in the UK

Contents

Preface to Second Edition

It is now seven years since the first edition of *Essential Microbiology* was published, so it is high time the contents were updated, and I have taken the opportunity to revise the layout in the hope that it will better serve its target readership. The main change to the book from its original incarnation is the inclusion of a chapter on microbial disease in humans. When preparing the content of the first edition, the one major area of doubt I had was whether or not to include a chapter or section on medical microbiology. I was urged to do so by a number of colleagues, but in the end I resisted, feeling it to be too large a topic for inclusion in a general introductory text. The invitation to prepare a second edition, however, has given me an opportunity to reconsider the matter, and comments from several reviewers, together with further reflection on my own part, have persuaded me to change my mind. I have therefore introduced a new chapter on microbial disease in humans, supplementing new material with some expanded and repackaged from other chapters in the first edition. This has resulted in a shuffling and reordering of the second half of the book, which I hope leads to a more logical structure. The new edition no longer features end-of-chapter quizzes; however, these and other forms of self-assessment can now be found on the book's dedicated website. The other major change that will be noticed by anyone familiar with the original book is the introduction of colour. I feel strongly that a book such as this should be visually attractive as well as instructive, and am grateful to my editorial team at Wiley for allowing me this indulgence, in spite of the additional pressure it creates in trying to keep the selling price to a minimum, which was always one of the principal aims of *Essential Microbiology*.

As always, I should be grateful to receive any comments and suggestions for improvement from students or their tutors.

Stuart Hogg
September 2012

Preface to First Edition

Every year, in UK universities alone, many hundreds of students study microbiology as part of an undergraduate course. For some, the subject will form the major part of their studies, leading to a BSc degree in Microbiology, or a related subject such as Bacteriology or Biotechnology. For the majority, however, the study of microbiology will be a brief encounter, forming only a minor part of their course content.

A number of excellent and well-established textbooks are available to support the study of microbiology; such titles are mostly over 1000 pages in length, beautifully illustrated in colour, and rather expensive. This book in no way seeks to replace or compete with such texts, which will serve specialist students well throughout their three years of study, and represent a sound investment. It is directed rather towards the second group of students, who require a text that is less detailed, less comprehensive, and less expensive! The majority of the students in my own classes are enrolled on BSc degrees in Biology, Human Biology and Forensic Science; I have felt increasingly uncomfortable about recommending that they invest a substantial sum of money on a book much of whose content is irrelevant to their needs. Alternative recommendations, however, are not thick on the ground. This, then, was my initial stimulus to write a book of 'microbiology for the non-microbiologist'.

The facts and principles you will find here are no different from those described elsewhere, but I have tried to select those topics that one might expect to encounter in years 1 and 2 of a typical non-specialist degree in the life sciences or related disciplines. Above all, I have tried to *explain* concepts or mechanisms; one thing my research for this book has taught me is that textbooks are not always right, and they certainly don't always explain things as clearly as they might. It is my wish that the present text will give the attentive reader a clear understanding of sometimes complex issues, whilst avoiding over-simplification.

The book is arranged into seven sections, the fourth of which, Microbial Genetics, acts as a pivot, leading from principles to applications of microbiology. Depending on their starting knowledge, readers may 'dip into' the book at specific topics, but those whose biological and chemical knowledge is limited are strongly recommended to read Chapters 2 and 3 for the foundation necessary for the understanding of later chapters. Occasional boxes are inserted into the text, which provide some further enlightenment on the topic being discussed, or offer supplementary information for the inquisitive reader. As far as possible, diagrams are limited to simple line drawings, most of which could be memorised for reproduction in an examination setting. Although a Glossary is provided at the end of the book, new words are also defined in the text at the point of their first introduction, to facilitate uninterrupted reading. All chapters except the first are followed by a self-test section in which readers may review their knowledge and understanding by 'filling in the gaps' in incomplete sentences; the answers are all to be found in the text, and so are not provided separately. The only exceptions to this are two numerical questions, the solutions to which are to be found at the back of the book. By completing the self-test questions, the reader effectively provides a summary for the chapter.

A book such as this stands or falls by the reception it receives from its target readership. I should be pleased to receive any comments on the content and style of *Essential Microbiology* from students and their tutors, all of which will be given serious consideration for inclusion in any further editions.

Stuart Hogg
January 2005

Acknowledgements

I would like to thank those colleagues who took the time to read over individual chapters of this book, and those who reviewed the entire manuscript. Their comments have been gratefully received, and in some cases spared me from the embarrassment of seeing my mistakes perpetuated in print.

Thanks are also due to my editorial team at John Wiley, Rachael Ballard and Fiona Seymour, and production editor Jasmine Chang for ensuring smooth production of this book.

I am grateful to those publishers and individuals who have granted permission to reproduce diagrams. Every effort has been made to trace holders of copyright; any inadvertent omissions will gladly be rectified in any future editions of this book.

Finally, I would like to express my gratitude to my family for allowing me to devote so many weekends to 'the book'.

About the Companion Website

This book is accompanied by a companion website:

www.wiley.com/go/hogg/essentialmicrobiology

The website includes:
- Powerpoints of all figures from the book for downloading
- PDFs of tables from the book
- Short Answer Questions
- Key concepts
- Links for further reading

I
Introduction

1
Microbiology: What, Why and How?

Microorganisms (or microbes) inhabit every corner of the globe, and are essential for the maintenance of the world's ecosystems. They include organisms responsible for some of the most deadly human diseases, and others that form the basis of important industrial processes. Yet until a few hundred years ago, nobody knew they existed! This book offers an introduction to the world of microorganisms, and in this opening chapter, we offer some answers to three questions:

- *What* is microbiology?
- *Why* is it such an important subject?
- *How* have we gained our present knowledge of microbiology?

1.1   What is microbiology?

Things aren't always the way they seem. On the face of it, 'microbiology' should be an easy word to define: the science (*logos*) of small (*micro*) life (*bios*), or to put it another way, the study of living things so small that they can't be seen with the naked eye. Bacteria neatly fit this definition, but what about fungi and algae? These two groups each contain members that are far from microscopic. On the other hand, certain animals, such as nematode worms, can be microscopic, yet are not considered to be the domain of the microbiologist. Viruses represent another special case; they are most certainly microscopic; indeed, most are submicroscopic, but by most accepted definitions they are not living (why? – see Chapter 10 for an explanation). Nevertheless, these too fall within the remit of the microbiologist.

Essential Microbiology, Second Edition. Stuart Hogg.
© 2013 John Wiley & Sons, Ltd. Published 2013 by John Wiley & Sons, Ltd.

   In the central section of this book you can read about the thorny issue of microbial classification and gain some understanding of just what is and what is not regarded as a microorganism.

1.2   Why is microbiology important?

To the lay person, microbiology means the study of sinister, invisible 'bugs' that cause disease. As a subject, it generally tends to impinge on the popular consciousness in news coverage concerning the latest 'health scare'. It may come as something of a surprise therefore to learn that the vast majority of microorganisms coexist alongside us without causing any harm; indeed, at least a thousand different species of bacteria are to be found on human skin! In addition, many microorganisms are positively beneficial, performing vital tasks such as the recycling of essential elements, without which life on our planet could not continue, as we'll examine in Chapter 14. Other microorganisms have been exploited by humans for our own benefit, for instance in the manufacture of antibiotics (Chapter 17) and foodstuffs (Chapter 18). To get some idea of the importance of microbiology in the world today, just consider the following list of some of the general areas in which the expertise of a microbiologist might be used:

- medicine
- environmental science
- food and drink production
- fundamental research
- agriculture
- pharmaceutical industry
- genetic engineering

The popular perception among the general public, however, remains one of infections and plagues. Think back to the first time you ever heard about microorganisms; almost certainly, it was when you were a child and your parents impressed on you the dangers of ingesting 'germs' from dirty hands or putting things in your mouth after they'd been on the floor. In reality, only a couple of hundred out of the half million or so known bacterial species give rise to infections in humans; these are termed *pathogens*, and have tended to dominate our view of the microbial world.

   In the next few pages we shall review some of the landmark developments in the history of microbiology, and see how the main driving force throughout this time, but particularly in the early days, has been the desire to understand the nature and cause of infectious diseases in humans.

> A *pathogen* is an organism with the potential to cause disease.

1.3 How do we know? Microbiology in perspective: to the Golden Age and beyond

We have learnt an astonishing amount about the invisible world of microorganisms, particularly over the last century and a half. How has this happened? The penetrating insights of brilliant individuals are rightly celebrated, but a great many 'breakthroughs' or 'discoveries' have only been made possible thanks to some (frequently unsung) development in microbiological methodology. For example, on the basis that 'seeing is believing', it was only when we had the means to *see* microorganisms under a microscope that we could prove their existence.

Microorganisms had been on the Earth for some 4000 million years when Antoni van Leeuwenhoek started his pioneering microscope work in 1673. Leeuwenhoek was an amateur scientist who spent much of his spare time grinding glass lenses to produce simple microscopes (Figure 1.1). His detailed drawings make it clear that the 'animalcules' he observed from a variety of sources included representatives of what later became known as protozoa, bacteria and fungi. Where did these creatures come from? Arguments about the origin of living things revolved around the long-held belief in *spontaneous generation*, the idea that living organisms could arise from non-living matter. In an elegant experiment, the Italian Francesco Redi (1626–1697) showed

Figure 1.1 Leeuwenhoek's microscope. The lens (a) was held between two brass plates and used to view the specimen, which was placed on the mounting pin (b). Focusing was achieved by means of two screws (c) and (d). Some of Leeuwenhoek's microscopes could magnify up to 300 times. Original source: antoni van Leeuwenhoek and his little animals by CE Dobell (1932).

that the larvae found on putrefying meat arose from eggs deposited by flies, and not spontaneously as a result of the decay process. This can be seen as the beginning of the end for the spontaneous generation theory, but many still clung to the idea, claiming that while it may not have been true for larger organisms, it must surely be so for minute creatures such as those demonstrated by Leeuwenhoek. Despite mounting evidence against the theory, as late as 1859 fresh 'proof' was still being brought forward in its support. Enter onto the scene Louis Pasteur (1822–95), still arguably the most famous figure in the history of microbiology. Pasteur trained as a chemist, and made a lasting contribution to the science of stereochemistry before turning his attention to spoilage problems in the wine industry. He noticed that when lactic acid was produced instead of alcohol in wine, rod-shaped bacteria were always present as well as the expected yeast cells. This led him to believe that while the yeast produced the alcohol, the bacteria were responsible for the spoilage, and must have originated in the environment. Exasperated by continued efforts to substantiate the theory of spontaneous generation, he set out to disprove it once and for all. In response to a call from the French Academy of Science, he carried out a series of experiments that led to the acceptance of *biogenesis*, the idea that life arises only from already existing life. Using his famous swan-necked flasks (Figure 1.2), he demonstrated that as long as dust particles (and the microorganisms carried on them) were excluded, the contents would remain sterile. This also disproved the idea held by many that there was some element in the air itself that was capable of initiating microbial growth. In Pasteur's words '....the doctrine of spontaneous generation will never recover from this mortal blow. *There is no known circumstance in which it can be affirmed that microscopic beings came into the world without germs, without parents similar to themselves*' [author's italics]. Pasteur's findings on the role of microorganisms in wine contamination led inevitably to the idea that they may also be responsible for diseases in humans, animals and plants.

The notion that some invisible (and therefore presumably extremely small) living creatures were responsible for certain diseases was not a new one. Long before microorganisms had been shown to exist, the Roman philosopher Lucretius (~98–55 BC) and much later the physician Girolamo Fracastoro (1478–1553) had supported the idea. Fracastoro wrote 'Contagion is an infection that passes from one thing to another' and recognised three forms of transmission: by direct contact, through inanimate objects and via the air; we still class transmissibility of infectious disease in much the same way today (see Chapter 15). The prevailing belief at the time, however, was that an infectious disease was due to something called a *miasma*, a poisonous vapour arising from dead or diseased bodies, or to an imbalance between the four humours of the body (blood, phlegm, yellow bile and black bile).

During the nineteenth century, many diseases were shown, one by one, to be caused by microorganisms. In 1835, Agostino Bassi showed that a disease

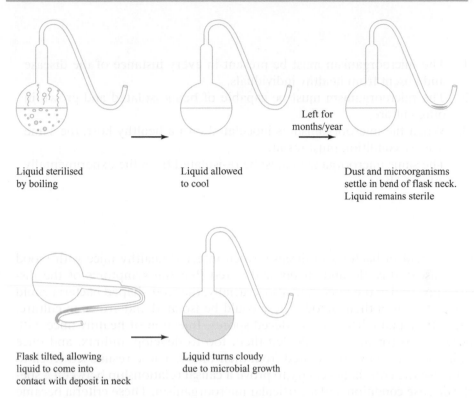

Liquid sterilised
by boiling

Liquid allowed
to cool

Left for
months/year

Dust and microorganisms
settle in bend of flask neck.
Liquid remains sterile

Flask tilted, allowing
liquid to come into
contact with deposit in neck

Liquid turns cloudy
due to microbial growth

Figure 1.2   Pasteur's swan-necked flasks. Broth solutions rich in nutrients were placed in flasks and boiled. The necks of the flasks were heated and drawn out into a curve, but kept open to the atmosphere. Pasteur showed that the broth remained sterile because any contaminating dust and microorganisms remained trapped in the neck of the flask as long as it remained upright.

of silkworms was due to a fungal infection, and 10 years later, Miles Berkeley demonstrated that a fungus was also responsible for the great Irish potato blight. Joseph Lister's pioneering work on antiseptic surgery provided strong, albeit indirect, evidence of the involvement of microorganisms in infections of humans. The use of heat-treated instruments and of phenol both on dressings and actually sprayed in a mist over the surgical area, was found greatly to reduce the number of fatalities following surgery. Around the same time, in the 1860s, the indefatigable Pasteur had shown that a parasitic protozoan was the cause of another disease of silkworms called 'pébrine', which had devastated the French silk industry.

The definitive proof of the germ theory of disease came from the German, Robert Koch, who in 1876 showed the relationship between the cattle disease anthrax and a bacillus we now know as *Bacillus anthracis*. This was also the first demonstration of the

> A *bacillus* is a rod-shaped bacterium.

Box 1.1   Koch's postulates

1. The microorganism must be present in every instance of the disease and absent from healthy individuals.
2. The microorganism must be capable of being isolated and grown in pure culture.
3. When the microorganism is inoculated into a healthy host, the same disease condition must result.
4. The same microorganism must be re-isolated from the experimentally infected host.

---

involvement of bacteria in disease. Koch infected healthy mice with blood from diseased cattle and sheep, and noted that the symptoms of the disease appeared in the mice, and also, crucially, that rod-shaped bacteria could be isolated from their blood. These could be isolated and grown in culture, where they multiplied and produced spores. Injection of healthy mice with these spores (or more bacilli) led them too to develop anthrax, and once again the bacteria were isolated from their blood. These results led Koch to formalise the criteria necessary to prove a causal relationship between a specific disease condition and a particular microorganism. These criteria became known as *Koch's postulates* (Box 1.1), and are still in use today.

Despite their value, it is now realised that Koch's postulates do have certain limitations. It is known for example that certain agents responsible for causing disease (e.g. viruses, prions: see Chapter 10) can't be grown *in vitro*, but only in host cells. Also, the healthy animal in Postulate 3 is seldom human, so a degree of extrapolation is necessary – if agent X doesn't

> The term *in vitro* (= 'in glass') is used to describe procedures performed outside of the living organism in test tubes, etc. (cf. *in vivo*).

cause disease in a laboratory animal, can we be sure it won't in humans? Furthermore, some diseases are caused by more than one organism, and some organisms are responsible for more than one disease. On the other hand, the value of Koch's postulates goes beyond just defining the causative agent of a particular disease, and allows us to ascribe a specific effect (of whatever kind) to a given microorganism.

Critical to the development of Koch's postulates was the advance in microbial culturing techniques, enabling the isolation of pure cultures of specific microorganisms. These are discussed in more detail in Chapter 4. The development of pure cultures revolutionised

> A *pure* or *axenic culture* contains one type of organism only, and is completely free from contaminants.

Table 1.1   The discovery of some major human pathogens

Year	Disease	Causative agent	Discoverer
1876	Anthrax	*Bacillus anthracis*	Koch
1879	Gonorrhoea	*Neisseria gonorrhoeae*	Neisser
1880	Typhoid fever	*Salmonella typhi*	Gaffky
1880	Malaria	*Plasmodium* spp.	Laveran
1882	Tuberculosis	*Mycobacterium tuberculosis*	Koch
1883	Cholera	*Vibrio cholerae*	Koch
1883/4	Diphtheria	*Corynebacterium diphtheriae*	Klebs and Loeffler
1885	Tetanus	*Clostridium tetani*	Nicolaier and Kitasato
1886	Pneumonia (bacterial)	*Streptococcus pneumoniae*	Fraenkel
1892	Gas gangrene	*Clostridium perfringens*	Welch and Nuttall
1894	Plague	*Yersinia pestis*	Kitasato and Yersin
1896	Botulism	*Clostridium botulinum*	Van Ermengem
1898	Dysentery	*Shigella dysenteriae*	Shiga
1901	Yellow fever	Flavivirus	Reed
1905	Syphilis	*Treponema pallidum*	Schaudinn and Hoffman
1906	Whooping cough	*Bordetella pertussis*	Bordet and Gengou
1909	Rocky Mountain spotted fever	*Rickettsia rickettsii*	Ricketts

microbiology, and within 30 years or so of Koch's work on anthrax, the pathogens responsible for the majority of common human bacterial diseases had been isolated and identified. Not without just cause is this period known as the 'golden age' of microbiology! Table 1.1 summarises the discovery of some major human pathogens.

Koch's greatest achievement was in using the advances in methodology and the principles of his own postulates to demonstrate the identity of the causative agent of tuberculosis, which at the time was responsible for around one in every seven human deaths in Europe. Although it was believed by many to have a microbial

> *Charles Chamberland, a pupil of Pasteur's, invented the autoclave, contributing greatly to the development of pure cultures.*

cause, the causative agent had never been observed, either in culture or in the affected tissues. We now know this is because *Mycobacterium tuberculosis* (the tubercle bacillus) is very difficult to stain by conventional methods due to the high lipid content of the cell wall surface. Koch developed a staining technique that enabled it to be seen, but realised that in order to satisfy his own postulates, he must isolate the organism and grow it in culture. Again, there were technical difficulties, since even under favourable conditions, *M. tuberculosis* grows slowly, but eventually Koch was able to demonstrate the infectivity of the cultured organisms towards guinea pigs. He was then able to isolate them again from the diseased animal and use them to cause disease in uninfected animals, thus satisfying the remainder of his postulates.

Although most bacterial diseases of humans and their aetiological agents have now been identified, important variants continue to

> *Aetiology* is the cause or origin of a disease.

evolve and sometimes emerge; examples in recent decades include Lyme disease and legionellosis (legionnaire's disease); the latter is an acute respiratory infection caused by the previously unrecognised genus, *Legionella*. Also, *Helicobacter pylori,* only discovered in the 1980s, has been shown to play an important (and previously unsuspected) role in the development of stomach ulcers. There still remain a few diseases that some investigators suspect are caused by bacteria, but for which no pathogen has been identified.

Another cause of infectious diseases are viruses, and following their discovery during the last decade of the nineteenth century, it was soon established that many diseases of plants, animals and humans were caused by these minute, non-cellular agents.

The major achievement of the first half of the twentieth century was the development of antibiotics and other antimicrobial agents, a topic discussed in some detail in Chapter 17. Infectious diseases that previously accounted for millions of deaths became treatable by a simple course of therapy, at least in the affluent West, where such medications were readily available.

If the decades either side of 1900 have become known as the golden age of microbiology, the second half of the twentieth century will surely be remembered as the golden age of molecular genetics. Following on from the achievements of others such as Griffith and Avery, the publication of Watson and Crick's structure for DNA in 1953 heralded an extraordinary period of achievement in this area, culminating at the turn of the twenty-first century in the completion of the Human Genome Project.

You may ask, what has this genetic revolution to do with microbiology? Well, all the early work in molecular genetics was carried out on bacteria and viruses, as you'll learn in Chapter 11, and microbial systems have also been absolutely central to the development of the techniques of genetic engineering. In addition, as part of the Human Genome Project, the genomes of many microorganisms have been

> The *Human Genome Project* is an international effort to map and sequence all the DNA in the human genome. The project has also sequenced the genomes of many other organisms.

decoded, something that has now become almost routine, thanks to methodological advances made during the project. Having this information will help us to understand in greater detail the disease strategies of microorganisms, and to devise ways of countering them.

As we have seen, a recurring theme in the history of microbiology has been the way that advances in knowledge have followed on from methodological or technological developments, and we shall refer to a number of such developments during the course of this book. To conclude this introduction to microbiology, we shall return to the instrument that, in some respects, started it all. In any microbiology course, you are sure to spend some time

Eyepiece tube

Main focus knob

Objectives

Stage

Substage condenser

Base with built-in
illumination

Figure 1.3 The compound light microscope. Modern microscopes have a built-in light source. The light is focused onto the specimen by the condenser lens, and then passes into the body of the microscope via the objective lens. Rotating the objective nosepiece allows different magnifications to be selected. The amount of light entering the microscope is controlled by an iris diaphragm. Light microscopy allows meaningful magnification of up to around 1000 × .

looking down a microscope, and to get the most out of the instrument it is essential that you understand the principles of how it works. The following pages attempt to explain these principles.

1.4   Light microscopy

Try this simple experiment. Fill a glass with water, then partly immerse a pencil and observe from above; what do you see? The apparent 'bending' of the pencil is due to rays of light being slowed down as they enter the water, because air and water have different *refractive indices*. Light rays are similarly retarded as they enter glass, and all optical instruments are based on this phenomenon of *refraction*.

The compound light microscope consists of three sets of lenses (Figure 1.3):

> The *refractive index* of a substance is the ratio between the velocity of light as it passes through that substance and its velocity in a vacuum. It is a measure of how much the substance slows down and therefore refracts the light.

- the *condenser* focuses light onto the specimen to give optimum illumination;
- the *objective* provides a magnified and inverted image of the specimen;
- the *eyepiece* adds further magnification.

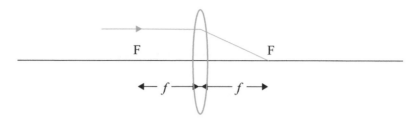

Figure 1.4   Light rays parallel to the axis of a convex lens pass through the focal point. The distance from the centre of the lens to the focal point is called the *focal length* of the lens (f).

Most microscopes have three or four different objectives, giving a range of magnifications, typically from $10 \times$ to $100 \times$. The total magnification is obtained by multiplying this by the eyepiece value (usually $10 \times$), thus giving a maximum magnification of $1000 \times$.

In order to appreciate how this magnification is achieved, we need to understand the behaviour of light passing through a convex lens:

- rays parallel to the axis of the lens are brought to a focus at the *focal point* of the lens (Figure 1.4);
- similarly, rays entering the lens from the focal point emerge parallel to the axis;
- rays passing through the centre of the lens from any angle are undeviated.

Because the condenser is not involved in magnification, it need not concern us here. Consider now what happens when light passes through an objective lens from an object AB situated slightly beyond its focal point (Figure 1.5a). Starting at the tip of the object, a ray parallel to the axis will leave the lens and pass through the focal point; a ray leaving the same point and passing through the centre of the lens will be undeviated. The point at which the two rays converge is an image of the original point formed by the lens. The same thing happens at an infinite number of points along the object's length, resulting in a *primary image* of the specimen, A′B′. What can we say about this image, compared to the original specimen AB? It is *magnified* and it is *inverted* (i.e. it appears upside down).

This primary image now serves as an object for a second lens, the eyepiece, and is magnified further (Figure 1.5b); this time the object is situated within the focal length. Using the same principles as before, we can construct a ray diagram, but this time we find that the two lines emerging from a point don't converge on the other side of the lens, but actually get further apart. The point at which the lines do

> A *real image* is one that can be projected onto a flat surface such as a screen. A *virtual image* does not exist in space and cannot be projected in this way. A familiar example is the image seen in a mirror.

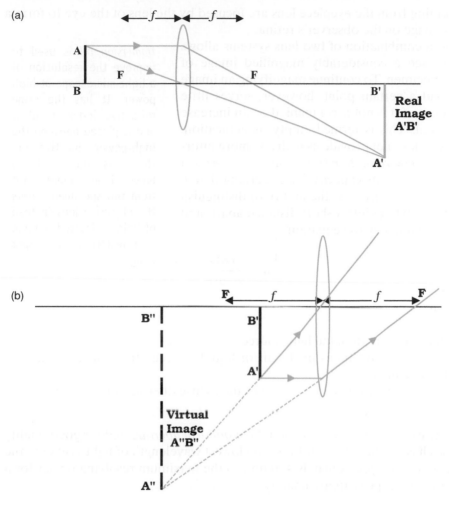

Figure 1.5   The objective lens and eyepiece lens combine to produce a magnified image of the specimen. (a) Light rays from the specimen AB pass through the objective lens to give a *magnified*, *inverted* and *real* primary image, A′B′. (b) The eyepiece lens magnifies this further to produce a *virtual* image of the specimen, A″B″.

eventually converge is actually 'further back' than the original object! What does this mean? The *secondary image* only *appears* to be coming from A″B″, and isn't actually there. An image such as this is called a *virtual image*. Today's readers, familiar with the concept of virtual reality, will probably find it easier to come to terms with this than some of their predecessors! The primary image A′B′, on the other hand, is a *real image*; if a screen was placed at that position, the image would be projected onto it. If we compare A″B″ with A′B′, we can see that it has been further magnified, but not further inverted, so it is still upside down compared with the original. The rays of light emerging from the eyepiece lens are focused by the lens of the eye to form a real image on the observer's retina.

So a combination of two lens systems allows us to see a considerably magnified image of our specimen. To continue magnifying an image beyond a certain point, however, serves little purpose, if it is not accompanied by an increase in detail. This is termed empty magnification, since it does not provide us with any more information. The *resolution* (resolving power, *d*) of a microscope is its capacity for discerning detail. More specifically, it is the ability to distinguish between two points a short distance apart, and is determined by the equation:

Immersion oil is used to improve the resolution of a light microscope at high power. It has the same refractive index as glass and is placed between the high-power objective and the glass slide. With no layer of air, more light from the specimen enters the objective lens instead of being refracted outside of it, resulting in a sharper image.

$$d = \frac{0.61\lambda}{n \sin \theta}$$

where:

λ = the wavelength of the light source;
n = the refractive index of the air or liquid between the objective lens and the specimen;
θ = the aperture angle (a measure of the light-gathering ability of the lens).

The expression $n \sin \theta$ is called the *numerical aperture* and for good quality lenses has a value of around 1.4. The lowest wavelength of light visible to the human eye is approximately 400 nm, so the maximum resolving power for a light microscope is approximately:

$$d = \frac{0.61 \times 400}{1.4} = 0.17 \ \mu m$$

that is, it cannot distinguish between two points closer together than about 0.2 microns. For comparison, the naked eye is unable to resolve two points more than about 0.2 mm apart.

For us to be able to discern detail in a specimen, it must have *contrast*; most biological specimens, however, are more or less colourless, so unless a structure is appreciably denser than its surroundings, it will not stand out using conventional light microscopy. This is why preparations are commonly subjected to *staining*

A *nanometre* (nm) is one-millionth of a millimetre. There are 1000 nanometres in 1 *micron* (μm), which is therefore one-thousandth of a millimetre.
$1 \ mm = 10^{-3}$ metre
$1 \ \mu m = 10^{-6}$ metre
$1 \ nm = 10^{-9}$ metre

procedures prior to viewing. The introduction of coloured dyes, which bind to certain structures, enables the viewer to discern more detail.

Since staining procedures involve the addition and washing off of liquid stains, the sample must clearly be immobilised or *fixed* to the slide if it is not to end up down the drain. The commonest way of doing this is to make a *heat fixed smear*; this kills the cells and attaches them to the glass microscope slide. A thin aqueous suspension of the cells is spread across the slide, allowed to dry, then passed (sample side up!) through a flame a few times. Excessive heating must be avoided, as it would distort the natural structure of the cells.

Using simple stains, such as methylene blue, we can see the size and shape of bacterial cells, for example, and their arrangement, while the binding properties of *differential stains* react with specific structures, helping us to distinguish between bacterial types. Probably the most widely used bacterial stain is the *Gram stain* (see Box 1.2), which for more than 100 years has been an invaluable first step in the identification of unknown bacteria.

The Gram stain is a differential stain, which only takes a few minutes to carry out, and which enables us to place a bacterial specimen into one of two groups – Gram-positive or Gram-negative. The reason for this differential reaction to the stain was not understood for many years, but is now seen to be a reflection of differences in cell wall structure, discussed in more detail in Chapter 3.

Box 1.2 The Gram stain

The Gram stain involves the sequential use of two stains. The critical stage is step 3; some cells will resist the alcohol treatment and retain the crystal violet, while others become decolourised. The counterstain (safranin or neutral red) is weaker than the crystal violet, and will only be apparent in those cells that have been decolourised.

| Add crystal violet (primary stain) | Add iodine (mordant) | Alocohol wash (decolourisation) | Add counterstain |

Specialised forms of microscopy have been developed to allow the viewer to discern detail in living, unstained specimens; these include *phase-contrast* and *dark-field* microscopy. We can also gain an estimate of the number of microorganisms in a sample by directly counting them under the microscope. This is discussed along with other enumeration methods in Chapter 5.

> *Phase-contrast microscopy* exploits differences in thickness and refractive index of transparent objects such as living cells to give improved contrast.

1.5   Electron microscopy

From the equation shown in the previous section, you can see that if it were possible to use a shorter wavelength of light, we could improve the resolving power of a microscope. However,

> *Dark-field microscopy* employs a modified condenser. It works by blocking out direct light, and viewing the object only by the light it diffracts.

because we are limited by the wavelength of light visible to the human eye we are not able to do this with the light microscope. The *electron microscope*, however, is able to achieve greater magnification and resolution because it uses a high-voltage beam of electrons, whose wavelength is very much shorter than that of visible light. Consequently we are able to resolve points that are much closer together than is possible even with the very best light microscope. The resolving power of an electron microscope may be as low as 1–2 nm, enabling us to see viruses, for example, or the internal structure of cells in considerable detail. The greatly improved resolution means that specimens can be meaningfully magnified over 100 000 × .

Electron microscopes, which were first developed in the 1930s and 40s, use ring-shaped electromagnets as 'lenses' to focus the beam of electrons onto the specimen. Because the electrons would collide with, and be deflected by, molecules in the air, electron microscopes require a pump to maintain a vacuum in the column of the instrument. There are two principal types of electron microscope: the *transmission electron microscope* (TEM) and the *scanning electron microscope* (SEM).

Figure 1.6 shows the main features of a TEM. As the name suggests, in TEM, the electron beam passes *through* the specimen and is scattered according to the density of the different parts. Due to the limited penetrating power of the electrons, extremely thin sections (<100 nm, or less than one-tenth of the diameter of a bacterial cell) must be cut, using a diamond knife. To allow this, the specimen must be fixed and dehydrated, a process that can introduce shrinkage and distortion to its structure if not correctly performed.

After being magnified by an objective 'lens', an image of the specimen is projected onto a fluorescent screen or photographic plate. Denser areas, which scatter the beam, appear dark, and those which allow it to pass through are light. It is often necessary to enhance contrast artificially, by means of

Figure 1.6 The transmission electron microscope (TEM). Electrons from a tungsten filament pass through a vacuum chamber and are focused by powerful electromagnets. Passage through the specimen causes a scattering of the electrons to form an image that is captured on a fluorescent screen. Reproduced from Black, JG (1999) Microbiology: Principles and Explorations, 4th edn, with permission from John Wiley & Sons.

'staining' techniques that involve coating the specimen with a thin layer of a compound containing a heavy metal, such as osmium or palladium. It will be evident from the foregoing description of sample preparation and use of a vacuum that electron microscopy cannot be used to study living specimens.

The TEM has been invaluable in advancing our knowledge of the fine structure of cells, microbial or otherwise. The resulting image is, however, a flat, two-dimensional one, and of limited use if we wish to learn about the surface of a cell or a virus. For this, we turn to SEM. The scanning electron microscope was developed in the 1960s and provides vivid, sometimes startling, three-dimensional images of surface structure. Samples are dehydrated and coated with gold to give a layer a few nanometres thick. A fine beam of electrons probes back and forth across the surface of the specimen and causes secondary electrons to be given off. The number of these, and the angle at which they are emitted, depends on the topography of the specimen's surface. SEM does not have quite the resolving power of the TEM, and therefore does not operate at such high magnifications.

Between them, SEM and TEM have opened up a whole new world to microbiologists, allowing us to put advances in our knowledge of microbial biochemistry and genetics into a structural context.

2
Biochemical Principles

All matter, whether living or non-living, is made up of *atoms*; the atom is the smallest unit of matter capable of entering into a chemical reaction. Atoms can combine together by bonding, to form *molecules*, which range from the small and simple to the large and complex. The latter are known as *macromolecules*; major cellular constituents such as carbohydrates and proteins belong to this group, and it is with these that this chapter is mainly concerned (Table 2.1). In order to appreciate how these macromolecules operate in the structure and function of microbial cells, however, you need to appreciate the basic principles of how atoms are constructed and how they interact with one other.

2.1 Atomic structure

All atoms have a central, positively charged *nucleus*, which is very dense and makes up most of the mass of the atom. The nucleus is made up of two types of particle, *protons* and *neutrons*. Protons carry a positive charge, and neutrons are uncharged, hence the nucleus overall is positively charged. It is surrounded by much lighter, and rapidly orbiting, *electrons* (Figure 2.1). These are negatively charged, with the charge on each electron being equal (but of course opposite) to that of the protons; however, the electrons have only 1/1840 of the mass of either protons or neutrons. It is the attractive force between the positively charged protons and the negatively charged electrons that holds the atom together.

The number of protons in the nucleus is called the *atomic number*, and ranges from one to over one hundred. The combined total of protons and neutrons is known as the *mass number*. All atoms have an equal number of protons and electrons, so regardless of the atomic number, the overall charge on the atom will always be zero.

Essential Microbiology, Second Edition. Stuart Hogg.
© 2013 John Wiley & Sons, Ltd. Published 2013 by John Wiley & Sons, Ltd.

Table 2.1   Biological macromolecules. Important examples of each of the four major classes of macromolecule found in biological systems

Proteins	Carbohydrates	Lipids	Nucleic acids
Enzymes	Sugars	Triacylglycerols	DNA
Receptors	Cellulose	(fats)	RNA
Antibodies	Starch	Phospholipids	
Structural		Waxes	
proteins		Sterols	

Atoms having the same atomic number have the same chemical properties; such atoms all belong to the same *element*. An element is made up of one type of atom only and cannot be chemically broken down into simpler substances; thus pure copper, for example, is made up entirely of copper atoms. There are 92 of these elements, 26 of which commonly occur in living things. Each element has been given a universally agreed symbol; examples that we shall encounter in biological macromolecules include carbon (C), hydrogen (H) and oxygen (O). The atomic numbers of selected elements are shown in Table 2.2.

The relationship between neutrons, protons, atomic number and mass number is illustrated in Table 2.3. We have used carbon as an example, since all living matter is based upon this element. The carbon represented can be expressed in the form:

$$\begin{matrix} 12 \\ \text{C} \\ 6 \end{matrix} = \begin{matrix} \text{(mass number)} \\ \text{(element symbol)} \\ \text{(atomic number)} \end{matrix}$$

The number of neutrons in an atom can be deduced by subtracting the atomic number from the mass number. In the case of carbon, this is the same

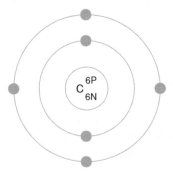

Figure 2.1   Atomic structure. The nucleus of a carbon atom contains six protons and six neutrons, surrounded by six electrons. Note how these are distributed between inner (2) and outer (4) electron shells.

Table 2.2 Symbols and atomic numbers of some elements occurring in living systems

Element	Symbol	Atomic no.
Hydrogen	H	1
Carbon	C	6
Nitrogen	N	7
Oxygen	O	8
Sodium	Na	11
Magnesium	Mg	12
Phosphorus	P	15
Sulphur	S	16
Chlorine	Cl	17
Potassium	K	19
Iron	Fe	26

as the number of protons (6), but this is not always so; phosphorus for example has 15 protons and 16 neutrons, giving it an atomic number of 15 and a mass number of 31.

2.1.1 Isotopes

Although the number of protons in the nucleus of a given element is always the same, the number of neutrons can vary, giving different forms, or *isotopes*, of that element. Carbon-14 (^{14}C) is a naturally occurring but rare isotope of carbon that has eight neutrons instead of six, hence the atomic mass of 14. Carbon-13 (^{13}C) is a rather more common isotope, making up around 1% of naturally occurring carbon; it has seven neutrons per atomic nucleus. The *atomic mass* (or atomic weight) of an element is the average of the mass numbers of an element's different isotopes, taking into account the proportions in which they occur. Carbon-12 is by far the predominant form of the element in nature, but the existence of small amounts of the other (slightly heavier) forms means that the atomic mass is 12.011 rather than exactly 12. Some isotopes are stable, while others decay spontaneously, with the release of subatomic particles. The latter are called *radioisotopes*; ^{14}C is a radioisotope, while the other two forms of carbon are stable isotopes. Radioisotopes have been an extremely useful research tool in a number of areas of molecular biology.

Table 2.3 The vital statistics of carbon

No. of protons	No. of neutrons	Atomic number	Mass number	Atomic mass
6	6	6	12	12.011

Box 2.1   How heavy is a mole?

When you work in a laboratory, something you'll need to come to grips with sooner or later is the matter of quantifying the amounts and concentrations of substances used. Central to this is the *mole*, so before we go any further, let's define this:

A mole is the molecular weight of a compound expressed in grams.

(The *molecular weight* is simply the sum of the atomic weights of all the atoms in a compound.)

So, to take sodium chloride as an example:

Molecular formula   =   NaCl (one atom each of sodium
                                     and chlorine)
Atomic weight of sodium   =   22.99
Atomic weight of chlorine =   35.45
∴ Molecular weight   =   58.44

Thus one mole of sodium chloride equals 58.44 grams (58.44 g)

Concentrations are expressed in terms of mass per volume, so here we introduce the idea of the *molar solution*. This is a solution containing one mole dissolved in a final volume of one litre of an appropriate solvent (usually water).

Molar solution = one mole per litre

A one molar (1 M) solution of sodium chloride therefore contains 58.44 g dissolved in water and made up to one litre. A 2 M solution would contain 116.88 g in 1 litre, and so on.

In biological systems, a molar solution of anything is actually rather concentrated, so we tend to deal in solutions that are so many millimolar (mM, one-thousandth of a mole per litre) or micromolar (μM, one-millionth of a mole per litre).

Why bother with moles?

So far, so good, but why can't we just deal in grams, or grams per litre?

Consider the following example:

You've been let loose in the laboratory, and been asked to compare the effects of supplementing the growth medium of a bacterial culture with several different amino acids. 'Easy', you think. 'Add X milligrams of each to the normal growth medium, and see which stimulates growth the most'.

The problem is that although you may be adding the same *weight* of each amino acid, you're not adding the same number of *molecules*, because each has a different molecular weight. If you add the same number of moles (or millimoles or micromoles) of each instead, you would be comparing the effect of the same number of molecules of each, and thus obtain a much more meaningful comparison. This is because *a mole of one compound contains the same number of molecules as a mole of any other compound.* This number is called *Avogadro's number*, and is 6.023×10^{23} molecules per mole.

The electrons that orbit around the nucleus do not do so randomly, but are arranged in a series of *electron shells*, radiating out from the nucleus (see Figure 2.1). These layers correspond to different energy levels, with the highest energy levels being located furthest away from the nucleus. Each shell can accommodate a maximum number of electrons, and they always fill up starting at the innermost one, that is, the one with the lowest energy level. In our example, carbon has filled the first shell with two electrons, and occupied four of the eight available spaces on the second.

The chemical properties of atoms are determined by the number of electrons in the outermost occupied shell. Neon, one of the so-called 'noble' gases, has an atomic number of 10, completely filling the first two shells, and is chemically unreactive or *inert*. Atoms that achieve a similar configuration are all stable. If, on the other hand, such an arrangement is not achieved, the atom is unstable, or reactive. Reactions take place between atoms that attempt to achieve stability by attaining a full outer shell. These reactions may involve atoms of the same element or ones of different elements; the result in either case is a molecule or ion. Figure 2.2 shows how atoms combine to form a molecule. A substance made up of molecules containing two or more different elements is called a *compound*. In each example, the

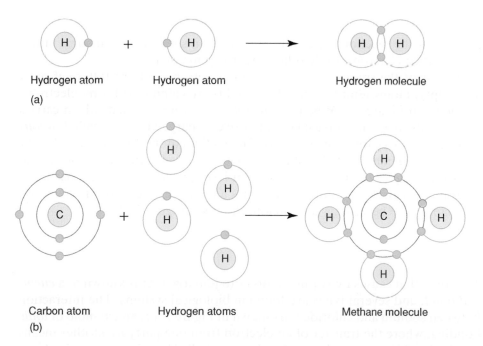

Hydrogen atom   Hydrogen atom   Hydrogen molecule

(a)

Carbon atom   Hydrogen atoms   Methane molecule

(b)

Figure 2.2   Atoms join to form molecules. The formation of (a) hydrogen and (b) methane by covalent bonding. Each atom achieves a full set of electrons in its outer shell by sharing with another atom. A shared pair of electrons constitutes a covalent bond.

Figure 2.3 Ion formation. Sodium achieves stability by losing the lone electron from its outermost shell. The resulting sodium ion Na$^+$ has 11 protons and 10 electrons, hence it carries a single positive charge. Chlorine becomes ionised to chloride (Cl$^-$) when it gains an electron to complete its outer shell.

product of the reaction has a full outer electron shell; note that some atoms are donating electrons, while others are accepting them.

If most of the spaces in the outermost electron shell are full, or if most are empty, atoms tend to strive for stability by gaining or losing electrons, as shown in Figure 2.3. When this happens, an *ion* is formed, which carries either a positive or negative charge. Positively charged ions are called *cations* and negatively charged ones *anions*. The sodium atom, for example, has 11 electrons, meaning that the inner two electron shells are filled and a lone electron occupies the third shell. When it loses this last electron, it has more protons than electrons, and therefore has a net positive charge of one; when this happens, it becomes a sodium ion, Na$^+$ (see Figure 2.3).

2.1.2 Chemical bonds

The force that causes two or more atoms to join together is known as a *chemical bond*, and several types are found in biological systems. The interaction between sodium and chloride ions shown in Figure 2.4 is an example of *ionic* bonding, where the transfer of an electron from one party to another means that both achieve a complete outer electron shell. There is an attractive force between positively and negatively charged ions, called an ionic bond. Certain elements form ions with more than a single charge, by gaining or losing two

Formation of
ionic bond

Sodium chloride
NaCl

Figure 2.4 Ionic bonding. A positively charged Na$^+$ and negatively charged Cl$^-$ attract each other, and an ionic bond is formed. The result is a molecule of sodium chloride.

or more electrons in order to achieve a full outer electron shell; thus calcium ions (Ca^{2+}) are formed by the loss of two electrons from a calcium atom.

The goal of stability through a full complement of outer shell electrons may also be achieved by means of *sharing* one or more pairs of electrons. Consider the formation of methane (see Figure 2.2); a carbon atom, which has four spaces in its outer shell, can achieve a full complement by sharing electrons from four separate hydrogen atoms. This type of bond is a *covalent* bond.

Sometimes, a pair of atoms share not one but two pairs of electrons (Figure 2.5). This involves the formation of a *double bond*. Triple bonding, through the sharing of three pairs of electrons, is also possible, but rare.

In the examples of covalent bonding we've looked at so far, the sharing of the electrons has been equal, but this is not always the case because sometimes the electrons may be drawn closer to one atom than another (Figure 2.6a). This has the effect of making one atom slightly negative and another slightly positive. Molecules like this are called *polar* molecules and the bonds are polar bonds. Sometimes a large molecule may have both polar and nonpolar areas. Polar molecules are attracted to each other, with the

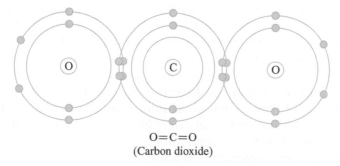

O=C=O
(Carbon dioxide)

Figure 2.5 Double bond formation. In the formation of carbon dioxide, the carbon atom shares *two* pairs of electrons with each oxygen atom.

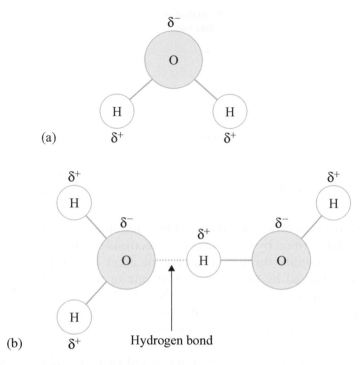

Figure 2.6 Water is a polar molecule. (a) The electrons of the hydrogen atoms are strongly attracted to the oxygen atom, causing this part of the water molecule to carry a slightly negative charge, and the hydrogen part a slightly positive one. (b) Because of their polar nature, water molecules are attracted to each other by hydrogen bonding. Hydrogen bonding is much weaker than ionic or covalent bonding, but plays an important role in the structure of macromolecules such as proteins and nucleic acids.

negative areas of one molecule attracted to the positive areas of another (Figure 2.6b). In water, hydrogen atoms bearing a positive charge are drawn to the negatively charged oxygens.

This attraction between polar atoms is called *hydrogen bonding*, and can take place between covalently bonded hydrogen and any electronegative atom, most commonly oxygen or nitrogen. Hydrogen bonds are much weaker than either ionic or covalent bonds; however, if sufficient of them form in a compound, the overall bonding force can be appreciable. Each water molecule can form hydrogen bonds with others of its kind in four places. In order to break all these bonds, a substantial input of energy is required, explaining why water has such a relatively high boiling point, and why most of the water on our planet is in liquid form.

Another weak form of interaction is brought about by *van der Waals forces*, which occur briefly when two nonpolar molecules (or parts of molecules) come into very close contact with one another. Although transient, and generally even weaker than hydrogen bonds, they occur in great numbers

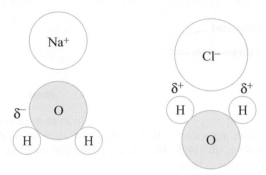

Figure 2.7   Ions dissolve in water. An ionic compound such as sodium chloride dissociates in water to its constituent ions. Water molecules form hydration shells around both Na⁺ and Cl⁻ ions.

in certain macromolecules and play an important role in holding proteins together (see Section 2.3.3).

Water is essential for living things, both in the composition of their cells and in the environment surrounding them. Organisms are made up of 60–95% water by weight, and even inert dormant forms like spores and seeds have a significant water component. This dependence on water is a function of its unique properties, which in turn derive from its polar nature.

Water is the medium in which most biochemical reactions take place; it is a highly efficient solvent; indeed, more substances will dissolve in water than in any other solvent. Substances held together by ionic bonds tend to dissociate into anions and cations in water, because as individual solute molecules become surrounded by molecules of water, *hydration shells* are formed, in which the negatively charged parts of the solute attract the positive region of the water molecule, and the positive parts the negative region (Figure 2.7). The attractive forces that allow the solute to dissolve are called *hydrophilic* forces, and substances that are water-soluble are hydrophilic (water-loving). Other polar substances such as sugars and proteins are also soluble in water by forming hydrophilic interactions.

Molecules such as oils and fats are nonpolar, and because of their non-reactivity with water are termed *hydrophobic* ('water-fearing'). If such a molecule is mixed with water, it will be excluded, as water molecules 'stick together'. This very exclusion by water can act as a cohesive force among hydrophobic molecules (or hydrophobic areas of large molecules). This is often called hydrophobic bonding, but it isn't really bonding as such, rather a shared avoidance of water. All living cells have a hydrophilic interior surrounded by a hydrophobic membrane, as we'll see in Chapter 3.

An *amphipathic* substance is one that is part polar and part nonpolar. When such a substance is mixed with water, *micelles* are formed (Figure 2.8); the nonpolar parts are excluded by the water and group together as described above, leaving the polar groups pointing outwards into the water, where they are attracted by hydrophilic forces. Detergents exert their action by trapping

Water

Figure 2.8 Amphipathic molecules form micelles in water. In an aqueous environment, amphipathic substances align their molecules so that the nonpolar parts are hidden away from the water. Reproduced from Black, JG (1999) Microbiology: Principles and Explorations, 4th edn, with permission from John Wiley & Sons.

insoluble grease inside the centre of a micelle, while interaction with water allows them to be rinsed away (see Chapter 16).

Water plays a role in many essential metabolic reactions, and its polar nature allows for the breakdown to hydrogen and hydroxyl ions (H^+ and OH^-), and resynthesis as water. Water acts as a reactant in hydrolysis reactions such as:

$$XY + H_2O \rightarrow XH + YOH$$

or as a product in certain synthetic reactions, like

$$XH + YOH \rightarrow XY + H_2O$$

2.2   Acids, bases and pH

Only a minute proportion of water molecules, something like one in every 500 million, are present in the dissociated form, but as we've already seen, the H^+ and OH^- ions play an important part in cellular reactions.

A solution becomes acid or alkaline if there is an imbalance in the amount of these ions present. If there is an excess of H^+, the solution becomes *acid*, whilst if OH^- predominates, it becomes *alkaline*. The *pH* of a solution is an expression of the molar concentration of hydrogen ions; it is expressed thus:

$$pH = -\log_{10}[H^+]$$

In pure water, hydrogen ions are present at a concentration of 10^{-7} M, thus the pH is 7.0. This is called *neutrality*, where the solution is neither acid nor alkaline. At higher concentrations of H^+, such as 10^{-3} M (1 millimolar), the pH value is *lower*, in this case 3.0, so acid solutions have a value *below* 7. Conversely, alkaline solutions have a pH *above* 7. You will see from this example that an increase of 10^4 (10 000)-fold in the $[H^+]$ leads to a change of only four points on the pH scale. This is because it is a *logarithmic scale*, thus a solution of pH 10 is 10 times more alkaline than one of pH 9, and 100 times more than one of pH 8. Figure 2.9 shows the pH value of a number of familiar substances.

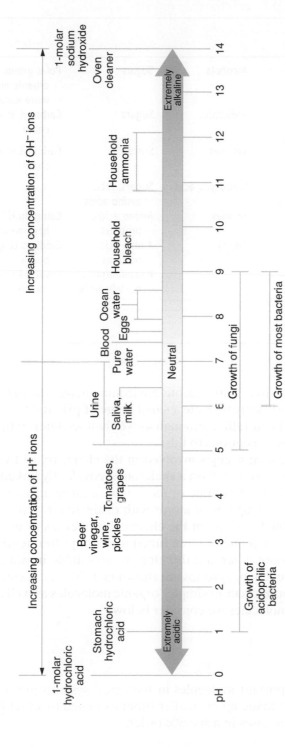

Figure 2.9 The pH value of some common substances. Most microorganisms exist at pH values around neutrality, but representatives are found at extremes of both acidity and alkalinity. Reproduced from Black, JG (1999) Microbiology: Principles and Explorations, 4th edn, with permission from John Wiley & Sons.

Table 2.4   Occurrence and characteristics of some functional groups

Functional group	Formula	Type of molecule	Found in:	Remarks
Hydroxyl	–OH	Alcohols	Sugars	Polar group, making organic molecules more water soluble
Carbonyl	$-C\underset{H}{\overset{O}{\lessgtr}}$	Aldehydes	Sugars	Carbonyl at end of chain
	$\overset{\|}{\underset{\|}{C}}=O$	Ketones	Sugars	Carbonyl elsewhere in chain
Carboxyl	–COOH	Carboxylic acids	Sugars, fats, amino acids	
Amino	$-NH_2$	Amines	Amino acids, proteins	Can gain H^+ to become $NH_3{}^+$
Sulphydryl	–SH	Thiols	Amino acids, proteins	Oxidises to give S=S bonds
Phosphate	$-O-\underset{\underset{O^-}{\|}}{\overset{\overset{O}{\|\|}}{P}}-O^-$		Phospholipids, nucleic acids	Involved in energy transfer

Most microorganisms live in an aqueous environment, and the pH of this is very important. Most will only tolerate a small range of pH, and the majority occupy a range around neutrality, although as we shall see later in this book, there are some startling exceptions to this.

Most of the important molecules involved in the chemistry of living cells are *organic*, that is, they are based on a skeleton of covalently linked carbon atoms. Biological molecules have one or more *functional groups* attached to this skeleton; these are groupings of atoms with distinctive reactive properties, and are responsible for many of the chemical properties of the organic molecule. The possession of one or more functional groups frequently makes an organic molecule more polar and therefore more soluble in water. Some of the most common functional groups are shown in Table 2.4. It can be seen that the functional groups occur in simpler organic molecules as well as in the more complex macromolecules we consider below.

2.3   Biomacromolecules

Many of the most important molecules in biological systems are *polymers*, that is, large molecules made up of smaller subunits joined together by covalent bonds, and in some cases in a specific order.

$$HC=O$$
$$HCOH$$
$$CH_2OH$$

$$CH_2OH$$
$$C=O$$
$$CH_2OH$$

(a)   (b)

Figure 2.10 Monosaccharides may be aldoses or ketoses. The three-carbon sugars (a) glyceraldehyde and (b) dihydroxyacetone share the same molecular formula, but have different functional groups. The two molecules are isomers (see Box 2.2).

2.3.1 Carbohydrates

Carbohydrates are made up of just three different elements: carbon, hydrogen and oxygen. The simplest carbohydrates are *monosaccharides*, or simple sugars; these have the general formula $(CH_2O)_n$. They are classed as either aldoses or ketoses, according to whether they contain an aldehyde group or a ketone group (Figure 2.10). Monosaccharides can furthermore be classified on the basis of the number of carbon atoms they contain. The simplest are trioses (three carbons) and the most important biologically are hexoses (six carbons) (see Boxes 2.2 and 2.3).

Glucose + Glucose   =   Maltose (α 1,4 linkage)

(a)

Galactose + Glucose   =   Lactose (β 1,4 linkage)

(b)

Figure 2.11 Monosaccharides can be joined by a glycosidic linkage. The result is a condensation reaction, in which a molecule of water is lost. α- and β- linkages result in different orientations in space.

Monosaccharides are generally crystalline solids that are soluble in water and have a sweet taste. They are all reducing sugars, so called because they are able to reduce alkaline solutions of copper ions from the cupric (Cu^{2+}) to the cuprous (Cu^+) form.

A *disaccharide* is formed when two monosaccharides (which may be of the same type or different), join together with a concomitant loss of a water molecule (Figure 2.11). Further monosaccharides can be added, giving chains of three, four, five or more units. These are termed *oligosaccharides* (*oligo* = a few), and chains with many units are *polysaccharides*. The chemical bond

---

Box 2.2   Isomers: same formula, different structure

The simplest monosaccharides are the trioses glyceraldehyde and dihydroxyacetone (see Figure 2.10). Look carefully at the structures, and you will see that although they both share the same number of carbons (3), hydrogens (6) and oxygens (3), the way in which these atoms are arranged is different in the two sugars. Molecules such as these, which have the same chemical formula but different structural formulas, are said to be structural *isomers*. The different groupings of atoms lead to structural isomers having different chemical properties. When we come to look at the *hexoses* (six-carbon sugars), we see that there are many structural possibilities for the general formula $C_6H_{12}O_6$; some of these are shown below.

CHO	CHO	CHO	CHO	CH$_2$OH
HO–C–H	H–C–OH	H–C–OH	HO–C–H	C=O
H–C–OH	HO–C–H	HO–C–H	HO–C–H	HO–C–H
HO–C–H	H–C–OH	HO–C–H	H–C–OH	H–C–OH
HO–C–H	H–C–OH	H–C–OH	H–C–OH	H–C–OH
CH$_2$OH	CH$_2$OH	CH$_2$OH	CH$_2$OH	CH$_2$OH
D-glucose	L-glucose	D-galactose	D-mannose	D–fructose

Note that some of these structures are identical apart from the orientation of groups around the central axis; D-glucose and L-glucose, for example, differ only in the way H atoms and –OH groups are arranged to the right or left. They are said to be *stereoisomers* or optical isomers, and are mirror images of each other, just like your right and left hands. (D- and L- are short for *dextro-* and *laevo*-rotatory, meaning that the plane of polarised light is turned to the right and left respectively when passed through a solution of these substances). Generally, living cells will only synthesise one or other stereoisomer, not both.

Box 2.3 Sugars are more accurately shown as ring structures

When dissolved in water, the aldehyde or ketone group reacts with a hydroxyl group on the fifth carbon to give a cyclic form. D-glucose is shown in both forms below. The cyclic form of the molecule is shown below as a *Haworth projection*. The idea is that the ring is orientated at 90° to the page, with the edge that is shown thicker towards you, and the top edge away from you. Notice that there are even two forms of D-glucose! Depending on whether the –OH on carbon-1 is below or above the plane of the ring, we have α- or β-D-glucose.

α-D-Glucose
(linear form)

α-D-Glucose

β-D-Glucose

joining the monosaccharide units together is called a *glycosidic linkage*. The bond between the two glucose molecules that make up maltose is called an α-glycosidic linkage; in lactose, formed from one glucose and one galactose, we have a β-glycosidic linkage. The two bonds are formed in the same way, with the elimination of water, but they have a different orientation in space. Thus disaccharides bound together by α- and β-glycosidic linkages have a different overall shape and as a result the molecules behave differently in cellular reactions.

Biologically important molecules such as starch, cellulose and glycogen are all polysaccharides. Another is dextran, a sticky substance produced by some bacteria to aid their adhesion. They differ from monosaccharides in being generally insoluble in water, not tasting sweet and not being able to reduce cupric ions. Most polysaccharides are made up from either pentose or hexose sugars, and, like di- and oligosaccharides, can be broken down into their constituent subunits by hydrolysis reactions.

2.3.2 Proteins

Of the macromolecules commonly found in living systems, proteins have the greatest structural diversity, a fact reflected in their wide range of biological functions.

The five elements found in most naturally occurring proteins are carbon, hydrogen, oxygen, nitrogen and sulphur. In addition, other elements may be essential components of certain specialised proteins such as haemoglobin (iron) and casein (phosphorus).

Proteins can be very large molecules, with molecular weights of tens or even hundreds of thousands. Whatever their size, and in spite of the diversity referred to above, all proteins are made up of a collection of 'building bricks' called *amino acids* joined together. Amino acids are thought to have been among the first organic molecules formed in the early history of the Earth, and many different types exist in nature. All these, including the 20 commonly found in proteins, are based on a common structure, shown in Figure 2.12. It comprises a central carbon atom (known as the α-carbon) covalently bonded to an amino (NH_2) group, a carboxyl (COOH) group and a hydrogen atom. It is the group attached to the fourth and final valency bond of the α-carbon that varies from one amino acid to another; this is known as the '*R*'-group.

The 20 amino acids found in proteins can be conveniently divided into five groups, on the basis of the chemical nature of their 'R'-group. These range from a single hydrogen atom to a variety of quite complex side chains (Figure 2.13). It is unlikely nowadays that you would need to memorise the precise structure of all 20 as the author was asked to do in days gone by, but it would be advisable to familiarise yourself with the groupings and examples from each of them. The groups differentiate on the basis of a polar/nonpolar nature and on the presence or absence of an ionisable 'R'-group.

$$H_2N - \overset{\overset{\displaystyle H}{|}}{\underset{\underset{\displaystyle R}{|}}{C}} - COOH \qquad H_3N^+ - \overset{\overset{\displaystyle R}{|}}{\underset{\underset{\displaystyle H}{|}}{C}} - COO^-$$

(a)   (b)

Figure 2.12   Amino acid structure. (a) The basic structure of an amino acid. (b) In solution, the amino and carboxyl groups become ionised, giving rise to a *zwitterion* (a molecule with spatially separated positive and negative charges). All the 20 amino acids commonly found in proteins are based on a common structure, differing only in the nature of their 'R' group (see Figure 2.13).

Box 2.4   Amino acid shorthand

It is sometimes necessary to express in print the sequence of amino acids that make up the primary structure of a particular protein; clearly it would be desperately tedious to express a sequence of hundreds of bases in the form 'glycine, phenylalanine, tryptophan, methionine, … etc.', so a system of abbreviations for each amino acid has been agreed. Each amino acid can be reduced to a three-letter code, thus you might see something like:

1	2	3	4	5	6	7	8	9	10	11
Gly	Phe	Tyr	Met	His	Lys	Gly	Ala	His	Val	Glu… and so on.

Note that each residue has a number; this *numbering always begins at the N-terminus*.

Each amino acid can also be represented by a single letter. The abbreviations using the two systems are shown below.

A	Ala	Alanine	M	Met	Methionine
B	Asx	Asparagine/aspartic acid	N	Asn	Asparagine
C	Cys	Cysteine	P	Pro	Proline
D	Asp	Aspartic acid	Q	Gln	Glutamine
E	Glu	Glutamic acid	R	Arg	Arginine
F	Phe	Phenylalanine	S	Ser	Serine
G	Gly	Glycine	T	Thr	Threonine
H	His	Histidine	V	Val	Valine
I	Ile	Isoleucine	W	Trp	Tryptophan
K	Lys	Lysine	Y	Tyr	Tyrosine
L	Leu	Leucine	Z	Glx	Glutamine/glutamic acid

Note that one amino acid, *proline*, falls outside the main groups. Closer inspection of its molecular structure reveals that this differs from the other amino acids by having one of its N–H linkages replaced by an N–C, which forms part of a cyclic structure (see Figure 2.13). This puts certain conformational constraints upon proteins containing proline residues.

As can be seen from Figure 2.13, the simplest amino acid is glycine, whose R-group is simply a hydrogen atom. This means that the glycine molecule is symmetrical, with a hydrogen atom on opposite valency bonds. All the other amino acids, however, are asymmetrical: the α-carbon acts as what is called a chiral centre, giving the molecule right or left 'handedness'. Thus two *stereoisomers*, known as the D- and L- forms, are possible for each of the amino acids except glycine. Note that all the amino acids found in naturally occurring proteins have the L-form; the D-form also occurs in nature but only in certain specific, nonprotein contexts.

Proteins, as we've seen, are made up of amino acids; these are joined together by means of *peptide bonds*, involving the $-NH_2$ group of one amino

Figure 2.13 The 20 amino acids found in proteins. The 'R' group of each amino acid is shown. These range from the simplest, glycine, to more complex representatives such as tryptophan.

Figure 2.14 Amino acids are joined by peptide linkage. The carboxyl group of one amino acid is joined to the amino group of another. This is another example of a condensation reaction (cf. Figure 2.11). No matter how many amino acids are added, the resulting structure always has a free carboxyl group at one end and a free amino group at the other.

acid and the –COOH group of another. The formation of a peptide bind is a form of *condensation reaction* in which water is lost (Figure 2.14). The resulting structure of two linked amino acids is called a *dipeptide*, and retains an –NH$_2$ at one end and a –COOH at the other. If we were to add another amino acid to form a tripeptide, this would still be so, and if we kept on adding them until we had a *polypeptide*, we would still have the same two groupings at the extremities of the molecule. These are referred to as the *N-terminus* and the *C-terminus* of the polypeptide. Since a water molecule has been removed at the formation of each peptide bond, we refer to the chain so formed as being composed of amino acid *residues*, rather than amino acids. Generally, when a polypeptide reaches 50 residues in length it is termed a protein, although the actual distinction is not clear-cut.

So far, we can think of proteins as long chains of many amino acid residues, rather like a string of beads. This is called the *primary structure* of the protein; it is determined by the relative proportions of each of the 20 amino acids, and also the order in which they are joined together. It is the basis of all the remaining levels of structural complexity, and it *ultimately determines the properties of a particular protein*. It is also what makes one protein different from another. Since the 20 types of amino acid can be linked together in any order, the number of possible sequences is astronomical, and it is this great variety of structural possibilities that gives proteins such diverse structures and functions.

> In theory, there are 20^{100} or some 10^{130} different ways in which 20 different amino acids could combine to give a protein 100 amino acid residues in length!

2.3.3 Higher levels of protein structure

The structure of proteins is a good deal more complicated than just a linear chain of amino acids. A long thin chain would be a very fragile structure;

Figure 2.15 Secondary structure in proteins. Hydrogen bonding occurs between the –CO and –NH groups of amino acids on the backbone of a polypeptide chain. The two amino acids may be on the same or different chains.

proteins that have undergone folding are more stable and compact. The results of this folding are the secondary and tertiary structures of a protein.

The *secondary structure* is due to hydrogen bonding between a carbonyl (–CO) group and an amido (–NH) group of amino acid residues on the peptide backbone (Figure 2.15). *Note that the 'R'-group plays no part in secondary protein structure.* Two regular patterns of folding result from this; the α-helix and the β-pleated sheet.

The *α-helix* occurs when hydrogen bonding takes place between amino acids close together in the primary structure. A stable helix is formed by the –NH group of an amino acid bonding to the –CO group of the amino acid four residues further along the chain (Figure 2.16a). This causes the chain to twist into the characteristic helical shape. One turn of the

> Very small distances within molecules are measured in Angstrom units (Å). One Angstrom unit is equal to one tenbillionth (10^{-10}) of a metre.

helix occurs every 3.6 amino acid residues, and results in a rise of 5.4 Å; this is called the *pitch height* of the helix. The ability to form a helix like this is dependent on the component amino acids; if there are too many with large R-groups, or R-groups carrying the same charge, a stable helix will not be formed. Because of its rigid structure, proline (see Figure 2.13) cannot be accommodated in an α-helix. Naturally occurring α-helices are always *right-handed*, that is, the chain of amino acids coils round the central axis in a clockwise direction. This is a much more stable configuration than a left-handed helix, due to the fact that there is less steric hindrance (overlapping of electron clouds) between the R-groups and the C=O group on the peptide backbone. (Note that if proteins were made up of the D-form of amino acids, we would have the reverse situation, with a left-handed form favoured). In the β-*pleated sheet,* the hydrogen bonding occurs between amino acids either on

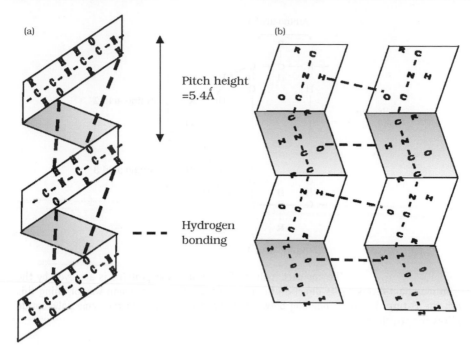

Pitch height =5.4Å

Hydrogen bonding

Figure 2.16 Secondary structure in proteins: the α-helix and β-pleated sheet. (a) Hydrogen bonding between amino acids four residues apart in the primary sequence results in the formation of an α-helix. (b) In the β-pleated sheet hydrogen bonding joins adjacent chains. Note how each chain is more fully extended than in the α-helix. In the example shown, the chains run in the same direction (parallel).

separate polypeptide chains or on residues on the same chain but far apart in the primary structure (Figure 2.16b). The chains in a β-pleated sheet are fully extended, with 3.5 Å between adjacent amino acid residues (cf. α-helix, 1.5 Å). When two or more of these chains lie next to each other, extensive hydrogen bonding occurs between the chains. Adjacent strands in a β-pleated sheet can either run in the same direction (e.g. N → C), giving rise to a parallel β-pleated sheet, or in opposite directions (antiparallel β-pleated sheet).

A common structural element in the secondary structure of proteins is the β-*turn*. This occurs when a chain doubles back on itself, such as in an antiparallel β-pleated sheet. The –CO group of one amino acid is hydrogen bonded to the –NH group of the residue three further along the chain. Frequently, it is called a *hairpin turn*, for obvious reasons (Figure 2.17). Numerous changes in direction of the polypeptide chains result in a compact, globular shape to the molecule.

Typically about 50% of a protein's secondary structure will have an irregular form. Although this is often referred to as random coiling, it is only random in the sense that there is no regular pattern; it still contributes towards

Amino acid 1

Figure 2.17 The β-turn. The compact folding of many globular proteins is achieved by the polypeptide chain reversing its direction in one or more places. A common way of doing this is with the β-turn. Hydrogen bonding between amino acid residues on the same polypeptide stabilises the structure.

the integrity of the molecule. The proportions and combinations in which α-helix, β-pleated sheet and random coiling occur vary from one protein to another. Keratin, a structural protein found in skin, horn and feathers, is an example of a protein entirely made up of α-helix, whilst the lectin (sugar-binding protein) concanavalin A is mostly made up of β-pleated sheets.

The *tertiary structure* of a protein is due to interactions between side chains, that is, R-groups of amino acid residues, resulting in the folding of the molecule to produce a thermodynamically more favourable structure. The structure is formed by a variety of weak, non-covalent forces; these include hydrogen bonding, ionic bonds, hydrophobic interactions, and van der Waals forces. The strength of these forces diminishes with distance, therefore the formation of a compact structure is encouraged. In addition, the –SH groups on separate cysteine residues can form a covalent –S–S– linkage. This is known as a *disulphide bridge* and may have the effect of bringing together two cysteine residues that are far apart in the primary sequence (Figure 2.18).

In globular proteins, the R-groups are distributed according to their polarities; nonpolar residues such as valine and leucine nearly always occur on the inside, away from the aqueous phase, while charged, polar residues including glutamic acid and histidine generally occur at the surface, in contact with the water.

The protein can be *denatured* by heating or treatment with certain chemicals; this causes the tertiary structure to break down and the molecule to

Figure 2.18 Disulphide bond formation. (a) Disulphide bonds formed by the oxidation of cysteine residues result in cross-linking of a polypeptide chain. (b) This can have the effect of bringing together residues that lie far apart in the primary amino acid sequence. Disulphide bonds are often found in proteins that are exported from the cell, but rarely in intracellular proteins.

unfold, resulting in a loss of the protein's biological properties. Cooling, or removal of the chemical agents, will lead to a restoration of both the tertiary structure and biological activity, showing that both are entirely dependent on the primary sequence of amino acids.

Even the tertiary structure may not be the ultimate level of organisation of a protein, because some are made up of two or more polypeptide chains, each with its own secondary and tertiary structure, combined together to give the

Figure 2.19 Polypeptide chains may join to form quaternary structure. The example shown comprises two identical polypeptide subunits. Coils indicate α-helical sequences, arrows are β-pleated sheets. Reproduced from Bolsover, SR, et al. (1997) From Genes to Cells, with permission from John Wiley & Sons.

quaternary structure (Figure 2.19). These chains may be identical or different, depending on the protein. As with the tertiary structure, non-covalent forces between R-groups are responsible, the difference being that this time they link amino acid residues on separate chains rather than on the same one.

Such proteins lose their functional properties if dissociated into their constituent units; the quaternary joining is essential for their activity. The enzyme phosphorylase A is an example of a protein with a quaternary structure. It has four subunits, which have no catalytic activity unless joined together as a tetramer.

Although all proteins are polymers of amino acids existing in various levels of structural complexity as we've seen above, some have additional, non-amino acid components. They may be organic, such as sugars (glycoproteins) or lipids (lipoproteins) or inorganic, including metals (metalloproteins) or phosphate groups (phosphoproteins). These components, which form an integral part of the protein's structure, are called *prosthetic groups*.

2.3.4 Nucleic acids

The third class of polymeric macromolecules are the *nucleic acids*. These are deoxyribonucleic acid (DNA) and ribonucleic acid (RNA), and both are polymers of smaller molecules called nucleotides.

Figure 2.20 A nucleotide has three parts: a pentose sugar, a phosphate group and a nitrogenous base (see Figure 2.21). Note the difference between the sugars (a) deoxyribose (DNA) and (b) ribose (RNA).

As we shall see, there are important differences both in the overall structures of RNA and DNA and in the nucleotides they contain, so we shall consider each of them in turn.

The structure of DNA The composition of a DNA nucleotide is shown in Figure 2.20(a). It has three parts: a five-carbon sugar called *deoxyribose*, a phosphate group and a base. This base can be any one of four molecules; as can be seen in Figure 2.21, these are all based on a cyclic structure containing nitrogen. Two of the bases, *cytosine* and *thymine*, have a single ring and are called *pyrimidines*. The other two, *guanine* and *adenine*, have a double ring structure; these are the *purines*. For convenience, the four bases are often referred to by their initial letter only: A, C, G and T.

One nucleotide differs from another by the identity of the base it contains; the rest of the molecule (sugar and phosphate) is identical. You will recall from the previous section that the properties of a protein depend on the order in which its constituent amino acids are linked together; we have exactly the same situation with nucleic acids, except that instead of an 'alphabet' of 20 'letters', here we have one of only four. Nevertheless, because nucleic acid molecules are extremely long, and the bases

Erwin Chargaff measured the proportions of the different nucleotides in a range of DNA samples. He found that T always = A and C always = G. Watson and Crick interpreted this as meaning that the bases always paired up in this way.

Figure 2.21 Bases belong to two classes. Nucleotides differ from each other in the identity of the nitrogenous base. (a) In DNA these are adenine (A), cytosine (C), guanine (G) or thymine (T). The purines (A and G) have a two-ring structure, while the pyrimidines (C and T) have only one ring. (b) In RNA, thymine is replaced by a similar molecule, uracil (U).

can occur in almost any order, an astronomically large number of different sequences is possible.

The nucleotides join together by means of a *phosphodiester bond* that links the phosphate group of one nucleotide to an –OH group on the 3-carbon of the deoxyribose sugar of another (Figure 2.22). The chain of nucleotides therefore has a free –OH group on a 3-carbon (the *3′ end*) and a phosphate

Figure 2.22 The phosphodiester bond. A chain of DNA is made longer by the addition of nucleotides containing not one but three phosphate groups; on joining the chain, two of these phosphates are removed. Nucleotides are joined to each other by a phosphodiester bond, linking the phosphate group on the 5-carbon of one deoxyribose to the –OH group on the 3-carbon of another. (These carbons are known as 5' and 3' to distinguish them from the 3- and 5-carbon on the nitrogenous base.) Note that the resulting chain, however many nucleotides it may comprise, always has a 5'(PO₄) group at one end and a 3'(OH) group at the other.

group on a 5-carbon (the *5' end*). This remains the case however long the chain becomes.

However, the structure of DNA is not just a single chain of linked nucleotides, but two chains wound around each other to give the famous *double helix* form proposed by James Watson and Francis Crick in 1953 (Figure 2.23; see also Chapter 11). If we compare this to an open spiral staircase, alternate sugar and phosphate groups make up the 'skeleton' of the staircase, while the inward-facing bases pair up by hydrogen bonding to form the steps. Notice that each nucleotide pair always comprises three rings, resulting from a combination of one purine and one pyrimidine base. This means that the two strands of the helix are always evenly spaced. The way in which the bases pair up is further governed by the phenomenon of complementary base

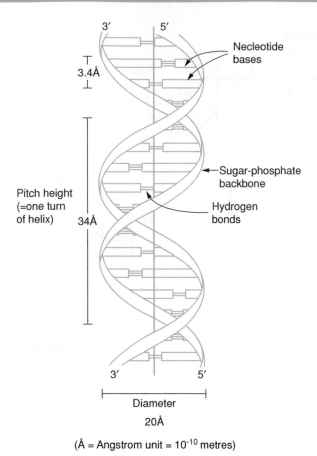

Figure 2.23   DNA is a double helix. The model proposed by Watson and Crick has two chains of nucleotides joined together by hydrogen-bonded base pairs pointing inwards towards the centre of the helix. The rules of complementary base pairing mean that the sequence of one chain can be predicted from the sequence of the other. Note how the chains run in opposite directions (antiparallel).

pairing. *A nucleotide containing thymine will only pair with one containing adenine, and likewise guanine always pairs with cytosine* (Figure 2.24). Thus the sequence of nucleotides on one strand of the double helix determines that of the other, as it has a complementary structure. Figure 2.23 shows how the two strands of the double helix are *antiparallel*, that is, they run in opposite directions, one $5' \rightarrow 3'$ and the other $3' \rightarrow 5'$. In Chapter 11 we shall look at how this structure was used to propose a mechanism for the way in which DNA replicates and genetic material is copied.

The structure of RNA   In view of the similarities in the structure of DNA and RNA, we shall confine ourselves here to a consideration of the major

Figure 2.24   DNA nucleotides obey strict rules of base-pairing. Adenine pairs only with thymine, and guanine with cytosine, thus if the sequence of bases in one strand of a DNA molecule is known, that of the other can be predicted. This critical feature of Watson and Crick's model offers an explanation for how DNA is able to replicate itself. Note that GC pairs are held together by three hydrogen bonds, while AT pairs only have two.

differences. There are two important differences in the composition of nucleotides of RNA and DNA. The central sugar molecule of RNA is not deoxyribose, but *ribose*; as shown in Figure 2.20, these differ only in the possession of an H atom or an –OH group attached to carbon-2. Second, although RNA shares three of DNA's nitrogenous bases (A, C and G), instead of thymine it has *uracil*. Like thymine, this can form pairs specifically with adenine.

The final main difference between RNA and DNA is the fact that RNA generally comprises only a single polynucleotide chain, although this may be subject to secondary and tertiary folding as a result of complementary base pairing within the same strand. The roles of the three different forms of RNA will be discussed in Chapter 11.

2.3.5   Lipids

Although lipids can be large molecules, they are not regarded as macro-molecules because unlike proteins, polysaccharides and nucleic acids, they are not polymers of a basic subunit. Moreover, lipids do not share any single structural characteristic; they are a diverse group structurally, but have in common the fact that they are *insoluble in water,* but soluble in a range of organic solvents. This nonpolar nature is due to the predominance of covalent bonding, mainly between atoms of carbon and hydrogen.

Fats are simple lipids, whose structure is based on *fatty acids*. Fatty acids are long hydrocarbon chains ending in a carboxyl (–COOH) group. They have the general formula:

$$CH_3 - (CH_2)_n - COOH$$

where '*n*' is usually an even number. They combine with glycerol according to the basic reaction:

$$Alcohol + Acid \rightarrow Ester$$

The bond so formed is called an ester linkage, and the result is an acylglycerol (Figure 2.25). One, two or all three of the –OH groups may be esterified with a fatty acid, to give respectively mono-, di- and triacylglycerols (or mono-, di- and triglycerides). Natural fats generally contain a mixture of two or three different fatty acids substituted at the three positions; consequently, a considerable diversity is possible among fats. Fats serve as energy stores; a higher proportion of C–C and C–H bonds in comparison with proteins or carbohydrates results in a greater energy-storing capacity.

The second main group of lipids to be found in living cells are *phospholipids*. These have a similar structure to triacylglycerols, except that instead of a third fatty acid chain, they have a phosphate group joined to the glycerol (Figure 2.26), introducing a hydrophilic element to an otherwise hydrophobic molecule. Thus phospholipids are an example of an amphipathic molecule, with a polar region at one end of the molecule and a nonpolar region at the other. This fact is essential for the formation of a bilayer when the phospholipid is introduced into an aqueous environment; the hydrophilic phosphate groups point outwards towards the water, while the hydrophobic hydrocarbon chains 'hide' inside (Figure 2.27, and cf. Figure 2.8, micelle formation).

This bilayer structure forms the basis of all biological membranes (see Chapter 3), forming a barrier around cells and certain organelles. Phospholipids generally have another polar group attached to the phosphate; Figure 2.26 shows the effect of substituting choline.

The structural diversity of lipids can be illustrated by comparing fats and phospholipids with the final group of lipids we need to consider, the steroids. As can be seen from Figure 2.28, these have a completely different form,

Figure 2.25 Acylglycerol formation. Fatty acids are linked to glycerol to form an acylglycerol. When all three –OH groups on the glycerol are esterified, the result is a triacylglycerol or triglyceride. The three fatty acids may or may not be the same. In the example shown, one of the fatty acids is unsaturated (see Box 2.5).

Figure 2.26 Phospholipids. Phospholipids introduce a polar element to acylglycerols by substituting a phosphate at one of the glycerol –OH groups. A second charged group may attach to the phosphate group; the phospholipid shown is phosphatidylcholine.

Box 2.5   Saturated or unsaturated?

You may well have heard of saturated and unsaturated fats in the context of the sorts of foods we should and shouldn't be eating. This terminology derives from the type of fatty acids that make up the different fats.

Each carbon atom in the hydrocarbon chain of a *saturated* fatty acid such as stearic acid is bonded to the maximum possible number of hydrogen atoms (i.e. it is saturated with them).

Fatty acids containing one or more double bonds have fewer hydrogen atoms and are said to be *unsaturated*.

Compare the structures of stearic acid and oleic acid below. Both have identical structures except that oleic acid has two fewer hydrogen atoms and in their place a C=C double bond. A *kink* or bend is introduced into the chain at the point of the double bond; this means that adjacent fatty acids do not pack together so neatly, leading to a drop in the melting point. The presence of unsaturated fatty acids in membrane phospholipids makes the membrane more fluid.

Stearic acid (18.0)
(saturated)

Oleic acid (18.1)
(monounsturated)

Linoleic acid (18.2)
(polyunsaturated)

but still share in common the property of hydrophobicity. The four-ring planar structure is common to all steroids, with the substitution of different side groups producing great differences in function. Cholesterol is an important component of many membranes.

Figure 2.27 Phospholipids can form a bilayer in aqueous surroundings. A 'sandwich' arrangement is achieved by the polar phosphate groups facing outwards and burying the fatty acid chains within. Water is thus excluded from the hydrophobic region, a key property of biological membranes (see Figure 3.5).

General steroid structure   Cholesterol

Figure 2.28 Steroids. All steroids are based on a four-ringed structure. The presence of an –OH group on the lower left ring makes the molecule a *sterol*. Cholesterol plays an important role in the fluidity of animal membranes by interposing itself among the fatty acid tails of phospholipids. The only bacterial group to contain sterols are the mycoplasmas; however, some other groups contain *hopanoids*, which have a similar structure and are thought to play a comparable role in membrane stability.

It would be wrong to gain the impression that living cells contain only molecules of the four groups outlined above. Smaller organic molecules play important roles as precursors or intermediates in metabolic pathways (see Chapter 6), and several inorganic ions such as potassium, sodium and chloride play essential roles in maintaining the living cell. Finally, some macromolecules comprise elements of more than one group, for example, lipopolysaccharides (carbohydrate and lipid) and glycoproteins (protein and carbohydrate).

3
Cell Structure and Organisation

The basic unit of all living things is the cell. The *cell theory* is one of the fundamental concepts of biology; it states that:

- all organisms are made up of cells;
- all cells derive from other, pre-existing cells.

As we shall see in this chapter, there may exist within a cell many smaller, subcellular structures, each with its own characteristics and function, but these are not capable of independent life.

An organism may comprise just a single cell (unicellular), a collection of cells that are not morphologically or functionally differentiated (colonial), or many distinct cell types organised into tissues or organs (multicellular). Among microorganisms, all bacteria and protozoans are unicellular; fungi may be unicellular or multicellular, while algae may exist in all three forms. There is, however, one way that organisms can be differentiated from each other that is even more fundamental than whether they are uni- or multicellular. It is a difference that is greater than that between a lion and a mushroom or an earthworm and an oak tree, and it exists at the level of the individual cell. All organisms are made up of one or other (but definitely not both!) of two very distinct cell types, which we call *prokaryotic* and *eukaryotic* cells, both of which exist in the microbial world. These differ from each other in many ways, including size, structural complexity and organisation of genetic material (Table 3.1).

> The names given to the two cell types derive from Greek words:
> *Prokaryotic* = 'before nucleus'
> *Eukaryotic* = 'true nucleus'

Essential Microbiology, Second Edition. Stuart Hogg.
© 2013 John Wiley & Sons, Ltd. Published 2013 by John Wiley & Sons, Ltd.

Table 3.1   Similarities and differences between prokaryotic and eukaryotic cell structure

Similarities
Cell contents bounded by a *plasma membrane*
Genetic information encoded on *DNA*
Ribosomes act as site of protein synthesis

Differences	
PROKARYOTIC	EUKARYOTIC
Size	
Typically 1–5 microns (μm)	Typically 10–100 μm
Genetic material	
Free in cytoplasm	Contained within a membrane-bound *nucleus*
Single circular chromosome or *nucleoid*	Multiple chromosomes, generally in pairs
Histones absent	DNA complexed with histone proteins
Internal features	
Membrane-bound organelles absent	Several membrane-bound organelles present, including *mitochondria, Golgi body, endoplasmic reticulum* and (in plants and algae) *chloroplasts*
Ribosomes smaller (70S), free in cytoplasm	Ribosomes larger (80S), free in cytoplasm or attached to membranes
Respiratory enzymes bound to plasma membrane	Respiratory enzymes located in mitochondria
Cell wall	
Usually based on *peptidoglycan* (not Archaea)	When present, based on cellulose or chitin
External features	
Cilia absent	Cilia may be present
Flagella, if present, composed of flagellin. Provide rotating motility	Flagella, if present, have complex $(9+2)$ structure. Provide 'whiplash' motility
Pili may be present	Pili absent
Outside layer (slime layer, capsule, glycocalyx) present in some types	Pellicle or test present in some types

The most fundamental difference between prokaryotic and eukaryotic cells is reflected in their names; eukaryotic cells possess a true *nucleus*, and several other distinct subcellular *organelles* that are bounded by a membrane. Prokaryotes have no such organelles. Most of these differences only became apparent after the development of the electron microscopy techniques described in Chapter 1.

Table 3.2 Principal groups of prokaryotic and eukaryotic organisms

Prokaryotes	Eukaryotes
Archaea	Fungi
Bacteria	Algae
	Protozoa
	Plants
	Animals

As can be seen from Table 3.2, the prokaryotes comprise the simpler and more primitive types of microorganisms; they are generally single-celled, and arose much earlier in evolutionary history than the eukaryotes. Indeed, as discussed later in this chapter, it is widely accepted that eukaryotic cells actually developed from their more primitive counterparts. Note that the viruses do not appear in Table 3.2, because they do not have a cellular structure at all, and are incapable of independent replication. They are not therefore considered to be living organisms. (See Chapter 10 for a fuller discussion of the viruses.)

Modern nucleic acid sequencing methods used to determine *phylogenetic* relationships between organisms have revealed that within the prokaryotes there is another fundamental division. One group of bacteria were shown to differ greatly from all the others; we now call these the *Archaea*, to differentiate them from the true Bacteria. These two groups, together with the eukaryotes, are thought to have evolved from a common ancestor, and represent the three *domains* of life (Figure 3.1). The Archaea comprise a wide range of mostly anaerobic bacteria, including many of those that inhabit extreme environments such as hot springs. In this book we shall largely confine our discussions to the Bacteria; however, in Chapter 7 there is a discussion of the principal features of the Archaea and their main taxonomic groupings.

Phylogenetic: relating to the evolutionary relationship between organisms.

Despite their differences, *Archaea* and *Bacteria* are both prokaryotes.

Taxonomy is the science of classifying living (and once-living) organisms.

3.1 The prokaryotic cell

Bacteria are much smaller than eukaryotic cells; most fall into a size range of about 1–5 μm, although some may be larger than this. Some of the smallest bacteria, such as the Mollicutes (mycoplasmas), measure less than 1 μm, and are too small to be resolved clearly by an ordinary light microscope. Because of their extremely small size, it was only with the advent

Figure 3.1 The three domains of life. All life forms can be assigned to one of three domains on the basis of their ribosomal RNA sequences. The Archaea are quite distinct from the true bacteria and are thought to have diverged from a common ancestral line at a very early stage, before the evolution of eukaryotic organisms. The scheme illustrated is the one most widely accepted by microbiologists, but alternative models have been proposed.

of the electron microscope that we were able to learn about the detailed structure of bacterial cells. Using the light microscope, however, it is possible to recognise differences in the shape and arrangement of bacteria. Although a good deal of variation is possible, most have one of three basic shapes (Figure 3.2):

> In recent years, square, triangular and star-shaped bacteria have all been discovered!

- rod-shaped (bacillus; pl. bacilli);
- spherical (coccus; pl. cocci);
- curved: these range from comma-shaped (vibrio) to corkscrew-shaped (spirochaete).

All these shapes confer certain advantages to their owners; rods, with a large surface area, are better able to take up nutrients from the environment, while the cocci are less prone to drying out. The spiral forms are usually motile; their shape aids their movement through an aqueous medium.

As well as these characteristic cell shapes, bacteria may also be found grouped together in particular formations. When they divide, they may remain attached to one another, and the shape the groups of cells assume reflects the way the cell divides. Cocci, for example, are frequently found as chains of cells, a reflection of repeated division in one plane (Figure 3.2). Other cocci may form regular sheets or packets of cells, as a result of division

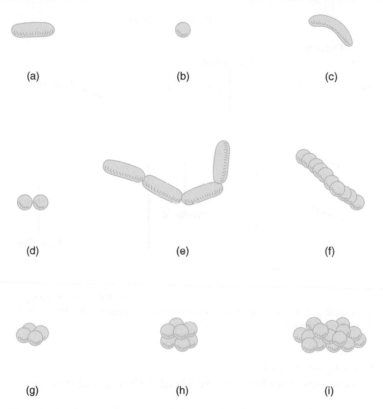

Figure 3.2 Bacterial shapes. Most bacteria are (a) rod-shaped, (b) spherical or (c) curved. These basic shapes may join to form (d) pairs, (e and f) chains, (g) sheets, (h) packets and (i) irregular aggregates.

in two or three planes. Yet others, such as the staphylococci, divide in several planes, producing the irregular and characteristic 'bunch of grapes' appearance. Rod-shaped bacteria only divide in a single plane and may therefore be found in chains, while spiral forms also divide in one plane, but tend not to stick together. Blue-greens form filaments; these are regarded as truly multicellular rather than as a loose association of individuals.

3.1.1 Prokaryotic cell structure

When compared with the profusion of elaborate organelles encountered inside a typical eukaryotic cell, the interior of a typical bacterium looks rather empty. The only internal structural features are:

- a bacterial chromosome or *nucleoid*, comprising a closed loop of double-stranded, supercoiled DNA; there may also be additional DNA in the form of *plasmids*;

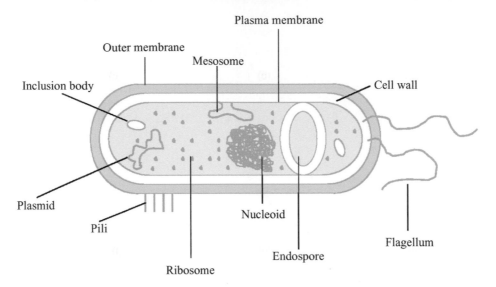

Figure 3.3 Structure of a generalised prokaryotic cell. Note the lack of complex internal organelles (cf. Figure 3.13). Gram-positive and Gram-negative bacteria differ in the details of their cell wall structure (see Figures 8.8 and 8.9).

- thousands of granular *ribosomes*;
- a variety of granular *inclusions* associated with nutrient storage.

All of these are contained in a thick aqueous soup of carbohydrates, proteins, lipids and inorganic salts known as the *cytoplasm*, which is surrounded by a *plasma membrane*. This in turn is wrapped in a *cell wall,* whose rigidity gives each bacterial cell its characteristic shape. Depending on the type of bacterium, there may be a further surrounding layer such as a *capsule* or *slime layer* and/or structures external to the cell associated with motility (*flagella*) or attachment (*pili/fimbriae*). Figure 3.3 shows these features in a generalised bacterial cell. In the following pages we shall examine these features in a little more detail, noting how each has a crucial role to play in the survival or reproduction of the cell.

3.1.2   Genetic material

Although it occupies a well-defined area within the cell, the genetic material of prokaryotes is not present as a true nucleus, as it lacks a surrounding nuclear membrane (cf. the eukaryotic nucleus, Figure 3.13). The *nucleoid* or bacterial chromosome comprises a closed circle of double-stranded DNA, many times the length

Not all bacteria conform to the model of a single circular chromosome; some have been shown to possess two with genes shared between them, while examples of linear chromosomes are also known.

of the cell and supercoiled to form a highly compact structure. (The common laboratory bacterium *E. coli* is around 3–4 μm in length, but contains a DNA molecule some 1400 μm in length!) The DNA may be associated with certain bacterial proteins, but these are not the same as the histones found in eukaryotic chromosomes.

Some bacteria contain additional DNA in the form of small, self-replicating extrachromosomal elements called *plasmids*. These do not carry any genes essential for growth and reproduction, and thus the cell may survive without them. They can be very important, however, as they may carry genes encoding toxins or resistance to antibiotics, and can be passed from cell to cell (see Chapter 12). Under certain environmental conditions, possession of a plasmid may confer an advantage to a bacterial cell.

3.1.3 Ribosomes

Apart from the nucleoid, the principal internal structures of prokaryotic cells are the ribosomes. There may be many thousands of these in an active cell, lending a speckled appearance to the cytoplasm. Ribosomes are composed of a complex of protein and RNA, and are the site of protein synthesis in the cell.

Although they carry out a similar function, the ribosomes of prokaryotic cells are smaller and lighter than their eukaryotic counterparts. Ribosomes are measured in Svedberg units (S), a function of their size and shape, and determined by their rate of sedimentation in a centrifuge; prokaryotic ribosomes are 70S, while those of eukaryotes are 80S. Some types of antibiotic exploit this difference by targeting the prokaryotic form and selectively disrupting bacterial protein synthesis (see Chapter 17).

All ribosomes comprise two unequal subunits (in prokaryotes, these are 50S and 30S, in eukaryotes 60S and 40S: Table 3.3). Each subunit contains its own RNA and a number of proteins (Figure 3.4). Many ribosomes may be attached simultaneously to a single mRNA molecule, forming a threadlike polysome. The role of ribosomes in bacterial protein synthesis is discussed in Chapter 11.

> A *polysome* is a chain of ribosomes attached to the same molecule of mRNA.

Table 3.3 Comparison of prokaryotic and eukaryotic ribosomes

	Prokaryotic	Eukaryotic
Overall size	70S	80S
Large subunit size	50S	60S
Large subunit RNA	23S and 5S	28S, 5.8S and 5S
Small subunit size	30S	40S
Small subunit RNA	16S	18S

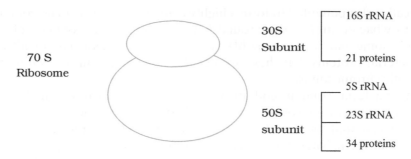

Figure 3.4   The bacterial ribosome. Each subunit comprises ribosomal RNA (rRNA) and proteins. The nucleotide sequence of small subunit (16S) rRNA is widely used in determining the phylogenetic (evolutionary) relationship between bacteria (see Chapter 7).

3.1.4   Inclusion bodies

Certain bacteria may contain within their cytoplasm granular structures known as inclusion bodies. These act as food reserves, and may contain organic compounds such as starch, glycogen or lipid. In addition, sulphur and polyphosphate can be stored as inclusion bodies, the latter being known as *volutin* or metachromatic granules. Two special types of inclusion body are worthy of mention. *Magnetosomes*, which contain a form of iron oxide, help some types of bacteria to orientate themselves downwards into favourable conditions, whilst *gas vacuoles* maintain buoyancy of the cell in blue-greens and some halobacteria.

3.1.5   Endospores

Certain bacteria such as *Bacillus* and *Clostridium* produce *endospores*. They are dormant forms of the cell that are highly resistant to extremes of temperature, pH and other environmental factors, and germinate into new bacterial cells when conditions become more favourable. The spore's resistance is due to the thick coat that surrounds it.

Endospores of pathogens such as *Clostridium botulinum* can resist boiling for several hours. It is this resistance that makes it necessary to autoclave at 121°C in order to ensure complete sterility.

3.1.6   The plasma membrane

The cytoplasm and its contents are surrounded by a plasma membrane, which can be thought of as a bilayer of phospholipid arranged like a sandwich, together with associated proteins (Figure 3.5).

Figure 3.5   The plasma membrane. Phospholipid molecules form a bilayer, with the hydrophobic hydrocarbon chains pointing in towards each other, leaving the hydrophilic phosphate groups to face outwards. Proteins embedded in the membrane are known as integral proteins, and may pass part of the way or all of the way through the phospholipid bilayer. The amino acid composition of such proteins reflects their location; the part actually embedded among the lipid component of the membrane comprises non-polar (hydrophobic) amino acids, while polar ones are found in the aqueous environment at either side. Reproduced from Singleton, P (1999) Bacteria in Biology, Biotechnology and Medicine, 5th edn, with permission from John Wiley & Sons.

The function of the plasma membrane is to keep the contents in, while at the same time allowing the selective passage of certain substances in and out of the cell (it is a semipermeable membrane).

Phospholipids comprise a compact, hydrophilic (water-loving) head and a long hydrophobic tail region (see Figure 2.27); this results in a highly ordered structure when the membrane is surrounded by water. The tails 'hide' from the water to form the inside of the membrane, while the heads project outwards. Also included in the membrane are a variety of proteins; these may pass right through the bilayer or be associated with the inner (cytoplasmic) or outer surface only. These proteins may play structural or functional roles in the life of the cell. Many enzymes associated with the metabolism of nutrients and the production of energy are associated with the plasma membrane in prokaryotes. As we'll see later in this chapter, this is fundamentally different from eukaryotic cells, where these reactions are carried out at specialised internal organelles. Proteins involved in the active transport of nutrients (see Chapter 4) are also to be found associated with the plasma membrane. The model of membrane structure as depicted in Figure 3.5 must not be thought of as static; in the widely accepted *fluid mosaic model*, the lipid is seen as a fluid state, in which proteins float around, rather like icebergs in an ocean.

The majority of bacterial membranes do not contain sterols (cf. eukaryotes: see later); however, many do contain molecules called hopanoids that are derived from the same precursors. Like sterols, they are thought to assist in maintaining membrane stability.

A comparison of the lipid components of plasma membranes reveals a distinct difference between members of the Archaea and the Bacteria.

3.1.7   The bacterial cell wall

Bacteria have a thick, rigid cell wall, which maintains the integrity of the cell and determines its characteristic shape. Since their cytoplasm contains high concentrations of dissolved substances, bacteria generally live in a hypotonic environment (i.e. one that is more dilute than their own cytoplasm). There is therefore a natural tendency for water to flow into the cell, and without the

> A *protoplast* is a cell that has had its cell wall removed.

cell wall the cell would fill and burst (this can be demonstrated by using enzymes to strip off the cell wall, leaving the naked *protoplast*).

The major component of the cell wall, which is responsible for its rigidity, is a substance unique to bacteria, called *peptidoglycan* (murein). This is a high molecular weight polymer whose basic subunit is made up of three parts: *N*-acetylglucosamine, *N*-acetylmuramic acid and a short peptide chain (Figure 3.6). The latter comprises the amino acids L-alanine, D-alanine, D-glutamic acid and either L-lysine or diaminopimelic acid (DAP). DAP is a rare amino acid, only found in the cell walls of prokaryotes. Note that some of the amino acids of peptidoglycan are found in the D-configuration. This is contrary to the situation in proteins, as you may recall from Chapter 2, and confers protection against proteases whose action is specifically directed against L-amino acids.

Precursor molecules for peptidoglycan are synthesised inside the cell, and transported across the plasma membrane by a carrier called bactoprenol phosphate before being incorporated into the cell wall structure. Enzymes called transpeptidases then covalently bond the tetrapeptide chains to one another, giving rise to a complex network (Figure 3.7); it is this cross-linking that gives the wall its mechanical strength. A number of antimicrobial agents exert their effect by inhibiting cell wall synthesis; these will be discussed further in Chapter 17.

Although all bacteria (with a few exceptions) have a cell wall containing peptidoglycan, there are two distinct structural types of cell wall, known as *Gram-positive* and *Gram-negative*. The names derive from the Danish scientist Christian Gram, who, in the 1880s, developed a rapid staining technique that could differentiate bacteria as belonging to one of two basic types (see Box 1.2). Although the usefulness of the Gram stain was recognised for many years, it was only with the advent of electron microscopy that the underlying molecular basis of the test could be explained, in terms of cell wall structure.

Gram-positive cell walls are relatively simple in structure, comprising several layers of peptidoglycan connected to each other by cross-linkages to form a strong, rigid scaffolding. In addition, they contain acidic polysaccharides

Figure 3.6 Peptidoglycan structure. **Peptidoglycan is a polymer made up of alternating molecules of *N*-acetylglucosamine (NAG) and *N*-acetylmuramic acid (NAM). A short peptide chain is linked to the NAM residues (see text for details). This is important in the cross-linking of the straight chain polymers to form a rigid network (see Figure 3.7). The composition of *E. coli* peptidoglycan is shown; the peptide chain may contain different amino acids in other bacteria. Partly reproduced from Hardy, SP (2002) Human Microbiology, with permission from Taylor & Francis Group.**

Figure 3.7 Cross-linking of peptidoglycan chains in *E. coli*. (a) The D-alanine on the short peptide chain attached to the *N*-acetylmuramic acid cross-links to a diaminopimelic acid residue on another chain. In other bacteria, the precise nature of the cross-linking may differ. (b) Further cross-linking produces a rigid network of peptidoglycan. The antibiotic penicillin acts by inhibiting the transpeptidase enzymes responsible for the cross-linking reaction (see Chapter 17). Reproduced from Hardy, SP (2002) Human Microbiology, with permission from Taylor & Francis Group.

called teichoic acids; these contain phosphate groups, which impart an overall negative charge to the cell surface. A diagram of the Gram-positive cell wall is shown in Figure 3.8.

   Gram-negative cells have a much thinner layer of peptidoglycan, making the wall less sturdy; however, the structure is made more complex by the presence of a layer of lipoprotein, polysaccharide and phospholipid known as the *outer membrane* (Figure 3.9). This misleading name derives from the fact that it superficially resembles the bilayer of the plasma membrane; however, instead of two layers of phospholipid, it has only one, the outer layer being made up of *lipopolysaccharide*. This has three parts: lipid A, core

Figure 3.8 The Gram-positive cell wall. Peptidoglycan is many layers thick in the Gram-positive cell wall and may account for 30–70% of its dry weight. Teichoic acids are negatively charged polysaccharides; they are polymers of ribitol phosphate and cross-link to peptidoglycan. *Lipoteichoic acids* are teichoic acids found in association with glycolipids. Reproduced from Henderson, B, et al. (1999) Cellular Microbiology: Bacteria-Host Interactions in Health and Disease, with permission from John Wiley & Sons.

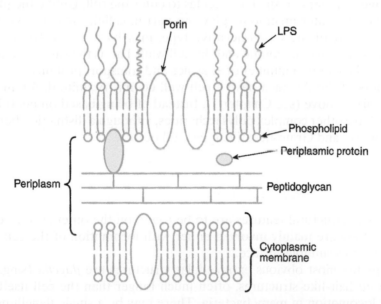

Figure 3.9 The Gram-negative cell wall. Note the thinner layer of peptidoglycan compared to the Gram-positive cell wall (Figure 3.8). It accounts for <10% of the dry weight. Beyond this lies the outer membrane, with its high lipopolysaccharide (LPS) content. Channels made of *porins* allow the passage of certain solutes into the cell. Reproduced from Henderson, B, et al. (1999) Cellular Microbiology: Bacteria-Host Interactions in Health and Disease, with permission from John Wiley & Sons.

Box 3.1   Mesosomes – the structures that never were?

When looked at under the electron microscope, Gram-positive bacteria often contained localised infoldings of the plasma membrane. These were given the name *mesosomes*, and were thought by some to act as attachment points for DNA during cell division, or to play a role in the formation of cross-walls. Others thought they were nothing more than artefacts produced by the rather elaborate sample preparation procedures necessary for electron microscopy. It is now generally accepted that this is the case.

---

polysaccharide and an O-specific side chain. The lipid A component may act as an *endotoxin*, which, if released into the bloodstream, can lead to serious conditions such as fever and toxic shock. The O-specific antigens are carbohydrate chains whose composition often varies between strains of the same species. Serological methods can distinguish between these, making them a valuable tool in the investigation, for example, of the origin of an outbreak of an infectious disease. Proteins called *porins* incorporated into the outer membrane and penetrating its entire thickness, form channels, which allow the passage of water and small molecules to enter the cell. Unlike the plasma membrane, the outer membrane plays no part in cellular respiration. Some bacteria, mostly of the Gram-negative type, may have a *periplasmic space* between the plasma membrane and the cell wall. This is the site of metabolic activity, and contains a number of enzymes and transport proteins.

Members of the Archaea have a cell wall chemistry quite different from that described above (see Chapter 7). Instead of being based on peptidoglycan, they have other complex polysaccharides, although a distinction between Gram-positive and Gram-negative types still occurs.

3.1.8   Beyond the cell wall

A number of structural features are to be found on the outer surface of the cell; these are mainly involved either with locomotion of the cell or its attachment to a suitable surface.

Perhaps the most obvious extracellular structures are *flagella* (sing. *flagellum*), thin hair-like structures, often much longer than the cell itself, and used for locomotion in many bacteria. There may be a single flagellum, one at each end, or many, depending on the bacteria concerned (Figure 3.10). Each flagellum is a hollow but rigid cylindrical filament made of the protein flagellin, attached via a hook to a basal body, which secures it to the cell wall and plasma membrane (Figure 3.11). The basal body comprises a

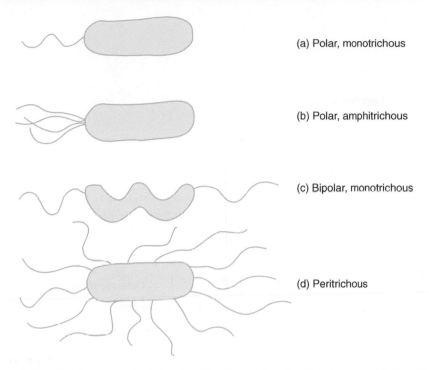

(a) Polar, monotrichous

(b) Polar, amphitrichous

(c) Bipolar, monotrichous

(d) Peritrichous

Figure 3.10   The arrangement of flagella. Flagella may be situated at one end (a and b), at both ends (c) or all over the cell surface (d).

series of rings, and is more complex in Gram-negative than in Gram-positive bacteria. Rotation of the flagellum, which may reach several hundred rpm, is an energy-dependent process driven by the basal body, and the direction of rotation determines the nature of the resulting cellular movement. Clockwise rotation of a single flagellum results in a directionless 'tumbling', but if it rotates anticlockwise, the bacterium will 'run' in a straight line (Figure 3.12a). Likewise, anticlockwise rotation causes bunched flagella to 'run' by winding around each other and acting as a single structure, whilst spinning in the opposite direction gives rise to multiple independent rotations and results in tumbling (Figure 3.12b).

Pili (sing. *pilus*) are structures that superficially resemble short flagella. They differ from flagella, however, in that they do not penetrate to the plasma membrane, and they are not associated with motility. Their function, rather, is to anchor the bacterium to an appropriate surface. Pathogenic bacteria have proteins called *adhesins* on their pili, which adhere to specific receptors on host tissues. Attachment pili are sometimes called *fimbriae*, to distinguish them from another distinct type of pilus, the *sex pilus*, which as its name suggests, is involved in the transfer of genetic information by *conjugation*. This is discussed in more detail in Chapter 11.

Flagellum

H+   H+

Switch complex

Outer membrane

Cell wall

Inner membrane

Figure 3.11   Bacterial flagella are anchored in the cell wall and plasma membrane. The filament of the flagellum is anchored by a basal body. In Gram-positive organisms, this comprises two rings inserted in the plasma membrane. In Gram-negative organisms (as shown), there are additional rings associated with the outer membrane and the peptidoglycan layer. Energy for rotation of the flagellum is derived from the proton motive force generated by the movement of protons across a membrane (see Chapter 6). Reproduced from Bolsover, SR, et al. (1997) From Genes to Cells, with permission from John Wiley & Sons.

Outside the cell wall, most bacteria have a polysaccharide layer called a glycocalyx. This may be a diffuse and loosely bound *slime layer*, or a better defined and generally thicker *capsule*. The slime layer helps protect against desiccation, and is instrumental in the attachment of certain bacteria to a substratum (the bacteria that stick to your teeth are a good example of this). Capsules offer protection to certain pathogenic bacteria against the phago-cytic cells of the immune system. Both capsules and slime layers are key com-ponents of *biofilms*, which form at liquid–solid interfaces, and can be highly significant in such varied settings as wastewater treatment systems, indwelling catheters and the inside of your mouth!

3.2   The eukaryotic cell

We have already seen that eukaryotic cells are, for the most part, larger and much more complex than prokaryotes, containing a range of specialised subcellular organelles (Figure 3.13). Within the microbial world, the major

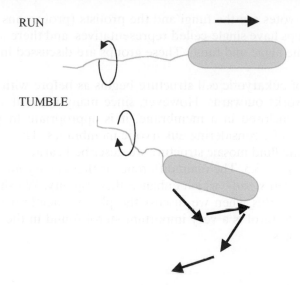

RUN

TUMBLE

Figure 3.12 Running and tumbling. **Anticlockwise rotation of bunched flagella gives rise to 'running' in a set direction. Reversing the direction of rotation causes 'tumbling', and allows the bacterial cell to change direction.**

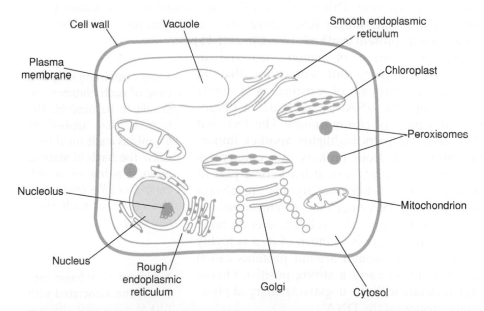

Figure 3.13 The eukaryotic cell. **This example of eukaryotic cell structure shows a plant cell. Other eukaryotic cells may differ with respect to the cell wall and the possession of chloroplasts. Note the much more elaborate internal structure compared to a typical prokaryotic cell (cf. Figure 3.3), in particular the presence of membrane-bounded organelles such as mitochondria, chloroplasts, endoplasmic reticulum and a true nucleus. Reproduced from Nicklin, J, et al. (2002) Instant Notes in Microbiology, 2nd edn, with permission from Taylor & Francis Group.**

groups of eukaryotes are the fungi and the protists (protozoans and algae); all of these groups have single-celled representatives, and there are multicellular forms in the algae and fungi. These groups are discussed in Chapters 8 and 9.

Our survey of eukaryotic cell structure begins as before with the genetic material, and works outwards. However, since many internal structures in eukaryotes are enclosed in a membrane, it is appropriate to preface our description by briefly considering eukaryotic membranes. These are, in fact, very similar to the fluid mosaic structure we described earlier in this chapter, as depicted in Figure 3.5. The main difference is that eukaryotic membranes contain lipids called *sterols*, which enhance their rigidity. We shall consider the significance of this when we discuss the plasma membrane of eukaryotes below. Cholesterol is a very important sterol found in the membranes of many eukaryotes.

3.2.1   The nucleus

The principal difference between prokaryotic and eukaryotic cells, and the one that gives the two forms their names, lies in the accommodation of their genetic material. Eukaryotic cells have a true nucleus, surrounded by a *nuclear membrane*. This is in fact a double membrane; it contains pores, through which messenger RNA leaves the nucleus on its way to the ribosomes during protein synthesis (see Chapter 11).

The organisation of genetic material in eukaryotes is very different from that in prokaryotes. Instead of existing as a single closed loop, the DNA of eukaryotes is organised into pairs of *chromosomes*. The fact that they occur in pairs highlights another important difference from prokaryotes: eukaryotes are genetically *diploid* in at least some part of their life cycle, while prokaryotes are always *haploid*. The DNA of eukaryotic chromosomes is linear in the sense that it has free ends; however, because there is so much of it, it is highly condensed and wound around proteins called *histones*. These carry a strong positive charge and associate with the negatively charged phosphate groups on the DNA.

> A cell containing only one copy of each chromosome is said to be *haploid*. The term is also applied to organisms made up of such cells. The haploid state is often denoted as *N* (cf. *diploid,* or *2N* – containing two copies of each chromosome).

> A *histone* is a basic protein found associated with DNA in eukaryotic chromosomes.

As well as the chromosomes, the nucleus also contains the *nucleolus*, a discrete structure rich in RNA, where ribosomes are assembled. The ribosomes themselves have the same function as their prokaryotic counterparts; the differences in size have already been discussed (see Table 3.3). They may be found free in the cytoplasm or associated with the endoplasmic reticulum (see below), depending on the type of protein they synthesise.

3.2.2 Endoplasmic reticulum

Running throughout the cell and taking up much of its volume, the endoplasmic reticulum (ER) is a complex membrane system of tubes and flattened sacs (*cisternae*). The presence of numerous ribosomes on their surface gives those parts of the ER involved in protein synthesis a granular appearance when seen under the electron microscope, giving rise to the name *rough ER*. Areas of the ER not associated with ribosomes are known as *smooth ER*; this is where the synthesis of membrane lipids takes place. The ER also serves as a communications network, allowing the transport of materials between different parts of the cell.

3.2.3 Golgi apparatus

The Golgi apparatus is another membranous organelle, comprising a set of flattened vesicles, usually arranged in a stack called a *dictyosome*. The function of the Golgi apparatus is to package newly synthesised substances such as proteins and assist in their transport away from the cell. The substances are contained in vesicles that are released from the main part of the complex, and fuse with the cytoplasmic membrane. The Golgi apparatus is poorly defined in certain fungi and protozoans.

3.2.4 Lysosomes

Another function of the Golgi apparatus is to package certain hydrolytic (digestive) enzymes into membrane-bound packets called *lysosomes*. The enzymes, which are needed to digest nutrient molecules that enter the cell by *endocytosis* (Figure 3.14), would break down the fabric of the cell itself if they were not contained within the lysosomes.

Peroxisomes are similar to lysosomes, but smaller, and also contain enzymes such as catalase, which breaks down the potentially toxic hydrogen peroxide generated by other breakdown reactions within the peroxisome.

3.2.5 Mitochondria

Whereas in prokaryotes the reactions involved in energy generation are associated with the plasma membrane, in eukaryotes they take place in specialised organelles called mitochondria. These are generally cigar-shaped and may be present in large numbers. They are enclosed by a double membrane, the inner surface of which is folded into finger-like projections called *cristae*. Respiratory enzymes are located on the increased surface area this provides, while

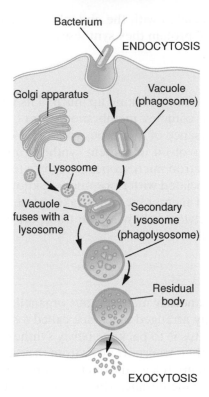

Figure 3.14   Endocytosis. Membrane-bound vacuoles surround a food particle and internalise it in the form of a *phagosome*. This fuses with a lysosome, which releases digestive enzymes, resulting in the breakdown of the contents. The process of endocytosis is unique to eukaryotic cells. The products of digestion are released from the cell by exocytosis. Reproduced from Black, JG (1999) Microbiology: Principles and Explorations, 4th edn, with permission from John Wiley & Sons.

other metabolic reactions take place in the semi-fluid matrix (Figure 3.15) (see also Chapter 6). The matrix of mitochondria contains a range of enzymes involved in oxidative metabolism as well as mitochondrial ribosomes. These are very similar to the ribosomes found in prokaryotic cells. The matrix also contains a small amount of DNA; this is typically circular, like the nucleoid of bacteria, and contains a limited number of genes. These mostly carry the code for proteins involved with the respiratory chain, and ribosomal and transfer RNAs.

   The mitochondrial cristae of algae, fungi and protozoans each have their own characteristic shapes. Until very recently, a few primitive protozoans, such as *Giardia*, were thought to lack mitochondria completely, and to represent an intermediate stage in the evolution of the eukaryotic condition. Recent research, however, has shown them to possess highly reduced remnants of mitochondria, which have been given the name *mitosomes*. It

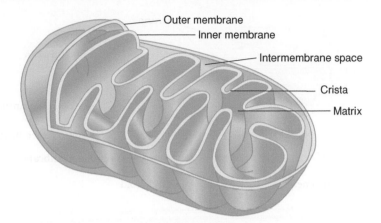

Figure 3.15 *Mitochondrial structure.* The inner membrane, the location of the electron transport chain in aerobic respiration, is formed by the invagination of the more permeable outer membrane. Mitochondria have similar dimensions to many bacteria (approx. 1–3 μm), but may vary in shape due to the plasticity of their membranes.

seems that such organisms did, after all, once possess mitochondria, but have subsequently lost much of their function – an example of so-called reductive evolution.

3.2.6 Chloroplasts

Chloroplasts are specialised organelles involved in the process of *photosyn thesis*, the conversion of light into cellular energy. As such, they are characteristic of green plants and algae. Like mitochondria, chloroplasts are surrounded by a double membrane, and serve as the location for energy-generating reactions. Inside the chloroplast are flattened membranous sacs known as *thylakoids*, which contain the photosynthetic pigment *chlorophyll.* Thylakoids are arranged in stacks called *grana* (Figure 3.16).

Mitochondria and chloroplasts both contain 70S ribosomes (similar to those found in prokaryotes), a limited amount of circular DNA and the means to replicate themselves. This is seen as key evidence for the *endosymbiotic theory* of eukaryotic evolution, which envisages that specialised organelles within eukaryotic cells arose from the ingestion of small prokaryotes, which over a long period of time lost their independent existence.

3.2.7 Vacuoles

Vacuoles are membrane-covered spaces within cells, and derive from the Golgi apparatus. They act as stores for various nutrients, and also for waste products. Some types of vacuole are important in regulating the water content of the cell.

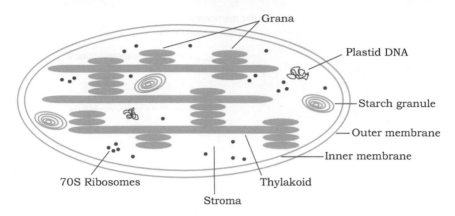

Figure 3.16 Chloroplast structure. Generation of ATP (adenosine triphosphate) from photosynthesis occurs on the thylakoid membranes. In green algae these take the form of discrete structures called grana. The enzyme ribulose bisphosphate carboxylase, responsible for fixing carbon dioxide via the Calvin cycle (see Chapter 6) is located in the stroma. The outer membrane of chloroplasts is relatively permeable, allowing the diffusion of the products of photosynthesis into the surrounding cytoplasm.

3.2.8   Plasma membrane

Many eukaryotes do not have cell walls, so the plasma membrane represents the outermost layer of the cell. The sterols mentioned earlier are important in helping these cells to resist the effects of osmotic pressure. The only prokaryotes to contain sterols are the Mollicutes, which are unusual in not possessing the typical bacterial cell wall. Although the eukaryotic plasma membrane does not have the role in cellular respiration associated with its prokaryotic counterpart, it does have additional functions. The process of *endocytosis* (and its reverse, *exocytosis*), by which particles or large soluble molecules are enveloped and brought into the cell, takes place at the plasma membrane. Also, carbohydrate residues in the membrane act as receptors for cell-to-cell recognition, and may be involved in cell adhesion.

3.2.9   Cell wall

As we have just noted, not all eukaryotes possess a cell wall; among those that do are fungi, algae and plants. Whilst the function, like that of prokaryotes, is to give strength to the cell, the chemical composition is very different, generally being a good deal simpler. The cell walls of plants, algae and lower members of the fungi are based on *cellulose* (Figure 3.17a), a repeating chain of glucose molecules joined by β-1,4 linkages, and may also include pectin and hemicellulose, both also polymers of simple sugars. Many fungi such as yeasts and mushrooms contain *chitin*, a polymer of *N*-acetylglucosamine (Figure 3.17b) We have encountered *N*-acetylglucosamine before, as a

(a)

(b)

Figure 3.17 Cellulose and chitin. The structures of (a) cellulose and (b) chitin. Cellulose is composed of repeating glucose units joined by β-1,4 linkages, and chitin is a polymer of N-acetylglucosamine.

component of peptidoglycan in bacterial walls. Chitin is also to be found as the major component of insect and crustacean exoskeletons, where the function is also to provide strength and rigidity. As in prokaryotes, the cell wall plays little part in the exchange of materials between the cell and its environment, a role fulfilled by the plasma membrane.

Some protozoans and unicellular algae are surrounded by a flexible *pellicle* made of protein.

3.2.10 Flagella and cilia

Motility in eukaryotic cells may be achieved by means of flagella or *cilia*; cilia can be thought of as, essentially, short flagella. Both are enclosed within the plasma membrane and anchored by means of a basal body. Flagellated cells generally have a single flagellum, whereas cilia are often present in very large numbers on each cell. In the microbial world, flagella are found in protozoans and motile algal forms, whilst cilia are mostly found in a class of protozoans called the Ciliophora. Flagella and cilia are not found in members of the Fungi.

Although they share the same thread-like gross morphology, eukaryotic flagella differ considerably from those of prokaryotes in their ultrastructure. Seen in cross-section, they have a very characteristic appearance, made up of two central *microtubules*, surrounded by a further nine pairs arranged in a circle (Figure 3.18). The microtubules are made of a protein called *tubulin*. Flagella in eukaryotes beat in waves, rather than rotating; cilia, which are present in large numbers, beat in a coordinated fashion so that some are in forward motion while others are in the recovery stroke (rather like a 'Mexican wave'!). In animals, ciliary motion has been adapted to move

Figure 3.18 Eukaryotic flagella have a characteristic '9 + 2' structure. **Although functionally analogous to their prokaryotic counterparts, eukaryotic flagella differ appreciably in their fine structure. A membrane surrounds an arrangement of proteinaceous microtubules, in which nine pairs surround a single central pair. Movement of eukaryotic flagella is by means of an ATP-driven whiplike motion.**

particulate matter across a tissue surface; ciliated cells of the respiratory tract, for example, act as a first line of defence in the removal of inhaled particles, such as bacteria from the airways.

3.3   Cell division in prokaryotes and eukaryotes

In unicellular prokaryotes, cell division by *binary fission* leads to the creation of new individuals. Growth occurs in individual cells until a maximum size is achieved, and a cross-wall is formed. Before cell division takes place, the genetic material must replicate itself (see Chapter 11), and one copy pass to each new daughter cell (Figure 3.19).

Cell division in eukaryotes also results in two identical daughter cells. In the case of unicellular eukaryotes, this results in two individual organisms (asexual reproduction), while in multicellular forms there is an increase in overall size. Cell division is preceded by a process of nuclear division called *mitosis*, which ensures that both daughter cells receive a full complement of chromosomes. The principal phases of mitosis are summarised in Figure 3.20a. In *interphase*, the chromosomes are not clearly visible under the microscope; DNA replication takes place during this period. The duplicated chromosomes, held together as sister *chromatids* by the centromere, move towards the centre of the cell during *prophase*. A series of microtubules form a spindle between the centrioles, and the chromosomes line up along this during *metaphase*. By this time, the nuclear membrane has broken down, allowing the paired sister chromatids to separate and migrate away from the centre to opposite ends of the spindle. This stage is called *anaphase*. Finally, in *telophase*, new nuclear membranes surround the two sets of chromosomes,

DNA replicates, making a
second copy of the chromosome
Origins of replication migrate to
ends of cell

Cell lengthens and new
cell wall is laid down
Plasma membrane
starts to grow inwards

Septum formation is
complete and daughter
cells separate

Figure 3.19   Binary fission in *E. coli.* Replication of the single circular chromosome is accompanied by an increase in cell size. The plasma membrane invaginates, and a new cross-wall is synthesised, resulting in two new daughter cells.

to form two nuclei. Mitosis is followed by cell division. Overall, the process of mitosis results in two identical nuclei containing the original (diploid) chromosome number.

At various stages of eukaryotic life cycles, a process of *meiosis* may occur, which halves the total number of chromosomes, so that each nucleus only contains one copy of each. In sexual reproduction, the haploid gametes are formed in this way, and the diploid condition is restored when two different gametes fuse. In some eukaryotes, not only the gametes but a substantial part of the life cycle may occur in the haploid form (see Chapters 8 and 9). Meiosis (Figure 3.20b) comprises two nuclear divisions, the second of which is very similar to the process of mitosis just described. In the first meiotic division, following DNA replication, homologous chromosomes (i.e. the two members of each pair) line up on the spindle together and eventually migrate to opposite poles. While they are together, it is possible for *crossing over* to occur, a process by which the two members of a chromosome pair swap homologous stretches of DNA (Figure 3.21). Since these may not be

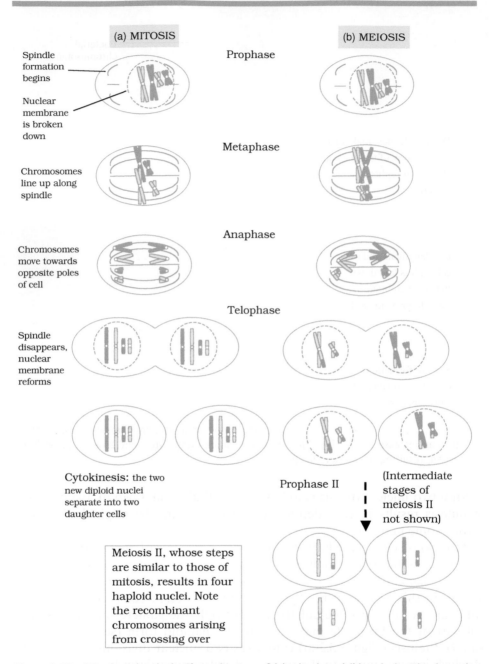

Figure 3.20 Mitosis and meiosis. The main steps of (a) mitosis and (b) meiosis. Mitosis results in two cells identical to the parent. Meiosis results in a reduction in the chromosome number and introduces genetic variation by means of crossing over. For details see the text.

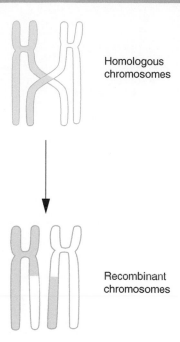

Homologous
chromosomes

Recombinant
chromosomes

Figure 3.21 Crossing over leads to recombination of genetic material. During crossing over, portions of homologous chromosomes are exchanged. This forms the basis of genetic recombination in eukaryotes, and ensures that offspring contain new combinations of genetic material.

identical in DNA sequence, crossing over serves to introduce genetic variation into the daughter nuclei. In the second meiotic division, sister chromatids separate as before, resulting in four haploid nuclei.

Meiosis ensures that all gametes are subtly different from each other and is a major contributor to the maintenance of genetic variation.

II
Microbial Nutrition, Growth and Metabolism

4
Microbial Nutrition and Cultivation

In Chapter 2 we introduced the major groups of macromolecules found in living cells; the raw materials from which these are synthesised are derived ultimately from the organism's environment in the form of nutrients (Table 4.1). These can be conveniently divided into those required in large quantities[1] (macronutrients) and those needed only in trace amounts (micronutrients or *trace elements*).

You will recall that carbon forms the central component of proteins, carbohydrates, nucleic acids and lipids; indeed, the living world is based on carbon, so it should come as no surprise that this is the most abundant element in all living cells, microbial or otherwise. Of the other macronutrients, nitrogen, oxygen, hydrogen, sulphur and phosphorus are also constituents of biological macromolecules, while the remainder (magnesium, potassium, sodium, calcium and iron in their ionised forms) are required in lesser quantities for a range of functions that will be described in due course. Micronutrients are all metal ions, and frequently serve as cofactors for enzymes. All microorganisms must have a supply of the nutrients described above, but they show great versatility in the means they use to satisfy these requirements.

The metabolic processes by which microorganisms assimilate nutrients to make cellular material and derive energy will be reviewed in Chapter 6. In the following section we briefly describe the role of each element, and the form in which it may be acquired.

Carbon is the central component of the biological macromolecules we discussed in Chapter 2. Carbon incorporated into biosynthetic pathways may be

---

[1] Everything is relative in the microbial world; a typical bacterial cell weighs around three ten-million-millionths (3×10^{-13}) of a gram!

Essential Microbiology, Second Edition. Stuart Hogg.
© 2013 John Wiley & Sons, Ltd. Published 2013 by John Wiley & Sons, Ltd.

Table 4.1   Elements found in living organisms

Element	Form in which usually supplied	Occurrence in biological systems
Macronutrients		
Carbon (C)	CO_2, organic compounds	Component of all organic molecules, CO_2
Hydrogen (H)	H_2O, organic compounds	Component of biological molecules, H^+ released by acids
Oxygen (O)	O_2, H_2O, organic compounds	Component of biological molecules; required for aerobic metabolism
Nitrogen (N)	NH_3, NO_3^-, N_2, organic N compounds	Component of proteins, nucleic acids
Sulphur (S)	H_2S, SO_4^{2-}, organic S compounds	Component of proteins; energy source for some bacteria
Phosphorus (P)	PO_4^{3-}	Found in nucleic acids, ATP, phospholipids
Potassium (K)	In solution as K^+	Important intracellular ion
Sodium (Na)	In solution as Na^+	Important extracellular ion
Chlorine (Cl)	In solution as Cl^-	Important extracellular ion
Calcium (Ca)	In solution as Ca^{2+}	Regulator of cellular processes
Magnesium (Mg)	In solution as Mg^{2+}	Coenzyme for many enzymes
Iron (Fe)	In solution as Fe^{2+} or Fe^{3+} or as FeS, $Fe(OH)_3$, etc.	Carries oxygen; energy source for some bacteria
Micronutrients (present as contaminants at very low concentrations)		
Copper (Cu)	Ionised in solution	Coenzyme; microbial growth inhibitor
Manganese (Mn)	Ionised in solution	Coenzyme
Cobalt (Co)	Ionised in solution	Vitamin B_{12}
Zinc (Zn)	Ionised in solution	Coenzyme; microbial growth inhibitor
Molybdenum (Mb)	Ionised in solution	Coenzyme
Nickel (Ni)	Ionised in solution	Coenzyme

derived from organic or inorganic sources (see below); some organisms can derive it from CO_2, while others require their carbon in 'ready-made', organic form.

Hydrogen is also a key component of macromolecules, and participates in energy generation processes in most microorganisms. In autotrophs (see 'Nutritional categories' later), hydrogen is required to reduce carbon dioxide in the synthesis of macromolecules.

Oxygen is of central importance to the respiration of many microorganisms, but in its molecular form (O_2), it can be toxic to some forms (see Chapter 5). These obtain from water the oxygen they need for the synthesis of macromolecules.

Nitrogen is needed for the synthesis of proteins and nucleic acids, as well as for important molecules such as ATP (adenosine triphosphate; you will learn more about ATP and its role in the cell's energy relations in Chapter 6). Microorganisms range in their demands for nitrogen from those that are able to assimilate ('fix') nitrogen (N_2) from the atmosphere to those that require all 20 amino acids to be provided preformed. Between these two extremes come species that are able to assimilate nitrogen from an inorganic source such as nitrate, and those that utilise ammonium salts or urea as a nitrogen source.

Sulphur is required for the synthesis of proteins and vitamins, and in some microorganisms is involved in cellular respiration and anoxygenic photosynthesis. It may be derived from sulphur-containing amino acids (methionine, cysteine), sulphates or sulphides.

Phosphorus is taken up as inorganic phosphate, and is incorporated in this form into nucleic acids and phospholipids, as well as other molecules such as ATP.

Metals such as *copper, iron* and *magnesium* are required as *cofactors* in enzyme reactions.

Many microorganisms are unable to synthesise certain organic compounds required for growth and must therefore acquire them from their growth medium. These compounds are termed *growth factors* (Table 4.2), of which three main groups can be identified: amino acids, purines and pyrimidines

> A *cofactor* is a non-protein component of an enzyme (often a metal ion) essential for its normal functioning.

Table 4.2 Microbial growth factors

Growth factor	Function
Amino acids	Components of proteins
p-Aminobenzoic acid	Precursor of folic acid, involved in nucleic acid synthesis
Cholesterol	Membrane component in mycobacteria and eukaryotes
Haem	Functional portion of cytochromes in electron transport
Nicotinamide adenine dinucleotide (reduced) (NADH)	Electron carrier
Niacin (nicotinic acid)	Precursor of NAD^+ and $NADP^+$
Pantothenic acid	Component of coenzyme A
Purines and pyrimidines	Components of nucleic acids
Pyridoxine (vitamin B_6)	Utilised in transamination syntheses of amino acids
Riboflavin (vitamin B_2)	Precursor of flavin adenine dinucelotide (FAD)
Thiamine (vitamin B_1)	Utilised in some decarboxylation reactions

(required for nucleic acid synthesis), and vitamins. You will already have read about the first two of these groups in Chapter 2. *Vitamins* are complex organic compounds required in very small amounts for the cell's normal functioning. They are often either *coenzymes* or their precursors (see Chapter 6). Microorganisms vary greatly in their vitamin requirements. Many bacteria are completely self-sufficient, while protozoans, for example, generally need to be supplied with a wide range of these dietary supplements. A vitamin requirement may be absolute or partial; an organism may be able, for example, to synthesise enough of a vitamin to survive, but grow more vigorously if an additional supply is made available to it.

4.1   Nutritional categories

Microorganisms can be categorised according to how they obtain their carbon and energy. As we've seen, carbon is the most abundant component of the microbial cell, and most microorganisms obtain it in the form of organic

> A *heterotroph* must use one or more organic molecules as its source of carbon.

molecules, derived directly or indirectly from other organisms. This mode of nutrition is the one that is familiar to us as humans (and all other animals); all the food we eat is derived as complex organic molecules from plants or animals (and selected representatives of the microbial world such as mushrooms). Microorganisms that obtain their carbon in this way are described as *heterotrophs*, and include all the fungi and protozoans as well as most types of bacteria. Microorganisms as a group are able to incorporate into cellular material the carbon from an incredibly wide range of organic compounds. In fact there is hardly any such compound occurring in nature that cannot be metabolised by some microorganism or other, explaining in part why microbial life is to be found thriving in the most unlikely habitats. Many synthetic materials can also serve as carbon sources for some microorganisms, which can have considerable economic and environmental significance.

A significant number of bacteria and all of the algae do not, however, take up their carbon preformed as organic molecules in this way, but derive it instead from carbon dioxide. These

> An *autotroph* can derive its carbon from carbon dioxide.

organisms are called *autotrophs*, and again we can draw a parallel with higher organisms, where all members of the plant kingdom obtain their carbon in a similar fashion.

We can also categorise microorganisms nutritionally by the way they derive the energy they require to carry out essential cellular reactions. *Autotrophs* thus fall into two categories. *Chemoautotrophs* obtain their energy as well as their carbon from inorganic sources; they do

> A *chemotroph* obtains its energy from chemical compounds.
> A *phototroph* uses light as its source of energy.

this by the oxidation of inorganic molecules such as sulphur or nitrite. *Photoautotrophs* have photosynthetic pigments, enabling them to convert light energy into chemical energy. The mechanisms by which this is achieved will be discussed in Chapter 6.

The great majority of heterotrophs obtain energy as well as carbon from the same organic source. Such organisms release energy by the chemical oxidation of organic nutrient molecules, and are therefore termed *chemoheterotrophs*. Those few heterotrophs that do not follow this mode of nutrition include the green and purple non-sulphur bacteria. These are able to carry out a form of photosynthesis and are known as *photoheterotrophs*.

There is one final subdivision of nutritional categories in microorganisms! Whether organisms are chemotrophs or phototrophs, they need a molecule to act as a source of electrons (reducing power) to drive their energy-generating systems (see Chapter 6). Those able to use an inorganic electron donor such as H_2O, H_2S or ammonia are called *lithotrophs*, while those requiring an organic molecule to fulfil the role are *organotrophs*. Most (but not all) microorganisms are either photolithotrophic autotrophs (algae, blue-greens) or chemoorganotrophic heterotrophs (most bacteria). For the latter category, a single organic compound can often act as the provider of carbon, energy and reducing power. The substance used by chemotrophs as an energy source may be organic (chemoorganotrophs) or inorganic (chemolithotrophs).

> A *lithotroph* is an organism that uses inorganic molecules as a source of electrons. An *organotroph* uses organic molecules for the same purpose.

4.2 How do nutrients get into the microbial cell?

Having found a source of a given nutrient, a microorganism must:

- have some means of taking it up from the environment;
- possess the appropriate enzyme systems to utilise it.

The plasma membrane represents a selective barrier, allowing into the cell only those substances it is able to utilise. This selectivity is due in large part to the hydrophobic nature of the lipid bilayer. A substance can be transported across the cell membrane in one of three ways: simple diffusion, facilitated diffusion or active transport.

In *simple diffusion*, small molecules move across the membrane in response to a concentration gradient (from high to low), until concentrations on either side of the membrane are in equilibrium. The ability to do this depends on being small (H_2O, Na^+, Cl^-) or soluble in the lipid component of the membrane (nonpolar gases such as O_2 and CO_2).

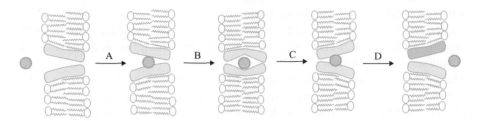

Figure 4.1   Facilitated diffusion and active transport. In facilitated diffusion, substances can move across the plasma membrane by binding to an embedded transport protein (shown shaded). No energy input is required, but the diffusion can only occur from an area of high concentration to one of low concentration. Reproduced from Thomas, G (2000) Medicinal Chemistry, an Introduction, with permission from John Wiley & Sons.

Larger, polar molecules such as glucose and amino acids are unable to enter the cell unless assisted by membrane-spanning *transport proteins* in the process of *facilitated diffusion* (Figure 4.1). Like enzymes, these proteins are specific for a single or small number of related solutes; another parallel is that they too can become saturated by too much 'substrate'. As with simple diffusion, there is no expenditure of cellular energy, and an inward concentration gradient is required. The transported substance tends to be metabolised rapidly once inside the cell, thus maintaining the concentration gradient from outside to inside.

Diffusion is only an effective method of internalising substances when their concentrations are greater outside the cell than inside. Generally, however, microorganisms find themselves in very dilute environments, hence the concentration gradient runs in the other direction, and diffusion into the cell is not possible. *Active transport* enables the cell to overcome this unfavourable gradient. Here, regardless of the direction of the gradient, transport takes place in one direction only, *into* the cell. Energy, derived from hydrolysis of ATP (see Chapter 6), is required to achieve this, and again specific transmembrane proteins are involved. They bind the solute molecules with high affinity outside of the cell, and then undergo a conformational change, which causes them to be released into the interior. Prokaryotic cells can carry out a specialised form of active transport called group translocation, whereby the solute is chemically modified as it crosses the membrane, preventing its escape. A well-studied example of this is the phosphorylation of glucose in *E. coli* by the phosphotransferase system. Glucose present in very low concentrations outside the cell can be concentrated within it by this mechanism. Glucose is unable to pass back across the membrane in its phosphorylated form (glucose-6-phosphate), but it can be utilised in metabolic pathways in this form.

Often it may be necessary to employ extracellular enzymes to break down large molecules before any of these mechanisms can be used to transport nutrients into the cell.

4.3 Laboratory cultivation of microorganisms

Critical to the development of microbiology during its 'golden age' were the advances in culturing techniques, which enabled the isolation and pure culture of specific microorganisms. The study of pure cultures made it possible to determine the properties of a specific organism such as its metabolic characteristics or its ability to cause a particular disease. It also opened up the possibility of classifying microorganisms, on the basis of the characteristics they display in pure culture.

The artificial culture of any organism requires a supply of the necessary nutrients, together with the provision of appropriate conditions such as temperature, pH and oxygen concentration. The nutrients and conditions provided in the laboratory are usually a reflection of those found in the organism's natural habitat. It is also essential that appropriate steps are taken to avoid contamination (Box 4.1). In the next section we shall describe the techniques used to isolate and propagate microorganisms in the laboratory.

4.3.1 Obtaining a pure culture

Microorganisms in the natural world do not live in pure cultures, but exist as part of complex ecosystems comprising numerous other organisms. The first step in the cultivation of specific microorganisms is therefore the creation of a *pure culture*. A key development for the production of pure cultures was the ability to grow microorganisms on a solid medium. Koch had noticed

Box 4.1 Aseptic technique

Most commonly used culture media will support the growth of a number of different bacteria. It is therefore essential when working in the microbiology laboratory that suitable precautions are taken to prevent the growth of unwanted *contaminants* in our cultures. These simple practical measures are termed *aseptic technique*, and it is essential to master them if reliable experimental results are to be obtained. Any glassware and equipment used is sterilised before work begins. Containers such as tubes, flasks and plates are kept open for the minimum amount of time, and the necks of bottles and tubes are passed through a flame to maintain their sterility. The wire loops and needles used to transfer small volumes of microbial cultures are sterilised by heating them to redness in a flame. Increasingly, these are being replaced by pre-sterilised, disposable plastic instruments. Your instructor will normally demonstrate aseptic technique to you in an early practical session.

that when a nutrient surface such as cut potato was exposed to air, individual microbial *colonies* grew up, and he inferred from this that these had arisen from the numerous divisions of single cells.

It soon became apparent that a number of organisms would not grow on potatoes, so Koch and his colleagues turned to gelatin as a means of solidifying a synthetic nutrient growth medium. Horizontal slabs were cut, and covered to help keep them free from atmospheric contaminants. Gelatin was a convenient means of solidifying media, as it could be boiled and then allowed to set in the desired vessel. There were two main drawbacks to its use, however; many organisms needed to be incubated at around body temperature (37°C), and gelatin melted before this temperature was reached. In addition, it was found that a number of bacteria were capable of utilising gelatin as a nutrient source, resulting in the liquefaction of the gel.

> Bacteria may be cultured using either liquid or solid media. *Solid* media are particularly useful in the isolation of bacteria; they are also used for their long-term storage. *Liquid* (broth) cultures are used for rapid and large-scale production of bacteria.

> A culture consisting entirely of one strain of organism is called a *pure* or *axenic culture*. In theory, such a culture represents the descendants of a single cell.

A more satisfactory alternative was soon found in the form of *agar*. This is a complex polysaccharide derived from seaweeds, and was suggested by the wife of one of Koch's colleagues, who had used it as a setting agent in jam making. Agar does not melt until near boiling point; this means that cultures can be incubated at 37°C or above without the medium melting. Moreover, when it cools, agar remains molten until just over 40°C· allowing heat-sensitive media components such as blood to be added. In addition, most bacteria can tolerate a short exposure to temperatures in this range, so they too can be inoculated into molten agar (see 'pour plate method' later). Crucially, agar is *more or less inert nutritionally*; only a very few organisms are known that can use agar as a food source, consequently it is the near ideal setting agent, resisting both thermal and microbial breakdown. Agar soon became the setting agent of choice, and has remained so ever since. Shortly afterwards, Richard Petri developed the two-part culture dish that was named after him; this could be sterilised separately from the medium and provided protection from contamination by means of its lid. This again is still standard equipment today, although the original glass has been largely replaced by pre-sterilised, disposable plastic.

> A *Petri dish* is the standard vessel for short-term growth of solid medium cultures in the laboratory. It comprises a circular dish with an overlapping lid.

The standard method of obtaining a pure bacterial culture is the creation of a *streak plate* (Figure 4.2). A sterilised wire inoculating loop is used to

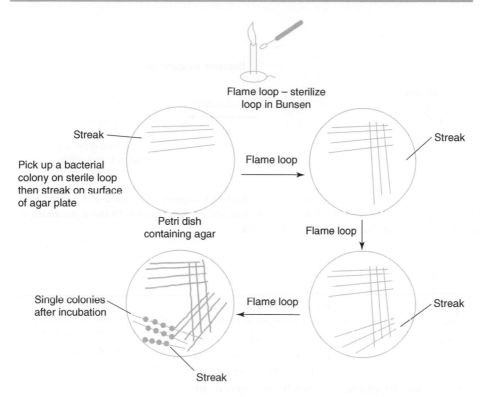

Flame loop – sterilize loop in Bunsen

Streak

Pick up a bacterial colony on sterile loop then streak on surface of agar plate

Petri dish containing agar

Flame loop

Streak

Flame loop

Single colonies after incubation

Flame loop

Streak

Streak

Figure 4.2 The streak plate. Streaking the sample across the agar surface eventually results in individual cells being deposited. Repeated cycles of cell division lead to the production of visible, isolated colonies. Reproduced from Nicklin, J, et al. (2002) Instant Notes in Microbiology, 2nd edn, with permission from Taylor & Francis Group.

spread out a drop of bacterial suspension on an agar plate in such a way that it becomes progressively more dilute; eventually, individual cells will be deposited on the agar surface. Following incubation at an appropriate temperature, a succession of cell divisions occurs, resulting in the formation of a bacterial colony, visible to the naked eye. Colonies arise because movement is not possible on the solid surface and all the progeny stay in the same place. A colony represents, in theory at least, the offspring of a single cell and its members are therefore genetically identical. (In reality, a clump of cells may be deposited together and give rise to a colony; this problem can be overcome by repeated isolation and re-streaking of single colonies.)

An alternative method for the isolation of pure cultures is the *pour plate* (Figure 4.3). In this method, a dilute suspension of bacteria is mixed with warm molten agar, and poured into an empty Petri plate. As the agar sets, cells are immobilised, and once again their progeny are all kept together, often within, as well as on, the agar. This method is especially useful for the isolation of bacteria that are unable to tolerate atmospheric levels of oxygen.

Figure 4.3   The pour plate. A sample of diluted bacterial suspension is mixed with molten agar and poured into a Petri dish. Most bacteria can tolerate a short exposure to the agar, which is held at a temperature just above its setting point.

4.3.2   Growth media for the cultivation of bacteria

A synthetic growth medium may be *defined*, that is, its exact chemical composition is known, or *undefined*. A defined growth medium may have few or many constituents, depending on the nutritional requirements of the organism in question. Examples of each are given in

> An *undefined* or *complex* *medium* is one whose precise chemical composition is not known.

Table 4.3. An undefined, or complex, medium may have a variable composition due to the inclusion of a component such as blood, yeast extract or tap water (Table 4.4). Peptones are also commonly found in complex media; these are the products of partially digesting protein sources such as beef or casein. The exact composition of a complex medium is neither known nor critically important. A medium of this type would generally be chosen for the cultivation of fastidious bacteria such as *Neisseria gonorrhoeae* (the causative agent of gonorrhoea); it is easier and less expensive to supply the many nutrients required by such an organism in this form rather than supplying them all

> A *fastidious* organism is unable to synthesise a range of nutrients and therefore has complex requirements in culture.

individually. Bacteria whose specific nutrient requirements are not known are also grown on complex media.

Whilst media such as nutrient agar are used to support the growth of a wide range of organisms, others are specifically designed for the isolation and identification of particular types. *Selective* media, such as bismuth sulphite medium, preferentially support the growth of particular bacteria. The bismuth ion inhibits the

> A *selective medium* is one that favours the growth of a particular organism or group of organisms, often by suppressing the growth of others.

Table 4.3   Defined growth media

Examples of defined (synthetic) media for (a) the iron-oxidising bacterium *Acidithiobacillus ferrooxidans* and (b) the lactic acid bacterium *Leuconostoc mesenteroides*. *L. mesenteroides* must be provided with numerous amino acids, nucleotides and vitamins as well as glucose as a carbon source, whereas *A. ferrooxidans* requires only mineral salts, including reduced iron to act as an energy source.

(a) Medium for *Acidithiobacillus ferrooxidans*

$FeSO_4.7H_2O$	40 g
$(NH_4)_2SO_4$	2 g
KH_2PO_4	0.5 g
$MgSO_4.7H_2O$	0.5 g
KCl	0.1 g
$Ca(NO_3)_2$	0.01 g
Distilled H_2O (pH 3.0)to 1 litre	

(b) Medium for *Leuconostoc mesenteroides*

Glucose	25 g	Phenylalanine	100 mg
Sodium acetate	20 g	Proline	100 mg
NH_4Cl	3 g	Serine	50 mg
KH_2PO_4	0.6 g	Threonine	200 mg
K_2HPO_4	0.6 g	Tryptophan	40 mg
NaCl	3 g	Tyrosine	100 mg
$MgSO_4.7H_2O$	0.2 g	Valine	250 mg
$MnSO_4.4H_2O$	20 mg	Adenine	10 mg
$FeSO_4.7H_2O$	10 mg	Cytosine	10 mg
Alanine	200 mg	Guanine	10 mg
Arginine	242 mg	Uracil	10 mg
Aspartic acid	100 mg	Nicotinic acid	1 mg
Asparagine	400 mg	Pyridoxine	1 mg
Cysteine	50 mg	Riboflavin	0.5 mg
Glutamic acid	300 mg	Thiamine	0.5 mg
Glycine	100 mg	Calcium pantothenate	0.5 mg
Histidine	62 mg	Pyridoxamine	0.3 mg
Isoleucine	250 mg	Pyridoxal	0.3 mg
Leucine	250 mg	*p*-Aminobenzoic acid	0.1 mg
Lysine	250 mg	Biotin	1 µg
Methionine	100 mg	Folic acid	10 µg
Water	To 1 litre		

growth of Gram-positive organisms as well as many Gram-negative types; this medium is used for the isolation of the pathogenic bacterium *Salmonella typhi*, one of the few organisms that can tolerate the bismuth. Specific media called *differential* media can be used to distinguish

A *differential* medium allows colonies of a particular organism to be differentiated from others growing in the same culture.

Table 4.4   Composition of an undefined growth medium

Brain heart infusion broth contains three undefined components. It is used for the culture of a wide variety of fastidious species, both bacterial and fungal.

Calf brain infusion	200 g
Beef heart infusion	250 g
Proteose peptone	10 g
Glucose	2 g
NaCl	5 g
Na_2HPO_4	2.5 g
H_2O (pH 7.4)	To one litre

between organisms whose growth they support, usually by means of a coloured indicator. MacConkey agar contains lactose and a pH indicator, allowing the differentiation between lactose fermenters (red colonies) and non-lactose fermenters (white/pale-pink colonies). Many media act both selectively and differentially; MacConkey agar, for example, also contains bile salts and the dye crystal violet, both of which serve to inhibit the growth of unwanted Gram-positive bacteria. Mannitol salt agar is also both selective and differential. The high (7.5%) salt content suppresses growth of most bacteria, whilst a combination of mannitol and an indicator permits the detection of mannitol fermenters in a similar fashion to that just described. Sometimes it is desirable to isolate an organism that is present in small numbers in a large mixed population (e.g. a sample of faeces or soil). *Enrichment* media provide conditions that selectively encourage the growth of these organisms; the use of blood agar in the isolation of streptococci provides an example of such a medium. Blood agar can act as a differential medium, in allowing the user to distinguish between haemolytic and non-haemolytic bacteria (see Chapter 7).

> An *enrichment* culture uses a selective medium to encourage the growth of an organism present in low numbers.

If we are to culture microorganisms successfully in the laboratory, we must provide appropriate physical conditions as well as providing an appropriate nutrient medium. In the next chapter, we shall examine how physical factors such as pH and temperature influence the growth of microorganisms, and describe how these conditions are provided in the laboratory.

4.3.3   Preservation of microbial cultures

Microbial cultures are preserved by storage at low temperatures, in order to suspend growth processes. For short periods, most organisms can be kept at refrigerator temperature (around 4°C), but for longer-term storage, more specialised treatment is necessary. Using deep-freezing or freeze-drying, cultures can be kept for many years, and then resurrected and recultured. Deep-freezing requires rapid freezing to between $-70°C$ and $-95°C$, while freeze-drying (lyophilisation) involves freezing at slightly less extreme temperatures and removing the water content under vacuum. Long-term storage may be desirable to avoid the development of mutations or loss of cell viability.

5
Microbial Growth

When we consider growth as applied to a multicellular organism such as a tree, a fish or a human being, we think in terms of an ordered increase in the size of the individual. Growth in unicellular microorganisms such as bacteria, yeasts and protozoans, however, is more properly defined in terms of an increase in the size of a given *population*. This may be expressed as an increase in either the number of individuals or the total amount of biomass. Methods employed in the measurement of growth of unicellular microorganisms

> *Biomass* is the total amount of cellular material in a system.

may be based on either of these. In this chapter we shall describe some of these methods, before considering the dynamics of microbial growth and some of the factors that affect it.

5.1 Estimation of microbial numbers

Several methods exist for the measurement of bacterial numbers, most of which are also applicable to the enumeration of other unicellular forms such as yeasts. Such methods fall into two main categories: those that count total cell numbers, and those that count viable cells only.

Total cell counts are generally done by direct microscopic examination. A specialised glass slide is employed, which carries an etched grid of known area (Figure 5.1). The depth of the trapped liquid sample is also known, so by counting the number of cells visible in the field of view, the number of cells per unit volume can be determined. The method may be made more accurate by the use of a fluorescent dye such as acridine orange, which binds to DNA, and hence avoids confusion with non-cellular debris. However, such methods cannot differentiate between living and non-living cells. Their usefulness is

Essential Microbiology, Second Edition. Stuart Hogg.
© 2013 John Wiley & Sons, Ltd. Published 2013 by John Wiley & Sons, Ltd.

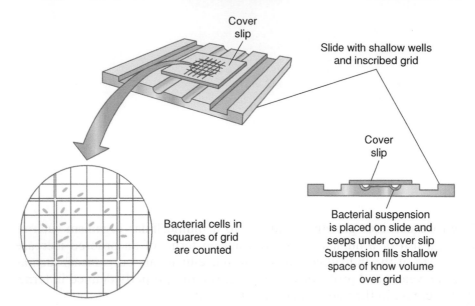

Figure 5.1 Estimation of total cell numbers by direct microscopic measurement. The Petroff–Hauser counting chamber is a specialised glass slide with an etched grid of known area. A droplet of cell suspension is placed on top, followed by a coverslip. Since the depth of liquid trapped is known, the volume covering the grid can be calculated. The number of cells present in several random squares is counted, and an average value obtained. The method does not distinguish between living and dead cells. Reproduced from Black, JG (1999) Microbiology: Principles and Explorations, 4th edn, with permission from John Wiley & Sons.

further limited by the fact that the smallest bacteria are difficult to resolve as individual cells by light microscopy. Other total cell count methods use cell-sorting devices, originally developed for separating blood cells in medical research. These pass the cell suspension through an extremely fine nozzle, and a detector registers the conductivity change each time a particle passes it. Again, no distinction can be made between viable and non-viable cells.

A *viable cell count*, on the other hand, is a measure of the number of *living* cells in a sample, or more specifically those capable of multiplying and producing a visible colony of cells. It is most commonly estimated by spreading a known volume of cell suspension onto an agar plate, and counting the number of colonies that arise after a period of incubation (Box 5.1). The method is based on the premise that each visible colony has derived from the repeated divisions of a single cell. In reality, it is accepted that this is not always the case, and so viable counts are expressed in *colony-forming units (cfu)*, rather than cells, per unit volume. It is generally necessary to dilute the suspension before plating out, otherwise the resulting colonies would be too numerous to count. In order to improve statistical reliability, plates are inoculated in duplicate or triplicate, and the mean value is taken.

Box 5.1   *Estimation of viable cell numbers*

In order that the sample to be plated out contains an appropriate number of cells, the original sample is subjected to *serial dilution*. In the example below, it is diluted by a factor of 10 at each stage to give a final dilution of 10^{-5} (one in a hundred thousand). Samples of 0.1 mL of each dilution are plated out on a suitable solid medium and colonies allowed to develop.

TNTC = too numerous to count

In the earlier tubes, the suspension of cells is too concentrated, resulting in too many colonies to count. In the final tube, the suspension is so dilute that there are no cells in the sample taken. The 10^{-3} dilution is used to calculate the concentration of cells in the original culture, as it falls within the range of 30–300 colonies regarded as being statistically reliable.

Colony count (10^{-3} dilution) = 65 (from 0.1 mL sample)
∴ Number of cfu per mL diluted suspension = $65 \times 10 = 650$
∴ Number of cfu per mL original suspension = $650 \times 10^{3} = 6.5 \times 10^{5}$

Box 5.2 The most probable number method

In the example below, three sets of five tubes of broth were inoculated with 10 mL, 1 mL and 0.1 mL of a water sample. The tubes were incubated to allow any bacteria present to multiply in number, and were scored as 'growth' (dark shading) or 'no growth' (no shading). The cell density statistically most likely to give rise to the result obtained (5-3-1) is then looked up on a set of most probable number (MPN) tables. The table (only part shown) indicates that there is a 95% probability that the sample fell within the range 40–300 cells/mL, with 110 cells/mL being the most likely value.

Number of tubes out of five showing growth			MPN /100 mL
10 mL	1 mL	0.1 mL	
4	3	1	33
4	4	0	34
5	0	0	23
5	0	1	30
5	0	2	40
5	1	0	30
5	1	1	50
5	1	2	60
5	2	0	50
5	2	1	70
5	2	2	90
5	3	0	80
5	3	1	110
5	3	2	140
5	3	3	170
5	4	0	130

Viable cell counts can also be made using liquid media, in the *most probable number* (*MPN*) technique (Box 5.2). Here, a series of tubes of broth are inoculated with progressively more dilute samples of a cell suspension, incubated, and examined for growth. The method is based on the statistical probability of each sample containing viable cells. It is well suited to the testing of drinking water, where low bacterial densities are to be expected.

Another method employed for the enumeration of bacteria in water is the *membrane filter test*. Here, a large volume of water is passed through a membrane filter with a pore size (0.45 μm) suitable for trapping bacteria (Figure 5.2). The filter is placed on an appropriate solid growth medium and colonies allowed to develop.

Figure 5.2 Estimation of cell numbers by membrane filtration. Microbial cells present in a water sample are trapped on a membrane, which is then placed on an agar medium to allow colony development. The technique is used to concentrate the cells present at low densities in a large volume. Reproduced from Prescott, LM, et al. (2002) Microbiology 5th edn, with permission from McGraw-Hill.

None of the methods described above provides a particularly rapid result, yet sometimes it is desirable to have an estimate of bacterial numbers immediately. A useful method for doing this is based on how cloudy or turbid the liquid growth medium becomes due to bacterial growth. *Turbidimetric* methods measure the change in optical density or absorbance of the medium, that is, how much a beam of light is scattered by the suspended particulate matter (Figure 5.3). They can be carried out very quickly by placing a sample in a spectrophotometer set at an appropriate wavelength. Values of optical density can be directly related to bacterial numbers or mass by reference to a standard calibration curve. Thus, an estimate of bacterial numbers, albeit a fairly approximate one, can be obtained almost instantaneously during an experimental procedure. Other indirect methods of measuring cell density include wet and dry weight estimations, and the measurement of cell components such as total nitrogen, protein or nucleic acid.

5.2 Factors affecting microbial growth

In Chapter 4 we discussed the nutrient requirements of microorganisms. Assuming these are present in an adequate supply, what other factors do we need to consider in order to provide favourable conditions for microbial growth? As the following section shows, growth may be profoundly affected by a number of physical factors.

5.2.1 Temperature

Microorganisms as a group are able to grow over a wide range of temperatures, from around freezing to above boiling point. For any organism, the minimum and maximum growth temperatures define the range over which

Figure 5.3 Indirect measurement of cell numbers by turbidimetric measurements. (a) Turbidimetry offers an immediate estimate of bacterial density by measuring the degree to which a culture scatters light shone through it in a spectrophotometer. (b) Within certain limits, there is a linear relationship between cell numbers and optical density (absorbance). By determining cell numbers or cell mass for samples of known optical density, a calibration graph can be produced.

growth is possible; this is typically about 25–30°C. Growth is slower at low temperatures because enzymes work less efficiently and also because lipids tend to harden and there is a loss of membrane fluidity. Growth rates increase with temperature until the *optimum temperature* is reached, and then the rate falls again (Figure 5.4). The optimum and limiting temperatures for an organism are a reflection of the temperature range of its enzyme systems, which in turn is determined by their three-dimensional protein structures (see Chapter 6). The optimum temperature is generally closer to the maximum growth temperature than the minimum. Once the optimum value is passed, the loss of activity caused by denaturation of enzymes causes the rate of growth to fall away sharply (see also Figures 5.6 and 5.7).

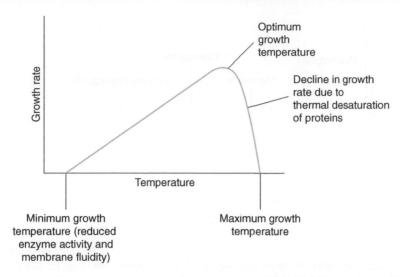

Figure 5.4 Effect of temperature on microbial growth rate. The factors governing the minimum, optimum and maximum temperatures for a particular organism are indicated. The curve is asymmetrical, with the optimum temperature being closer to the maximum than the minimum.

The majority of microorganisms achieve optimal growth at 'middling' temperatures of around 20–45°C; these are called *mesophiles* (Figure 5.5). Contrast these with *thermophiles*, which have become adapted not merely to surviving, but thriving at much higher temperatures. Typically, these would be capable of growth within a range of about 40–80°C, with an optimum around 50–65°C. *Extreme thermophiles* (*hyperthermophiles*) have optimum values in excess of this, and can tolerate temperatures in excess of 100°C. In 2003, a member of the primitive bacterial group called the Archaea (see Chapter 7) was reported as growing at a temperature of 121°C, a new world record! *Psychrophiles* occupy the other extreme of the temperature range; they can grow at 0°C, with optimal growth occurring at 15°C or below. Such organisms are not able to grow at temperatures above 25°C or so. *Psychrotrophs*, on the other hand, although they can also grow at 0°C, have much higher temperature optima (20–30°C). Members of this group are often economically significant due to their ability to grow on refrigerated foodstuffs.

> Extreme claims have been made for the temperature tolerance of the so-called 'red rain' cells that fell over India in 2001, which would represent new limits for life; however, these have yet to be substantiated.

In the laboratory, appropriate temperatures for growth are provided by culturing in an appropriate incubator. Incubators come in a variety of shapes and sizes, but all are thermostatically controlled and generally hold the temperature within a degree or two of the desired value.

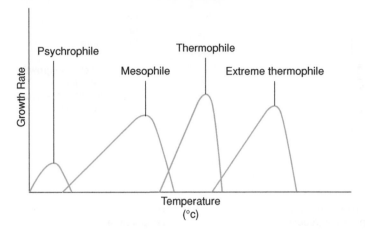

Figure 5.5 Different species occupy different temperature ranges. Microorganisms can be categorised according to the temperature range at which they grow. The range for two different thermophiles is shown. Virtually all species pathogenic in humans are mesophiles.

5.2.2 pH

Microorganisms are strongly influenced by the prevailing pH of their surroundings. As with temperature, we can define minimum, optimum and maximum values for growth of a particular type (Figure 5.6). The pH range (between minimum and maximum values) is greater in fungi than it is in

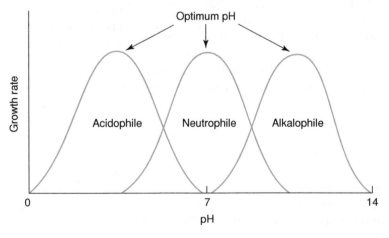

Figure 5.6 Effect of pH on microbial growth rate. Individual species of microorganism occupy a relatively narrow range of pH. Although for most species this is around neutrality, both acidophilic and alkalophilic forms exist. The shape of the curve reflects the properties of a particular organism's enzymes and other proteins.

> ## Box 5.3 Buffers
>
> The pH of a batch culture may change from its intended value as a result of an organism's metabolic activity. Buffers are added to a medium in order to minimise the effect of these changes. Such buffers must be non-toxic to the organism in question, and phosphate buffers are commonly used in microbial systems. Phosphate combines with hydrogen ions or hydroxyl ions to form respectively a weak acid or a weak base:
>
> $$H^+ + HPO_4^{2-} \rightarrow H_2PO_4^-$$
> $$OH^- + H_2PO_4^- \rightarrow HPO_4^{2-} + H_2O$$
>
> This reduces the impact on the pH of the medium.

bacteria. Most microorganisms grow best around neutrality (pH 7). Many bacteria prefer slightly alkaline conditions but relatively few are tolerant of acid conditions, and fewer still are acidophilic. Fungi, on the other hand, generally prefer slightly acid conditions and therefore tend to dominate over bacteria when

Acidophilic = 'acid-loving'; a term applied to organisms that show optimal growth in acid conditions (pH < 5.5).

these prevail. The reason for the growth rate falling away either side of the optimum value is again due to alterations in three-dimensional protein structure.

The pH value of a growth medium is adjusted to the desired value by the addition of acid or alkali during its preparation. The metabolic activities of microorganisms often mean that they change the pH of their environment as growth proceeds, so it is important that in a laboratory growth medium a desirable pH is not only set but maintained. This is achieved by the use of an appropriate buffer system. Phosphate buffers are widely used in the microbiology laboratory; they enable media to minimise changes in their pH when acid or alkali is produced (see Box 5.3).

5.2.3 Oxygen

Oxygen is present as a major constituent (20%) of our atmosphere, and most life forms are dependent upon it for survival and growth. Such organisms are termed *aerobes*. Not all organisms are aerobes, however; some *anaerobes* are able to survive in the absence of oxygen, and for some this is actually a necessity.

Aerobic organisms require oxygen to act as a terminal electron acceptor in their respiratory chains (see Chapter 6). Such organisms, when grown in laboratory culture, must therefore be provided with enough oxygen to satisfy their requirements. For a shallow layer of medium, such as that in a Petri dish, sufficient oxygen is available dissolved in surface moisture. In a deeper culture such as a flask of broth, however, aerobes will only grow beyond the surface layers if additional oxygen is provided (oxygen is poorly soluble in water). This is usually done by mechanical shaking or stirring.

> An *aerobe* is an organism that grows in the presence of molecular oxygen, which it uses as a terminal electron acceptor in aerobic respiration.
> An *anaerobe* is an organism that grows in the absence of molecular oxygen.

Obligate anaerobes cannot tolerate oxygen at all (see Box 5.4). They are cultured in special anaerobic chambers, and oxygen is excluded from all liquid and solid media. *Facultative anaerobes* are able to act like aerobes in the presence of oxygen, but have the added facility of being able to survive when conditions become anaerobic. *Aerotolerant anaerobes* are organisms that are basically anaerobic; although they are not inhibited by atmospheric oxygen, they do not utilise it. *Microaerophiles* require oxygen, but are only able to tolerate low concentrations of it (2–10%), and are harmed by higher concentrations. Organisms inoculated into a static culture medium will grow at positions that reflect their oxygen preferences (Figure 5.7).

5.2.4   Carbon dioxide

In Chapter 4 we saw that autotrophic organisms are able to use carbon dioxide as a carbon source; when grown in culture, these are provided with bicarbonate in their growth medium or incubated in a CO_2-enriched atmosphere. However, heterotrophic bacteria also require small amounts of carbon dioxide, which is incorporated into various metabolic intermediates. This dependency can be demonstrated by the failure of these organisms to grow if carbon dioxide is deliberately removed from the atmosphere.

Box 5.4   How can oxygen be toxic?

It seems strange to us to think of oxygen as a toxic substance; however, it can be converted by metabolic enzymes into highly reactive derivatives such as the superoxide free radical (O_2^-), which are very damaging to cells. Aerobes and most facultative anaerobes convert this to hydrogen peroxide, by means of the enzyme *superoxide dismutase*. This is further broken down by *catalase*. Obligate anaerobes do not possess either enzyme, and so cannot tolerate oxygen.

| (a) Obligate aerobes | (b) Facultative anaerobes | (c) Obligate anaerobes | (d) Aerotolerant anaerobes | (e) Microaerophiles |

Figure 5.7 Microorganisms have different oxygen requirements. In a static culture, microorganisms occupy different regions of the medium, reflecting their pattern of oxygen usage. (a) Obligate aerobes must grow at or near the surface, where oxygen is able to diffuse. (b) Facultative anaerobes are able to adjust their metabolism to the prevailing oxygen conditions. (c) Obligate anaerobes, in contrast, occupy those zones where no oxygen is present at all. (d) Aerotolerant anaerobes do not use oxygen, but neither are they inhibited by it. (e) Microaerophiles have specific oxygen requirements, and can only grow within a narrow range of oxygen tensions.

5.2.5 Osmotic pressure

Osmosis is the diffusion of water across a semipermeable membrane from a less concentrated solution to a more concentrated one, equalising concentrations. The pressure required to make this happen is called the *osmotic pressure*. If a cell were to be placed in a hypertonic solution (one whose solute concentration is higher),

> *Plasmolysis* is the shrinkage of the plasma membrane away from the cell wall, due to osmotic loss of water from the cell.

osmosis would lead to a loss of water from the cell (*plasmolysis*). This is the basis of using high concentrations of salt or other solutes in preserving foods against microbial attack. In the opposite situation, water would pass from a dilute (hypotonic) solution into the cell, causing it to swell and burst. The rigid cell walls of bacteria prevent them from bursting; this, together with their minute size, makes them less sensitive to variations in osmotic pressure than other types of cell. They are generally able to tolerate NaCl concentrations of between 0.5 and 3.0%. *Haloduric* ('salt-tolerant') bacteria are able to tolerate concentrations ten times as high, but prefer lower concentrations, whereas *halophilic* ('salt-loving') forms are adapted to grow best in conditions of high salinity such as those that prevail in the Dead Sea. In order to do this without plasmolysis occurring, they must build up a higher internal solute concentration, which they do by actively concentrating potassium ions inside the cell.

5.2.6 Light

Phototrophic organisms require light in order to carry out photosynthesis. In the laboratory, care must be taken that light of the correct wavelength is used, and that the source used does not also act as a heat source. Although fluorescent light produces little heat, it does not provide the wavelengths in excess of 750 nm needed by purple and green photosynthetic bacteria.

5.3 The kinetics of microbial growth

Unicellular organisms divide by *binary fission*; each cell grows to full size, replicates its genetic material then divides into two identical daughter cells. By identical means, two cells divide into four, four into eight and so on, leading to an exponential increase in cell numbers:

$$1 \rightarrow 2 \rightarrow 4 \rightarrow 8 \rightarrow 2^n$$

If we were to plot the number of cells in a population against time, we would get an exponential curve (Figure 5.8a). It is more convenient when plotting a growth curve to plot the logarithm of cell numbers against time, giving us a straight line (Figure 5.8b). Such exponential growth cannot continue indefinitely, however, and growth usually slows down either due to the supply of nutrients becoming exhausted, or because metabolism leads to an accumulation of harmful waste substances. Unicellular growth usually occurs in a series of different phases (Figure 5.9):

5.3.1 Lag phase

When an inoculum of bacteria is first introduced into some growth medium, it will probably require a period to adapt to its new surroundings – the less familiar these are, the longer the period of adaptation. If, for example, the carbon source in the medium is unfamiliar, the cells will need time to synthesise the necessary enzymes for its metabolism. The length of the lag phase will also depend on the age and general health of the cells in the inoculum. During this period, there is no net increase in bacterial numbers; however, the cells are metabolically active.

> *Inoculum* is the term given to the cells used to 'seed' a new culture.

5.3.2 Log (exponential) phase

When the bacteria have acclimatised to their new environment and synthesised the enzymes needed to utilise the available substrates, they are able to start regular division by binary fission. This leads to the exponential increase in numbers referred to above. Under optimal conditions, the population of cells will double in a constant and predictable length of time, known as the *generation (doubling) time*. The value for the widely used laboratory

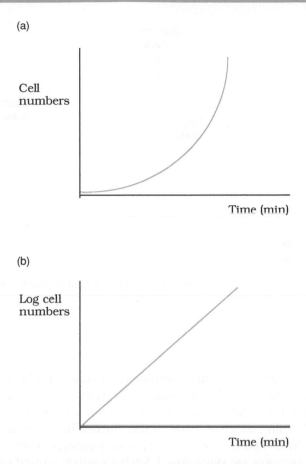

(a)

Cell numbers

Time (min)

(b)

Log cell numbers

Time (min)

Figure 5.8   Cell numbers increase exponentially. (a) Under ideal physico-chemical conditions, the number of cells in a population of a unicellular organism increases exponentially. (b) A plot of the log of cell numbers against time during exponential growth gives a straight line.

bacterium *E. coli* is 20 minutes, and for most organisms it is under 1 hour. There are some bacteria, however, whose generation time is many hours.

Thus, during exponential growth, the number of cells can be expressed as:

$$N_T = N_0 \times 2^n,$$

where N_0 is the number of cells at the start of exponential growth, N_T is the number of cells after time T, and n is the number of doubling times that have elapsed. Therefore n is equal to T/T_d, where T_d is the doubling time.

Substituting into the first equation:

$$N_T = N_0 \times 2^{T/T_d}$$

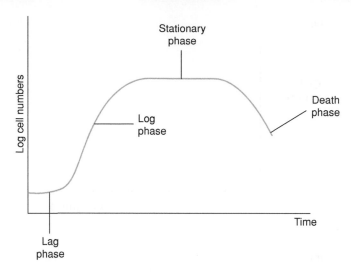

Figure 5.9   A microbial growth curve. The four phases of a typical growth curve are shown. See the text for details.

Box 5.5   Diauxic growth

When *E. coli* grows in a medium containing glucose and lactose, it preferentially metabolises the former since it is more energy efficient to do so. The cell has a regulatory mechanism that suppresses the synthesis of lactose-metabolising enzymes until all the glucose has been used up (see Chapter 11). At this point a second lag phase is entered, while the lactose metabolising enzymes are synthesised. Such growth is termed *diauxic*, and the resulting growth curve is characteristically biphasic.

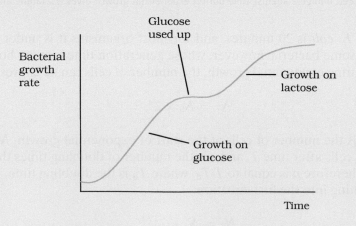

This can be expressed more conveniently by using logarithms to the base 2 (don't worry about how this is done!):

$$\log_2 N_T = \log_2 N_0 + T/T_d$$

Therefore:

$$\frac{\log_2 N_T - \log_2 N_0}{T} = \frac{1}{T_d}$$

Thus if we know the number of cells at the start and end of a period of exponential growth, we can calculate the doubling time. We can also determine the *mean growth rate constant* (*K*); this is a measure of the number of doublings of the population per unit time, and is equal to $1/T_d$. See Box 5.6 for a worked example.

Box 5.6   *Calculating an increase in microbial numbers*

Example:

An inoculum of 10^7 bacterial cells was introduced into a flask of culture medium and growth was monitored. No change was seen for 18 minutes (the lag phase), then growth occurred rapidly. After a further 76 minutes, the population had increased to 4.32×10^8 cells. What is the doubling time (T_d) of the culture?

  (*To obtain values of $\log_2$, multiply $\log_{10}$ values by 3.322*)

$$\frac{1}{T_d} = \frac{\log_2 N_T - \log_2 N_0}{T}$$

$$\frac{1}{T_d} = \frac{\log_2 (4.32 \times 10^8) - \log_2 10^7}{76}$$

$$\frac{1}{T_d} = \frac{(3.322 \times 8.6355) - (3.322 \times 7)}{76}$$

$$\frac{28.6871 - 23.254}{76} = \frac{1}{T_d}$$

$$T_d = \frac{76}{5.4331} = 14 \text{ minutes}$$

Box 5.7   See for yourself!

You can observe growth at a hyphal tip for yourself with a Petri dish culture of a fungus such as *Mucor* and a light microscope fitted with an eyepiece graticule. Focus on the outer visible edge of a colony, then identify the tip of one of the outwardly radiating hyphae. Align it with the graticule scale and note the position of the tip. Return after half an hour and note the new value, being careful not to disturb the microscope in the meantime. You may be surprised at how far the hypha has travelled in such a short time!

Many antibiotics such as penicillin (see Chapter 17) are only effective when cells are actively dividing, since they depend on disrupting new cell wall synthesis.

5.3.3   Stationary phase

As discussed above, the exponential phase is limited by environmental factors, and as the rate of growth slows down, the culture enters the next phase. The levelling out of the growth curve does not mean that cell division has ceased completely, but rather that the increase due to newly formed cells is cancelled out by a similar number of cell deaths. Eventually, however, as the death rate increases, the overall numbers fall and we enter the final phase of growth.

5.3.4   Death phase

As cells die off and the culture is unable to replace them, the total population of viable cells falls. This is the *death* (or *decline*) *phase*.

5.3.5   Batch culture and continuous culture

The phases of growth described above apply to a *batch culture*. In this form of culture, appropriate nutrients and other conditions are provided for growth, then an inoculum is added and the culture incubated. No further nutrients are added and no waste products are removed, thus conditions in the culture are continually changing. This results in active growth being of limited duration for the reasons outlined above. Sometimes it is desirable to keep the culture in the logarithmic phase, for example if the cells are being used to produce alcohol or antibiotics.

Figure 5.10 Continuous culture of microorganisms: the chemostat. By the continual addition of new nutrients and removal of waste, the culture is kept in a steady state, with both cell numbers and nutrient composition remaining constant.

In a *continuous culture*, nutrient concentrations and other conditions are held constant, and the cells are held in a state of exponential growth. This is achieved by continuously adding fresh culture medium and removing equal volumes of the old. Parameters such as pH can also be monitored and adjusted. The equipment used to do this is called a chemostat (Figure 5.10); it produces a steady-state culture whose population size is kept constant by careful control of flow rates and nutrient concentrations.

5.4 Growth in multicellular microorganisms

If uninterrupted, growth in fungi proceeds radially outwards from the initiating spore, allowing the fungal colony to colonise new regions potentially rich in nutrients. Actual growth occurs solely at the hyphal tip; as this happens, the terminal cell grows longer, until eventually a new cross wall or septum is formed. Cells away from the tip do not become any longer during hyphal extension; however, hyphae in this region may develop into aerial reproductive structures. Older hyphae at some distance from the tip may become completely empty of cytoplasm.

Cell counts and turbidometric measurements are not appropriate to estimate growth of fungi; however, total mycelial mass can be measured and its change plotted against time. A fungal growth cycle shows roughly the same phases of growth as described above for bacteria.

Box 5.8   Do you believe in fairies?

The well-known phenomenon of 'fairy rings' can be explained in terms of the radial growth of fungi. As the underground mycelium of certain members of the Basidiomycota extends outwards, it releases enzymes into the soil, degrading organic matter ahead of it and releasing nutrients including soluble nitrogen for the grass, whose growth becomes more lush at this point, and forms the familiar ring. Further back, the branching mycelium outcompetes the overlying grass and deprives it of minerals. Fairy rings are more likely to be found on cultivated land such as lawns and golf courses, because in order to spread uniformly they require a relatively homogeneous medium.

6

Microbial Metabolism

You may have wondered why it was necessary to learn all that biochemistry in Chapter 2; well, you're about to find out! In the following pages, you will learn about the processes by which microorganisms obtain and use *energy*.

6.1 Why is energy needed?

Like all other living things, microorganisms need to acquire energy in order to survive. Energy is required:

- to maintain the structural integrity of the cell by repairing any damage to its constituents;
- to synthesise new cellular components such as nucleic acids, polysaccharides and enzymes;
- to transport certain substances into the cell from its surroundings;
- for the cell to grow and multiply;
- for cellular movement.

Metabolism is the term used to describe all the biochemical reactions that take place inside a cell; it includes those reactions that release energy into the cell, and those that make use of that energy. Figure 6.1 summarises these processes.

As we saw in Chapter 4, most microorganisms obtain their energy from the nutrients they take into the cell; these may come from

> *Catabolism* is the term used to describe reactions that break down large molecules, usually coupled to a release of energy.

Essential Microbiology, Second Edition. Stuart Hogg.
© 2013 John Wiley & Sons, Ltd. Published 2013 by John Wiley & Sons, Ltd.

Figure 6.1   Microorganisms use a variety of processes to generate biochemical energy in the form of ATP. As ATP is broken down to ADP and inorganic phosphate, the energy released is used for the maintenance, reproduction and survival of the cell.

an organic or an inorganic source. Once inside the cell, these nutrients must then be biochemically processed by reactions that trap some of their chemical energy, at the same time breaking them down into smaller molecules. These then serve as building blocks for the synthesis of new cellular components. Chemical compounds contain potential energy within their molecular structure, and some of this can be released when they are broken down. In other metabolic types, energy is obtained from the sun by means of *photosynthesis*; once again, however, the energy is used for synthetic purposes.

> *Anabolism* is the term used to describe reactions involved in the synthesis of macromolecules, usually requiring an input of energy.

Central to the metabolic processes of any cell are *enzymes*. Without them, the many biochemical reactions referred to above simply wouldn't take place at a fast enough rate for living cells to maintain themselves. We shall start our consideration of metabolism by taking a look at enzymes: what they are, and how they work. In the later sections of the chapter, we shall consider in more detail those processes by which energy is acquired and spent.

> An *enzyme* is a cellular catalyst (usually a protein), specific to a particular reaction or group of reactions.

6.2 Enzymes

An *enzyme* is a cellular catalyst; it makes biochemical reactions proceed many times more rapidly than they would if uncatalysed. The participation of an enzyme can increase the rate of a reaction by a factor of millions, or even billions.

Traditionally, all enzymes have been thought of as globular proteins, but in the 1980s it was demonstrated that certain RNA molecules also have catalytic properties. These *ribozymes*, however, are very much in the minority, carrying out specific cut-and-splice reactions on RNA molecules, and in the present context can be ignored. In this book we shall confine our discussion of enzymes to the protein type.

Like any other catalyst, an enzyme remains unchanged at the end of a reaction. It must, however, at some point during the reaction bind to its *substrate* (the substance upon which it acts) to form an *enzyme–substrate complex* (Figure 6.2) by multiple weak forces such as electrostatic forces and hydrogen bonding. Only a small part of the enzyme's three-dimensional structure is involved in this binding; these few amino acids make up the *active site*, which forms a groove or dent in the enzyme's surface, into which the appropriate part of the substrate molecule fits (Figure 6.3). The amino acid residues that go to make up the active site may be widely separated in the enzyme's primary structure, but by means of the secondary and tertiary folding of the molecule, they are brought together to give a specific three-dimensional conformation, complementary to that of the substrate. It is this precise formation of the active site that accounts for one of the major characteristics of enzymes, their *specificity*. You should not think, however, that these few residues making up the active site are the only ones that matter; the enzyme can only fold in this way because the order and arrangement of the other amino acids allow it.

Figure 6.2 Enzyme–substrate interaction. An enzyme interacts with its substrate(s) to form an enzyme–substrate complex, leading to the formation of a product. In the example shown, two substrate molecules are held in position by the enzyme and joined together. Reproduced from Black, JG (1999) Microbiology: Principles and Explorations, 4th edn, with permission from John Wiley & Sons.

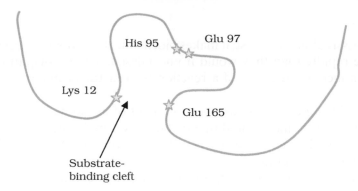

Figure 6.3 Catalytic activity occurs at the active site of an enzyme. A small number of amino acids form the active site of an enzyme, in which substrate molecules are held while bonds are formed or broken. Note how the amino acids that make up the active site may be far apart in the primary sequence, but brought together by subsequent protein folding.

6.2.1 Enzyme classification

Most enzymes have names that end in the suffix *-ase*. The first part of the name often gives an indication of the substrate; for example, urease and pyruvate decarboxylase. Other enzymes have names that are less helpful, such as trypsin, and others have several alternative names to confuse the issue further. To resolve such problems, an internationally agreed system of nomenclature has been devised. All enzymes are assigned initially to one of six broad groups according to the type of reaction they carry out, as shown in Table 6.1. Each enzyme is then placed into successively more specific groupings, each with a number. Thus regardless of any colloquial or alternative names, each

Table 6.1   Major classes of enzymes

Class	Name	Reaction type	Example
1	Oxidoreductases	Oxidation/reduction (electron transfer) reactions	Lactate dehydrogenase
2	Transferases	Transfer of functional groups, e.g. phosphate, amino	Glucokinase
3	Hydrolases	Cleavage of bonds with the addition of water (hydrolysis)	Glucose 6-phosphatase
4	Lyases	Cleavage of C–C, C–O or C–N bonds to form a double bond	Pyruvate decarboxylase
5	Isomerases	Rearrangement of atoms/groups within a molecule	Triose-phosphate isomerase
6	Ligases	Joining reactions, using energy from ATP	DNA ligase

enzyme has its own unique and unambiguous four-figure Enzyme Commission 'signature' (pyruvate decarboxylase, mentioned above, is EC 4.1.1.1).

6.2.2 Certain enzymes have a non-protein component

Many enzymes require the involvement of an additional, non-protein component in order to carry out their catalytic action. These 'extra' parts, or *cofactors*, are usually either metal ions (e.g. Mg^{2+}, Zn^{2+}) or complex organic molecules called *coenzymes*. Some of the most important coenzymes act by transferring electrons between substrate and product in redox reactions (see later).

> The purely protein component of an enzyme is known as the *apoenzyme*. The complex of apoenzyme and cofactor is called the *holoenzyme*. The apoenzyme on its own does not have biological activity.

6.2.3 How do enzymes speed up a reaction?

For any chemical reaction to take place there must be a small input of energy. This is called the *activation energy*, and is often likened to the small push that is needed to loosen a boulder and allow it to roll down a hill. It is the energy needed to convert the molecules at the start of a reaction into intermediate forms known as *transition states*, by the rearrangement of chemical bonds. The great gift of enzymes is that they can greatly *lower the activation energy* of a reaction, so that it requires a smaller energy input, and may therefore occur more readily (Figure 6.4).

Figure 6.4 An enzyme lowers the activation energy of a reaction. By lowering the amount of energy that must be expended in order for a reaction to commence, enzymes enable them to proceed much more quickly. Note that ΔG, the change in free energy, remains the same whether the reaction is catalysed or not.

(a)   (b)

Figure 6.5 Disruption of an enzyme's three-dimensional structure causes denaturation. Disruption of the bonds that form the secondary and tertiary protein structure of an enzyme (a) lead to a loss of catalytic activity, as the amino acids forming the active site are pulled apart (b). Reproduced from Bolsover, SR, et al. (1997) From Genes to Cells, with permission from John Wiley & Sons.

6.2.4 Environmental factors affect enzyme activity

The rate at which an enzyme converts its substrate into product is called its *velocity* (*v*), and is affected by a variety of factors.

Temperature   The rate of any chemical reaction increases with an increase in temperature due to the more rapid movement of molecules, and so it is with enzyme-catalysed reactions, until a peak is reached (the *optimum temperature*) after which the rate rapidly falls away. What causes this drop in the velocity? Recall from Chapter 2 that the very ordered secondary and tertiary structure of a protein molecule is due to the existence of numerous weak molecular bonds, such as hydrogen bonds. Disruption of these by excessive heat results in *denaturation* (Figure 6.5), that is, an unfolding of the three-dimensional structure. In the case of an enzyme, this leads to changes in the configuration of the active site, and a loss of catalytic properties. The effect of temperature on enzyme activity is shown in Figure 6.6. The graph can be thought of as a composite of two lines, one increasing with temperature due to the rise in thermal energy of the substrate molecules, and one falling due to denaturation of the protein structure. Before the optimum temperature it is the former that dominates, then the effect of the latter becomes more pronounced, and takes over completely.

pH   Enzyme velocity is similarly affected by the prevailing pH. Once again, this is due to alterations in three-dimensional protein structure. Changes in the pH affect the ionisation of charged 'R'-groups on amino acids at the active site and elsewhere, causing changes in the enzyme's precise shape, and a reduction in catalytic properties. As with temperature, enzymes have an optimum value at which they operate most effectively; when the pH deviates appreciably from this in either direction, denaturation occurs, leading to a reduction of enzyme activity (Figure 6.7).

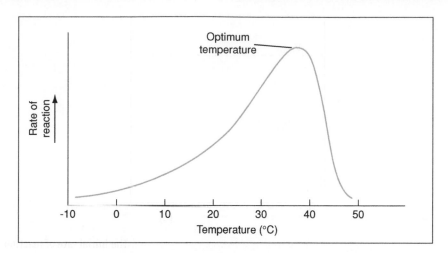

Figure 6.6 Effect of temperature on enzyme activity. Below the optimum temperature, the rate of reaction increases as the temperature rises. Above the optimum, there is a sharp falling off of reaction rate due to thermal denaturation of the enzyme's three-dimensional structure.

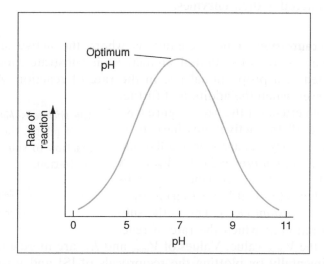

Figure 6.7 Effect of pH on enzyme activity. Either side of the optimum pH value, changes in ionisation of amino acid side chains lead to protein denaturation and a loss of enzyme activity. In the example shown, the optimum value is around pH7; some enzymes work best at acid or alkaline pH values, in which case the whole curve will be shifted to the left or right.

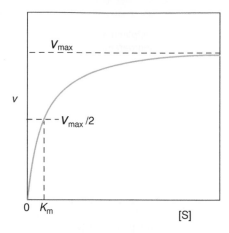

Figure 6.8   Enzyme activity is influenced by substrate concentration. The initial rate of reaction (V_0) is proportional to substrate concentration at low values of [S]. However, when the active sites of the enzyme molecules become saturated with substrate, a maximum rate of reaction (V_{max}) is reached. This cannot be exceeded, no matter how much the value of [S] increases. The curve of the graph fits the Michaelis–Menten equation.

Microorganisms are able to operate in a variety of physico-chemical environments, a fact reflected in the diversity of optimum values of temperature and pH encountered in their enzymes.

Substrate concentration   Under conditions where the active sites of an enzyme population are not saturated, an increase in substrate concentration will be reflected in a proportional rise in the rate of reaction. A point is reached, however, when the addition of further substrate has no effect on the rate (Figure 6.8). This is because all the active sites have been occupied and the enzymes are working flat out; this is called the *maximum velocity* (V_{max}). A measure of the affinity an enzyme has for its substrate (i.e. how closely it binds to it) is given by its *Michaelis constant* (K_m). This is the substrate concentration at which the rate of reaction is half of the V_{max} value. Values of V_{max} and K_m are more easily determined experimentally by plotting the reciprocals of [S] and v to obtain a straight line (Figure 6.9).

> The *Michaelis–Menten equation* relates the rate of a reaction to substrate concentration:
>
> $$v = \frac{V\,max\,[S]}{[S] + K_m}$$

Some enzymes do not obey Michaelis–Menton kinetics. The activity of *allosteric enzymes* is regulated by effector molecules, which bind at a position separate from the active site. By doing so, they induce a conformational change in the active site, which results in activation or inhibition of the enzyme. Thus effector molecules may be of two types, activators or inhibitors.

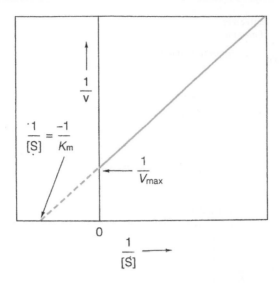

Figure 6.9 The Lineweaver-Burk plot. Plotting the reciprocal values of initial rate of reaction (V_o) and substrate concentration [S] enables values of the Michaelis constant (K_m) and maximum reaction rate (V_{max}) to be derived from the intercepts on a straight-line graph.

Enzyme inhibitors Many substances are able to interfere with an enzyme's ability to catalyse a reaction. As we shall see in Chapter 17, enzyme inhibition forms the basis of several methods of microbial control, so a consideration of the main types of inhibitor is appropriate here.

Perhaps the easiest form of enzyme inhibition to understand is *competitive inhibition*. Here, the inhibitory substance competes with the normal substrate for access to the enzyme's active site; if the active site is occupied by a molecule of inhibitor, it can't bind a molecule of substrate, thus the reaction will proceed less quickly (Figure 6.10a). The competitive inhibitor is able to act in this way because its molecular structure is sufficiently similar to that of the substrate for it to be able to fit into the active site. The effect is competitive because it depends on the relative concentrations of the substrate and inhibitor. If the inhibitor is only present at a low concentration, its effect will be minimal, since the number of enzyme–inhibitor interactions will be greatly outweighed by reactions with the 'correct' substrate. The V_{max} value for the enzyme is not reduced, but it is only reached more gradually. The apparent affinity of the enzyme for its substrate is thus decreased, reflected by an increase in the K_m (Figure 6.11).

Not all inhibitors act by competing for the active site, however. *Noncompetitive* inhibitors act by binding to a different part of the enzyme and in so doing alter its three-dimensional configuration (see Figure 6.10b). Although they do not affect substrate binding, they do reduce the rate at which product is formed. V_{max} thus cannot be reached; however, the value of K_m, is unchanged (Figure 6.12). Such inhibitors may bind to either the

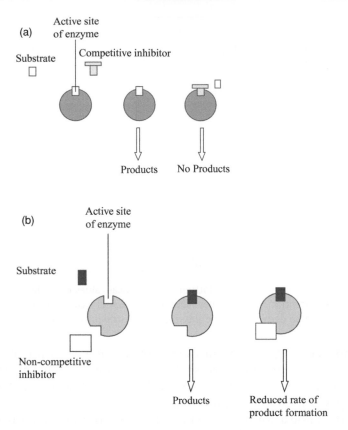

Figure 6.10   Enzyme inhibition. (a) A competitive inhibitor mimics the structure of the normal substrate molecule, enabling it to fit into the active site of the enzyme. Although it is not acted on by the enzyme and no products are formed, such an inhibitor prevents the normal substrate gaining access to the active site. (b) A non-competitive inhibitor binds to a second site on the enzyme and thus does not affect substrate binding; however, distortion of the enzyme molecule makes catalysis less efficient.

enzyme–substrate complex or to free enzyme. Both competitive and non-competitive forms of inhibition are *reversible*, since the inhibitor molecule is relatively weakly bound and can be displaced.

Irreversible inhibition, on the other hand, is due to the formation of a strong covalent linkage between the inhibitor and an amino acid residue on the enzyme. As a result of its binding, the inhibitor effectively makes a certain percentage of the enzyme population permanently unavailable to catalyse substrate conversion.

6.3   Principles of energy generation

In this section, we shall consider how enzyme-catalysed reactions are involved in the cellular capture and utilisation of energy.

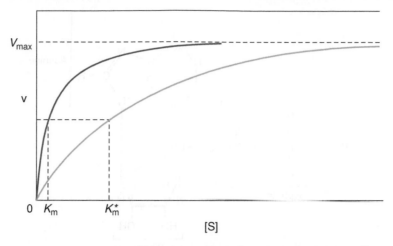

Figure 6.11   Competitive inhibition. In the presence of a competitive inhibitor (lower curve), the maximum rate of reaction (V_{max}) is reached eventually, but the apparent value of the Michaelis constant ($K_m{}^*$) is increased. Since there are fewer enzyme molecules in circulation, the apparent affinity of the enzyme for its substrate is diminished.

Figure 6.12   Non-competitive inhibition. Since product is not formed as efficiently in the presence of a non-competitive inhibitor (lower curve), the maximum rate of reaction achieved (V_{max}^*) is reduced. The Michaelis constant (K_m) is not altered.

Energy taken up by the cell, whether it be in the form of nutrients or sunlight, must be converted into a usable form. A simple analogy is selling goods for cash, which you can then use to buy whatever you want. The 'cash' of cellular metabolism is a compound called *adenosine triphosphate* (ATP). ATP is by far the most important of a class of compounds known as high-energy

Figure 6.13 Adenosine triphosphate (ATP). ATP is a nucleotide, similar to those depicted in Figures 2.20 and 2.21. Note the extra phosphate groups.

transfer compounds, which store the energy[1] from the breakdown of nutrients (or trapped by photosynthetic pigments) and release it when required by the cell. In *catabolic* reactions, in which molecules such as glucose are broken down, energy is released in the form of ATP, which can then be utilised in *anabolic* (synthetic) reactions.

ATP has a structure very similar to the nucleotides found in RNA, except it has two additional phosphate groups (Figure 6.13). The bond that links the third phosphate group requires a lot of energy for its formation, and is often referred to as a *'high-energy' phosphate bond*. Importantly for the cell, when this bond is broken, the same large amount of energy is released, so when ATP is broken down to ADP and a free phosphate group, energy is made available to the cell. It should be noted that 'high-energy' refers to the amount of energy needed to make or break the bond, and not to any intrinsic property of the bond itself. The process of adding or removing a phosphate group is called *phosphorylation* or *dephosphorylation* respectively.

6.3.1 Oxidation-reduction reactions

Many metabolic reactions involve the transfer of *electrons* from one molecule to another; these are called *oxidation-reduction*, or *redox*, reactions. When a molecule (or atom or ion) loses an electron, it is said to be *oxidised*. (Note that despite the terminology, oxygen does not necessarily take part in the reaction.) Conversely, when an electron is gained, the recipient is *reduced*

---

[1] Not all of the energy is converted into ATP. A proportion of it is lost as heat, some of which allows the enzyme-mediated reactions to proceed at a faster rate.

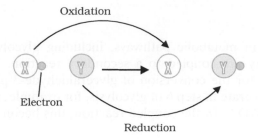

Figure 6.14 *Oxidation-reduction reactions.* When one molecule is oxidised, another is simultaneously reduced. In the example shown, 'X' loses an electron and thus becomes oxidised. By receiving the electron, 'Y' becomes reduced.

(Figure 6.14). Many metabolic reactions involve the loss of a hydrogen atom; since this contains one proton and one electron, the reaction is regarded as an oxidation, because an electron has been lost:

$$
\begin{array}{c}
\text{H} \\
| \\
\text{H}_3\text{C}-\text{C}-\text{COO} \\
| \\
\text{OH}
\end{array}
\longrightarrow
\begin{array}{c}
\\
\text{H}_3\text{C}-\text{C}-\text{COO}^- + 2\text{H}^+ + 2e \\
\| \\
\text{O}
\end{array}
$$

Lactate   Pyruvate

The lactate in the example above, by losing two hydrogen atoms, has automatically lost two electrons and thus become *oxidised* to pyruvate. Oxidation reactions are always associated with the *transfer of energy* from the oxidised substance to the reduced substance.

Two important molecules that we shall encounter a number of times in the following pages are the coenzymes nicotinamide adenine dinucleotide and nicotinamide adenine dinucleotide phosphate. You will be relieved to learn that they are nearly always referred to by their abbreviations, NAD^+ and $NADP^+$ respectively! Both are derivatives of the B vitamin niacin, and each can exist in an oxidised and a reduced form:

$$NAD^+ + H^+ + 2e^- \longleftrightarrow NADH$$
$$NADP^+ + H^+ + 2e^- \longleftrightarrow NADPH$$

Oxidised   Reduced

They are to be found associated with redox reactions (see Box 6.1), acting as carrier molecules for the transfer of electrons. In the oxidation of lactate shown above, the oxidising power is provided by the reduction of NAD^+, so the full story would be:

Lactate $\longrightarrow$ Pyruvate

NAD^+ $NADH + H^+$

Box 6.1   Coupled reactions

Many reactions in metabolic pathways, including glycolysis, can only take place if they are coupled to a secondary reaction. The oxidising power necessary for the conversion of glyceraldehyde-3-phosphate into 1,3-diphosphoglycerate in step 6 of glycolysis, for example, is provided by the coenzyme NAD^+. In the coupled reaction, this becomes reduced to NADH:

NAD⁺   NADH

Glyceraldehyde-3-phosphate ⟶ 1, 3-diphosphoglycerate

As the lactate is oxidised, so the NAD^+ in the coupled reaction is reduced. It is said to act as the *electron acceptor*. $NAD^+/NADH$ is generally involved in catabolic reactions, and $NADP^+/NADPH$ in anabolic ones.

As the equation above shows, there can be no oxidation without reduction, and vice versa; the two are irrevocably linked. The tendency of a compound to lose or gain electrons is termed its *redox potential* (E_o) (Box 6.2).

Box 6.2   Redox potentials

Substances vary in the affinity they have for binding electrons; this can be measured as their oxidation-reduction potential, or *redox potential*, relative to that of hydrogen. The flow of electrons in the electron transport chain occurs because the carriers are arranged in order of their redox potentials, with each having a greater electron affinity (more positive redox potential) than its predecessor. Thus electrons are donated to carriers with a more positive redox potential.

−	Strongly negative redox potential (good electron donors)
Reduction Potential	
0	
+	Strongly positive redox potential (good electron acceptors)

Figure 6.15 Catabolic pathways in heterotrophs. Pathways for the catabolism of proteins, nucleic acids and lipids as well as carbohydrates can all feed into the TCA cycle.

In the following section, we shall examine chemoheterotrophic metabolism, used by the majority of microorganisms to derive cellular energy from the oxidation of carbohydrates. Other groups have evolved their own systems of energy capture, and these will be considered later in the chapter.

Figure 6.15 provides a summary of the catabolic (breakdown) pathways used by heterotrophs. Complex nutrients such as proteins and polysaccharides must be enzymatically broken down and converted to substances that can then enter one of the degradative pathways that lead to energy production.

Glucose is the carbohydrate most widely used as an energy source by cells, and the processes by which it is broken down in the presence of oxygen to give carbon dioxide and water are common to many organisms. These have been very thoroughly studied and can be summarised:

$$C_6H_{12}O_6 + 6O_2 \rightarrow 6CO_2 + 6H_2O$$

What this equation, crucially, does *not* show is that as a result of this process, *energy* is released, and stored in the form of *38 molecules of ATP*, so for completeness, we need to add to the respective sides:

$$38ADP + 38Pi \rightarrow 38ATP$$
$$(Pi = \text{inorganic phosphate group})$$

The release of the energy contained within a molecule of glucose does not occur in a single reaction, but happens gradually, as the result of numerous reactions linked together in *biochemical pathways*, the first of which is *glycolysis* (Figure 6.16). Glycolysis can occur with or without oxygen, and is common to both aerobic and anaerobic organisms. Oxygen is essential, however, for *aerobic respiration*, by which ATP is generated from the products of

Figure 6.16 Glycolysis. Two molecules of ATP are 'spent' in the first stage of glycolysis, in which glucose is converted into the three-carbon compounds glyceraldehyde 3-phosphate and dihydroxyacetone phosphate. During the second stage, four ATPs are produced per molecule of glucose, so there is a net gain of two ATPs. In addition, reducing power is generated in the form of NADH (two molecules per molecule of glucose).

glycolysis. Anaerobes proceed down their own pathways following glycolysis as we shall see, but these lack the ATP-generating power of the aerobic process.

6.3.2   Why glucose?

By concentrating on glucose catabolism in this way, you may think we are ignoring the fate of other nutrient molecules. If you take another look at Figure 6.15, however, you will notice that the breakdown products of lipids, proteins and nucleic acids also find their way into our pathway sooner or later, having undergone transformations of their own.

6.3.3   Glycolysis

The initial sequence of reactions, in which a molecule of glucose is converted to two molecules of *pyruvate*,[2] is called *glycolysis* (Figures 6.16 and 6.17). In the first phase of glycolysis, glucose is phosphorylated and its six-carbon ring structure rearranged, before being cleaved into two three-carbon molecules. In the second phase, each of these undergoes oxidation, resulting in pyruvate.

---

[2] At physiological pH, carboxylic acids such as pyruvic acid and citric acid are found in their ionised form (pyruvate, citrate); however, long-established traditions persist, and you may well find reference elsewhere to the '-ic acid' forms.

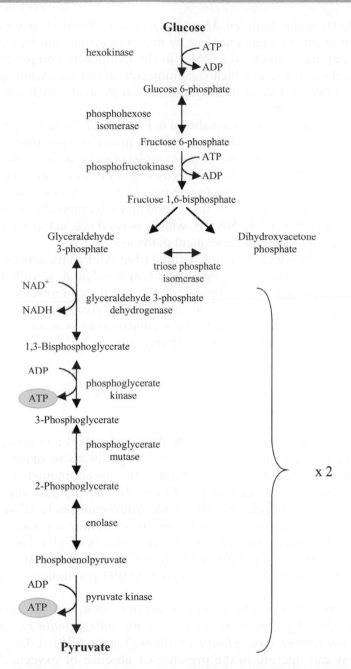

Figure 6.17 *Glycolysis: a more detailed look.* Glycolysis comprises ten separate enzyme-catalysed reactions. Of the two three-carbon compounds formed in the first stage, dihydroxyacetone phosphate cannot directly enter the later pathway, but must first be converted to its isomer glyceraldehyde 3-phosphate. For each molecule of glucose, two molecules of each compound are therefore produced from this point onwards, and the yield of ATP and NADH is likewise doubled.

Also known as the *Embden–Meyerhof pathway*, glycolysis is used for the metabolism of simple sugars not just by microorganisms, but by most living cells. The pathway, which takes place in the cytoplasm, comprises a series of ten linked reactions, in which each molecule of the six-carbon glucose is converted to two molecules of the three-carbon pyruvate, with a net gain of two molecules of ATP.

The full pathway of reactions is shown in Figure 6.17. Note how, in terms of energy, glycolysis can be divided conveniently into a 'sowing' phase, in which two molecules of ATP are *expended* per molecule of glucose, followed by a 'reaping' phase, which *yields* four molecules of ATP. The overall energy balance is therefore a *gain of two molecules of ATP* for each molecule of glucose oxidised to pyruvate. In addition, the second phase features the conversion of two molecules of NAD^+ to NADH, which, as we'll see, act as an important source of *reducing power* in subsequent pathways.

The reactions by which ATP is generated from ADP in the second phase of glycolysis are examples of *substrate-level phosphorylation*, so-called because the phosphate group is transferred directly from a donor molecule.

What happens next to the pyruvate produced by glycolysis depends on the organism concerned, and on whether the environment is aerobic or anaerobic; we shall look at these possibilities in due course.

6.3.4   Glycolysis is not the only way to metabolise glucose

Although glycolysis is widespread in both the microbial and non-microbial worlds, several bacterial types use alternative pathways to oxidise glucose. For certain Gram-negative groups, notably the pseudomonads (see Chapter 7), the main route used is the *Entner–Doudoroff pathway,* producing a mixture of pyruvate and glyceraldehyde-3-phosphate (Figure 6.18). The former, like that produced in glycolysis, can enter a number of pathways, while the latter can feed into the later stages of glycolysis. The net result of catabolism by the Entner–Doudoroff pathway is the production of one molecule each of ATP, NADH and NADPH per molecule of glucose degraded.

A secondary pathway, which may operate in tandem with glycolysis or the Entner–Doudoroff pathway, is the *pentose phosphate pathway,* sometimes known as the *hexose monophosphate shunt* (Figure 6.19). Like glycolysis, the pathway can operate in the presence or absence of oxygen. Although glyceraldehyde-3-phosphate can once again enter the glycolytic pathway and lead to ATP generation, for most organisms the pathway has a mainly anabolic (biosynthetic) function, acting as a source of precursor molecules for other metabolic pathways. The pentose phosphate pathway is a useful source of reducing power in the form of NADPH. In addition it acts as an

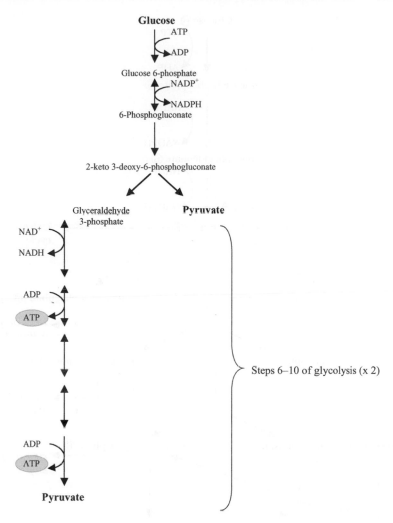

Figure 6.18   The Entner–Doudoroff pathway: an alternative way to metabolise glucose. **The** products of the pathway are pyruvate and glyceraldehyde 3-phosphate (G 3-P). There is a loss of one molecule of ATP in the opening reaction; however, when the G 3-P joins the later stages of glycolysis, two ATPs are generated, giving a net balance of +1. In addition, the pathway yields one molecule each of NADH and NADPH. The intermediate compound 6-phosphogluconate can enter the pentose phosphate pathway and be decarboxylated to the five-carbon compound ribulose 5-phosphate (see Figure 6.19).

important source of precursors in the synthesis of essential molecules; ribose 5-phosphate is an important precursor in the synthesis of nucleotides, while the four-carbon erythrose 4-phosphate is required for the synthesis of certain amino acids, and ribulose 5-phosphate is an intermediate in the Calvin cycle of carbon fixation (see Section 6.4.4).

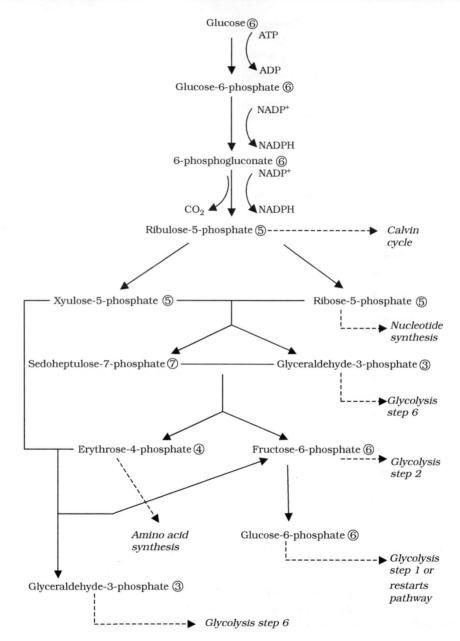

Figure 6.19 The pentose phosphate pathway. Operating simultaneously with glycolysis, the pathway serves as a source of precursors for other metabolic pathways. The metabolic fate of intermediates is indicated in italics. Circled numbers next to each molecule denote the number of carbons.

6.3.5 Aerobic respiration

We shall now examine the fate of the pyruvate produced as the end-product of glycolysis. As we have seen, this depends on whether the organism in question is aerobic or anaerobic.

You will recall that during glycolysis, NAD^+ was reduced to NADH. In order for glucose metabolism to continue, this supply of NAD^+ must be replenished; this is achieved either by *respiration* or *fermentation*. Respiration is the term used to describe those ATP-generating processes, aerobic or anaerobic, by which oxidation of a substrate occurs, with an inorganic substance acting as the final *electron acceptor*. In *aerobic respiration*, that substance is oxygen; in anaerobic respiration, a substance such as nitrate or sulphate can fulfil the role.

In most aerobic organisms, the pyruvate is completely oxidised to CO_2 and water by entering the *tricarboxylic acid (TCA) cycle*, also known as the *Krebs cycle* or simply the *citric acid cycle* (Figure 6.20). During this cycle, a series of redox reactions result in the gradual transfer of the energy contained in the pyruvate to coenzymes (mostly NADH). This energy is finally conserved in the form of ATP by a pro-

> The *TCA cycle* is a series of reactions that oxidise acetate to CO_2, generating reducing power in the form of NADH and $FADH_2$ for use in the electron transport chain.

cess of *oxidative phosphorylation*. We shall turn our attention to these important reactions in due course, but first let us examine the role of the TCA cycle in a little more detail.

Pyruvate does not itself directly participate in the TCA cycle, but must first be converted into the two-carbon compound *acetyl-coenzyme A*:

This is an important intermediate, as lipids and amino acids can also be metabolised into this form, and thereby feed into the TCA cycle. The main features of the cycle are as follows:

- each reaction is catalysed by a separate enzyme;
- four of the reactions involve substrate oxidation, with energy, in the form of electrons, passing to form NADH (mainly) and $FADH_2$;
- the two carbons present in acetyl-CoA are removed as CO_2;
- one reaction involves the generation of ATP by substrate-level phosphorylation.

Figure 6.20 The tricarboxylic acid (TCA) cycle. **Acetyl-CoA** may derive from the pyruvate of glycolysis or from lipid or amino acid metabolism. It joins with the four-carbon oxaloacetate to form the six-carbon citric acid. Two decarboxylation steps reduce the carbon number back to four and oxaloacetate re-enters the cycle once more. Although no ATP results directly from the cycle, the third phosphate on GTP can be easily transferred to ADP (GTP + ADP → GDP + ATP), thus generating one molecule of ATP per cycle. In addition, substantial reducing power is generated in the form of NADH and FADH$_2$. These carry electrons to the electron transport chain, where further ATPs are generated.

For each 'turn' of the citric acid cycle, *one molecule of ATP, three molecules of NADH and one molecule of FADH$_2$* are produced (FADH$_2$ is the reduced form of another coenzyme, flavin adenine dinucleotide, FAD). Since these derive from oxidation of a single acetyl-CoA molecule, we need to double these values per molecule of glucose originally entering glycolysis. Several of the intermediate molecules in the TCA cycle also act as precursors in other pathways, such as the synthesis of amino acids, fatty acids or purines and pyrimidines (see Section 6.5). Other pathways regenerate such intermediates for continued use in the TCA cycle (see Box 6.3).

Box 6.3   The glyoxylate cycle

The components of the TCA cycle may act as precursors for the biosynthesis of other molecules (e.g. both α-ketoglutarate and oxaloacetate can be used for the synthesis of amino acids). For the TCA cycle to continue, it must replace these compounds. Many microorganisms are able to do this by converting pyruvate to oxaloacetate via a carboxylation reaction. A pathway that replenishes intermediate compounds of another in this way is termed *anaplerotic*.

Organisms that use acetate (or molecules that give rise to it, e.g. fatty acids) as their sole carbon source regenerate TCA intermediates by means of the *glyoxylate cycle* (sometimes known as the glyoxylate shunt or bypass). This resembles the TCA cycle, but the two decarboxylation reactions (i.e. those where CO_2 is removed) are missed out (cf. Figure 6.20).

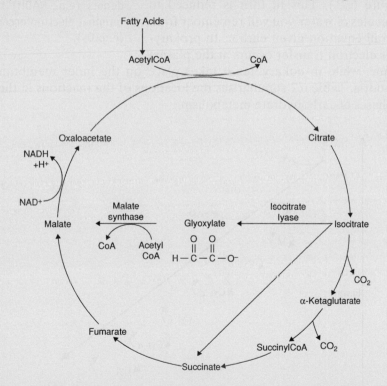

Thus isocitrate is converted directly to succinate and glyoxylate, and in another unique reaction, the glyoxylate is joined by acetyl-CoA to form malate. The result of this is that succinate can be removed to participate in a biosynthetic pathway, but oxaloacetate is still renewed via glyoxylate and malate.

So far, we're a long way short of the 38 molecules of ATP per molecule of glucose mentioned earlier; we've only managed two ATPs from glycolysis and a further two from the TCA cycle. Where do all the rest come from? Most of the energy originally stored in the glucose molecule is now held in the form of the reduced coenzymes (NADH and $FADH_2$) produced during glycolysis and the TCA cycle. This is now converted to no fewer than 34 molecules of ATP per glucose molecule by oxidative phosphorylation in the remaining steps in aerobic respiration (three from each molecule of NADH and two from each of $FADH_2$).

In the final phase of aerobic respiration, electrons are transferred from NAD and $FADH_2$ via a series of carrier molecules, known collectively as the *electron transport* (or *respiratory*) *chain*, to oxygen, the terminal electron acceptor (Figure 6.21). This in turn is reduced to the molecules of water you will remember from our overall equation given earlier. In prokaryotes, this electron transfer occurs at the plasma

> The *electron transport chain* is a series of donor/acceptor molecules that transfer electrons from donors (e.g. NADH) to a terminal electron acceptor (e.g. O_2).

membrane, while in eukaryotes it takes place on the inner membrane of mitochondria. Table 6.2 summarises the locations of the reactions in the different phases of carbohydrate metabolism.

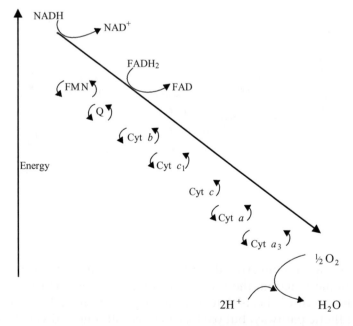

Figure 6.21 The electron transport chain. Electrons from NADH and $FADH_2$ pass from one electron carrier to another, with a gradual release of energy as ATP by chemiosmosis (see Figure 6.22). The electron carriers are arranged in order of their reduction potential (tendency to gain electrons). See text for further details.

Table 6.2 Location of respiratory enzymes

Reaction	Prokaryotes	Eukaryotes
Glycolysis	Cytoplasm	Cytoplasm
TCA cycle	Cytoplasm	Mitochondrial matrix
Electron transport	Plasma membrane	Mitochondrial inner membrane

6.3.6 Oxidative phosphorylation and the electron transport chain

The components of the electron transport chain differ between prokaryotes and eukaryotes, and even among bacterial systems, thus details may differ from the example outlined below. However, the purpose of the electron transport is the same for all systems, that is, the transfer of electrons from NADH/FADH$_2$ via a series of carriers to, ultimately, oxygen. Around half of the energy released during this process is conserved as ATP.

The carrier molecules, which act alternately as acceptors and donors of electrons, are mostly complex modified proteins such as flavoproteins and cytochromes, together with a class of lipid-soluble molecules called ubiquinones (also called coenzyme Q). The carriers are arranged in the chain such that each one has a more positive redox potential than the previous one. In the first step in the chain, NADH passes electrons to flavin mononucleotide (FMN), and in so doing becomes converted back to NAD$^+$, thereby ensuring a ready supply of the latter for the continuation of glycolysis (see Figure 6.21). From FMN, the electrons are transferred to coenzyme Q, and thence to a series of cytochromes. At each transfer of electrons the donor reverts to its oxidised form, ready to pick up more electrons. You may recall that FADH$_2$ yields only two, rather than three molecules of ATP per molecule; this is because it enters the electron transport chain at a later point than NADH, thereby missing one of the points where export of protons occurs. The final cytochrome in the chain transfers its electrons to molecular oxygen, which, as we've seen, acts as the terminal oxygen acceptor. The negatively charged oxygen combines with protons from its surroundings to form water. Four electrons and protons are required for the formation of each water molecule:

$$O_2 + 4e^- + 4H^+ \rightarrow 2H_2O$$

Since two electrons are released by the oxidation of each NADH, it follows that two NADH are needed for the oxidation of each oxygen.

How does this transfer of electrons lead to the formation of ATP? The *chemiosmotic theory* proposed by Peter Mitchell in 1961 offers an explanation. Although it was not immediately accepted, the validity of the chemiosmotic model is now widely recognised, and in 1978 Mitchell received a Nobel

Figure 6.22 Chemiosmosis. The active transport of protons across the membrane creates a gradient of charge and concentration (proton motive force). Special channels containing ATP synthetase allow the return of the protons; the energy released is captured as ATP. Reproduced from Hames, BD, et al. (1997) Instant Notes in Biochemistry, with permission from Taylor & Francis Group.

Prize for his work. As envisaged by Mitchell, sufficient energy is released at three points in the electron transport chain for the transfer of protons to the outside of the membrane, resulting in a gradient of both concentration and charge (*proton motive force*). The protons are able to return across the membrane and achieve an equilibrium through specific protein channels within the enzyme *ATP synthase*. The energy released by the protons as they return through these channels enables the ATP synthase to convert ADP to ATP (Figure 6.22).

Aerobic respiration in eukaryotes is slightly less efficient than in prokaryotes due to the fact that the three stages take place at separate locations (see Table 6.2). Thus the total number of ATPs generated is 36 rather than the 38 in prokaryotes (Table 6.3).

Table 6.3 Yield of ATP by aerobic respiration in prokaryotes

Process	ATP yield (per glucose molecule)
Glycolysis	2
TCA cycle	2
Electron transport chain	34*

*Derived from the oxidative phosphorylation of 2 × NADH from glycolysis, 8 × NADH and 2 × $FADH_2$ from the TCA cycle.

6.3.7   Fermentation

We turn now to the fate of pyruvate when oxygen is not available for aerobic respiration to take place. Microorganisms are able, by means of *fermentation*, to oxidise the pyruvate incompletely to a variety of end products.

It is worth spending a moment defining the word 'fermentation'. The term can cause some confusion to students, as it has come to have different meanings in different contexts. In everyday parlance, it is understood to mean simply alcohol production, while in an industrial context it generally means any large-scale microbial process such as beer or antibiotic production, which may be aerobic or anaerobic. To the microbiologist, the meaning is more precise:

A microbial process by which an organic substrate (usually a carbohydrate) is broken down without the involvement of oxygen or an electron transport chain, generating energy by substrate-level phosphorylation.

Two common fermentation pathways result in the production of lactic acid and ethanol respectively. Both are extremely important in the food and drink industries, and are discussed in more detail in Chapter 18. *Alcoholic fermentation*, which is more common in yeasts than in bacteria, results in pyruvate being oxidised via the intermediate compound acetaldehyde to ethanol.

$$\underset{\text{Pyruvate}}{CH_3COCOO^-} \xrightarrow[\text{pyruvate decarboxylase}]{CO_2} \underset{\text{Acetaldehyde}}{CH_3CHO} \xrightarrow[\text{alcohol dehydrogenase}]{NADH \quad NAD^+} \underset{\text{Ethanol}}{CH_3CH_2OH}$$

Electrons pass from the reduced coenzyme NADH to acetaldehyde, which acts as an electron acceptor, and NAD^+ is thereby regenerated for use in the glycolytic pathway. No further ATP is generated during these reactions, so the only ATP generated in the fermentation of a molecule such as glucose is that produced by the glycolysis steps. Thus, in contrast to aerobic respiration, which generates 38 molecules of ATP per molecule of glucose, fermentation is a very inefficient process, producing just two. Note that all the ATP produced by any fermentation is due to substrate-level phosphorylation, and does not involve an electron transport chain.

A variety of microorganisms carry out *lactic acid fermentation*. Some, such as *Streptococcus* and *Lactobacillus*, produce lactic acid as the only end product; this is referred to as *homolactic* fermentation:

$$\underset{\text{Pyruvate}}{CH_3COCOO^-} \xrightarrow[\text{Lactate dehydrogenase}]{NADH \quad NAD^+} \underset{\text{Lactate}}{CH_3CHOHCOO^-}$$

Certain other microorganisms, such as *Leuconostoc*, generate additional products such as alcohols and acids in a process called *heterolactic* fermentation.

In both alcohol and lactic acid fermentation, the two NADH molecules produced per molecule of glucose are reoxidised to NAD$^+$, ready to re-enter the glycolytic pathway.

6.3.8   Other types of fermentation

Members of the enteric bacteria metabolise pyruvate to a variety of organic compounds; two principal pathways are involved, both of them involving formic acid. In *mixed acid fermentation*, pyruvate is reduced by the NADH to give succinic, formic and acetic acids, together with ethanol (Figure 6.23a). *Escherichia*, *Shigella* and *Salmonella* belong to this group (see Chapter 7). Other enteric bacteria, such as *Klebsiella* and *Enterobacter*, carry out *2,3-butanediol fermentation*. In this, the products are not acidic, and include an intermediate called *acetoin*. Much more CO_2 is produced per molecule of glucose than in mixed acid fermentation (Figure 6.23b). The end products of the two types of fermentation provide a useful diagnostic test for the identification of unknown enteric bacteria.

6.3.9   Metabolism of lipids and proteins

We have concentrated so far on the metabolism of carbohydrates, but both lipids and proteins may also act as energy sources. Both are converted by a series of reactions to an intermediate compound that can then enter the pathways of metabolism we have discussed above.

Lipids often form an important energy source for microorganisms; they are plentiful in nature, as they form the major component of cell membranes, and may also exist as cellular storage structures. Lipids are hydrolysed to their constituent parts by a class of enzymes called *lipases*; the fatty acids so produced enter the cyclic *β-oxidation* pathway. In this, fatty acids are joined to coenzyme A to form an *acyl-CoA*, and shortened by two carbons in a series of reactions (Figure 6.24). Molecules of NADH and FADH$_2$ derived from β-oxidation can enter the electron transport chain to produce ATP. Acetyl-CoA, you will recall from earlier in this chapter, serves as the entry point into the TCA cycle. When you consider that a single turn of the TCA cycle gives rise to the production of 14 molecules of ATP, you can appreciate what a rich source of energy a 16- or 18-carbon fatty acid represents. The glycerol component of a lipid requires only slight modification in order to enter the glycolytic pathway as dihydroxyacetone phosphate (see Section 6.3.3 'Glycolysis').

Proteins are a less useful source of energy than lipids or carbohydrates, but may be utilised when these are in short supply. Like lipids, they are initially

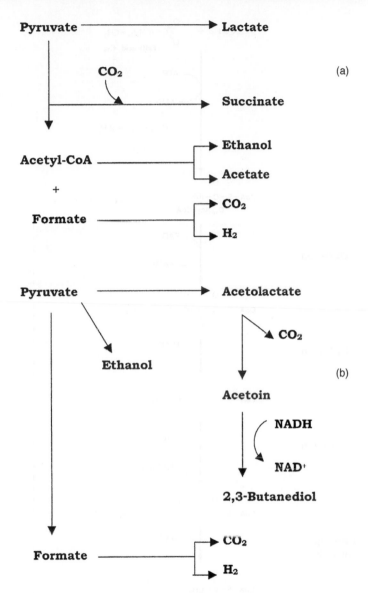

Figure 6.23 Patterns of fermentation. All members of the group produce pyruvate via the Embden–Meyerhof pathway (glycolysis), but subsequent reactions fall into one of two main types: (a) Mixed acid fermentation results in ethanol and a mixture of acids, mainly acetic, lactic, succinic and formic (e.g. *Escherichia, Salmonella*). (b) Butanediol fermentation involves the conversion of pyruvate to acetoin, then to 2,3-butanediol (e.g. *Enterobacter, Klebsiella*). The ratio of CO_2 to H_2 production is much greater in butanediol fermentation.

Figure 6.24 β-Oxidation of lipids. β-Oxidation comprises a series of four reactions, repeated for the removal of each two-carbon unit. Formation of the acyl-CoA ester requires the expenditure of ATP, but there is a net gain in reducing power (NADH + FADH₂), which can feed into the electron transport chain. The shortened acyl chain at the end of the process can re-enter the cycle and become further shortened, while acetyl CoA can enter directly into the TCA cycle. (CoA-SH = coenzyme A).

hydrolysed to their constituent 'building blocks', in this case, amino acids. These then undergo the loss of an amino group (deamination), resulting in a compound that is able to enter, either directly or indirectly, the TCA cycle.

$$
\underset{\text{Alanine}}{CH_3 - \overset{\displaystyle \overset{NH_2}{|} \, \overset{NH_2}{|}}{CH} - COOH} \longrightarrow \underset{\text{Pyruvate}}{CH_3 - \overset{\displaystyle \overset{O}{|} \, \overset{O}{|}}{C} - COO^- + NH_4^+}
$$

6.3.10 Anaerobic respiration

In the process of *anaerobic respiration*, carbohydrate can be metabolised by a process that utilises oxidative phosphorylation via an electron transport chain, but instead of oxygen serving as the terminal electron acceptor, a (usually) inorganic molecule such as nitrate or sulphate is used. These processes are referred to, respectively, as dissimilatory nitrate or sulphate reduction. *Obligate anaerobes* carry out this process, as they are unable to utilise oxygen; in addition, other organisms may turn to this form of respiration if oxygen is unavailable (*facultative anaerobes*). Other examples of inorganic electron acceptors for anaerobic respiration include Fe^{3+}, CO_2 and Mn^{4+}. In certain circumstances, an organic molecule such as fumarate may be used instead.

Anaerobic respiration is not as productive as its aerobic counterpart in terms of ATP production, because electron acceptors such as nitrate or sulphate have lower positive redox potentials than oxygen. Anaerobic respiration tends to occur in oxygen-depleted environments such as waterlogged soils. It must be stressed that anaerobic respiration is *not* the same as fermentation. The latter process does not involve the components of the electron transport chain (i.e. there is no oxidative phosphorylation), and generates much smaller amounts of energy.

6.3.11 Energy may be generated by the oxidation of inorganic molecules

In the previous pages, we have seen how electrons derived from a variety of organic sources can be channelled into the glycolytic pathway (or one of its alternatives), and how energy is generated by either oxygen or an organic/inorganic molecule acting as an electron acceptor. Certain bacteria, however, are able to derive their energy from the oxidation of inorganic substrates; these are termed *chemolithotrophs* (see Chapter 4). Molecules such as

NH_4^+, NO_2^-, Fe^{2+} and S^0 can be oxidised, with the concomitant generation of ATP.

$$NH_4^+ \rightarrow NO_2^-$$
$$NO_2^- \rightarrow NO_3^-$$
$$Fe^{2+} \rightarrow Fe^{3+}$$
$$S^0 \rightarrow SO_4^{2-}$$

The ΔG (change in free energy) for each of these reactions is much smaller than that for aerobic respiration. The value of ΔG is a measure of how much energy is released by a reaction. Thus bacteria using this form of metabolism need to oxidise a larger amount of their substrate in order to synthesise the same amount of cellular material.

In most cases, such bacteria are autotrophs, fixing carbon from carbon dioxide via the *Calvin cycle*. This is also used by phototrophic organisms, and is described at greater length in Section 6.4 on photosynthesis. If organic carbon is available, however, some organisms are able to act as heterotrophs, deriving their carbon, but not their energy, from such molecules. The overall energy yield from inorganic oxidation is much lower than that from aerobic respiration.

> The *Calvin cycle* is a pathway for the fixation of carbon dioxide used by photosynthetic organisms and some chemolithotrophs.

6.4 Photosynthesis

Photosynthetic organisms are differentiated from all other forms of life by their ability to derive their cellular energy not from chemical nutrients, but from the energy of the sun itself. A number of different microbial types are able to carry out photosynthesis, which we can regard as having two distinct forms:

* *oxygenic* photosynthesis, in which oxygen is produced; found in algae and cyanobacteria (blue greens), and in the non-microbial world, green plants;
* *anoxygenic* photosynthesis, in which oxygen is not generated; found in the purple and green photosynthetic bacteria.

Both forms of photosynthesis are dependent on a form of the pigment *chlorophyll*, similar, but not identical, to the chlorophyll found in green plants. We might at this point mention that there is a third method of phototrophic growth, quite distinct from the other two, in that chlorophyll plays no role. This form, which is found in halophilic members of the Archaea (see Chapter 7), uses instead a pigment called *bacteriorhodopsin*, similar in structure and function to the rhodopsin found in the retina of animals' eyes. In view of the fact that carbon dioxide is not fixed by this mechanism, and no form of chlorophyll is involved, it does not qualify to be described as photosynthesis by some definitions.

The reactions that make up photosynthesis fall into two distinct phases: in the *'light' reactions*, light energy is trapped and some of it conserved as ATP, and in the *'dark' reactions* the energy in the ATP is used to drive the synthesis of carbohydrate by the reduction of carbon dioxide. In the description of photosynthesis that follows, we shall discuss first the reactions of oxygenic photosynthesis, and then consider how these are modified in the anoxygenic form.

6.4.1 Oxygenic photosynthesis

The overall process of oxygenic photosynthesis can be summarised by the equation:

$$6CO_2 + 6H_2O \longrightarrow C_6H_{12}O_6 + 6O_2$$

You will perhaps have noticed that the equation is in essence the reverse of the one we saw earlier when discussing aerobic respiration. In fact, the equation above is not strictly accurate, since it is known that all the oxygen is derived from water. It is therefore necessary to rebalance the equation thus:

$$6CO_2 + 12H_2O \rightarrow C_6H_{12}O_6 + 6O_2 + 6H_2O$$

Unlike the metabolic reactions we have encountered so far, this reaction requires an input of energy, so the value of ΔG is positive rather than negative.

6.4.2 Where does photosynthesis take place?

In photosynthetic eukaryotes, photosynthesis takes place in specialised organelles, the *chloroplasts*. The light-gathering pigments are located on the stacks of flattened *thylakoid* membranes, while the dark reactions occur within the *stroma* (see Figure 3.16). The light reactions of cyanobacteria also take place on structures called thylakoids; however, since prokaryotic cells do not possess chloroplasts, these exist free in the cytoplasm. Their surface is studded

> *Thylakoids* are photosynthetic membranes found in chloroplasts or free in the cytoplasm (in cyanobacteria). They contain photosynthetic pigments and components of the electron transport chain.

with knob-like *phycobilisomes*, which contain unique accessory pigments called *phycobilins*.

> ## Box 6.4   How does a molecule become excited?
>
> When a photon of light strikes a pigment molecule, it is *absorbed*, instead of being reflected or going straight through, and the pigment molecule is raised to an excited (= higher energy) state. An electron in the pigment molecule moves to a higher electron shell (i.e. one further out); this is an unstable state, and the energy it has absorbed is either lost as light (fluorescence) or heat, or is transferred to an acceptor molecule. The pigment molecule then returns to its ground state. Chlorophyll in a test tube will re-emit light as fluorescence, but when present in intact chloroplasts or blue-green cells, it is able to pass on the energy to the next component of one or other photosystem.

6.4.3   'Light' reactions

The light reactions result in:

- splitting of water to release O_2 (photolysis);
- reduction of $NADP^+$ to NADPH;
- synthesis of ATP.

> **Over in a flash!**
> A molecule of chlorophyll may remain in its excited state for less than one picosecond (10^{-12} s).

   This first stage of photosynthesis is dependent on the ability of *chlorophyll* to absorb photons of light. Absorption of light causes a rearrangement of the electrons in the chlorophyll, so that the molecule attains an 'excited' state (see Box 6.4). Chlorophyll belongs to a class of organic compounds called tetrapyrroles, centred on a magnesium atom (Figure 6.25); a hydrophobic side chain allows the chlorophyll to embed itself in the thylakoid membrane. Several variants of the chlorophyll molecule exist, which differ slightly in their structure and the wavelengths of light they absorb. In organisms carrying out oxygenic photosynthesis, chlorophyll *a* and *b* predominate, while various *bacteriochlorophylls* operate in the anoxygenic phototrophs. Chlorophyll *a* absorbs light in the red and blue parts of the spectrum and reflects or transmits the green part (Figure 6.26); thus cells containing chlorophyll *a* appear green (unless masked by another pigment). Although other pigments are capable of absorbing light, only chlorophyll is able to pass the excited electrons via a series of electron acceptors/donors to convert $NADP^+$ to its reduced form, NADPH. Associated with chlorophyll are several *accessory pigments* such as *carotenoids* or (in cyanobacteria) phycobilins with their own absorption characteristics. They absorb light and transfer the energy to chlorophyll, enabling the organism to utilise light from a broader range of wavelengths.

Figure 6.25 Chlorophyll structure. All chlorophyll molecules are based on a tetrapyrrole ring centred on an atom of magnesium. Different types have a variety of side chains around the molecule, resulting in different light absorbance properties. The hydrophobic phytyl chain allows the chlorophyll to insert into thylakoid membranes, where photosynthetic reactions take place.

The light-gathering process occurs in collections of hundreds of pigment molecules known as *photosynthetic units*. Most of these act as antenna molecules, which absorb photons of radiant light. A large number of antenna molecules 'funnel' light towards a single reaction centre, allowing an organism to operate efficiently even at reduced light intensities. Following excitation of the chlorophyll molecule by a photon of light, an electron is transferred from one antenna to another, eventually reaching a specialised chlorophyll molecule in a pigment/protein complex called a *reaction centre* (Figure 6.27). There are two forms of this special chlorophyll, which absorb light maximally at different wavelengths, 680 nm and 700 nm. These are the starting points of two different pathways, known as *photosystem I* and *photosystem II*, which act in sequence together (as we shall see later in this chapter, anoxygenic photosynthesisers only have one of these systems).

Wavelengths of light are measured in *nanometres* (nm). A nanometre is one-millionth of a millimetre $(10^{-9}$ m).

Figure 6.26 Absorbance spectrum of chlorophyll *a*. Chlorophyll *a* absorbs light maximally in the blue and red regions of the spectrum. The green and yellow regions are reflected, giving the characteristic coloration to plants and algae.

The reactions of *photosystem I* are responsible for the reduction of NADP$^+$ (Figure 6.28). In its excited state, the chlorophyll molecule has a negative redox potential and sufficient energy to donate an excited electron to an acceptor molecule called *ferredoxin*. This is an electron carrier capable of existing in oxidised and reduced forms, like those we encountered in the electron transport chain of aerobic respiration. From here, the electron is passed through a series of successively less electronegative electron carriers

Figure 6.27 Light energy trapped by antenna chlorophyll passes to a reaction centre. Most of the chlorophyll molecules do not take part in converting light energy into ATP, but transfer electrons to specialised reaction centre chlorophylls.

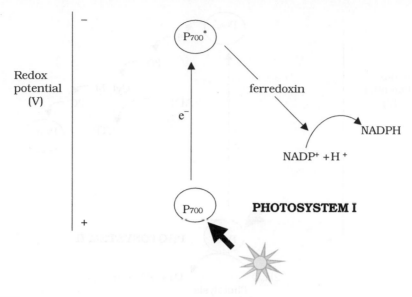

Redox potential (V)

Figure 6.28   Photosystem I. Electrons pass from excited chlorophyll P_{700} through a series of electron acceptors, eventually reducing $NADP^+$. Hydrogen ions are provided by the photolysis of water. The asterisk indicates the excited form of reaction centre chlorophyll.

until $NADP^+$ acts as the terminal electron acceptor, becoming reduced to NADPH. The hydrogen ions necessary for this are derived from the splitting of water (see later). In order for $NADP^+$ to become reduced as described above, the chlorophyll P_{700} must receive a constant supply of electrons. There are two ways in which this may be done: the first involves *photosystem II*. It is here that water is cleaved to produce oxygen in a process involving a light-sensitive enzyme:

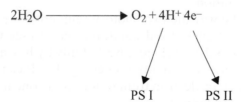

Chlorophyll P_{680} of photosystem II absorbs a photon of light, and is raised to an excited state, causing an electron to be released down an electrochemical gradient via carrier molecules as described above (Figure 6.29). The terminal electron acceptor for photosystem II is the chlorophyll P_{700} of photosystem I.

Photophosphorylation is the synthesis of ATP using light energy.

Figure 6.29   Photosystem II. Electrons released from excited molecules of chlorophyll P_{680} pass down an electrochemical gradient and replenish the supply for the chlorophyll P_{700} of photosystem I. The electron flow causes protons to be pumped across the photosynthetic membrane and drive the chemiosmotic synthesis of ATP. Electrons lost by chlorophyll P_{680} are replaced by electrons from the photolysis of water. Ph = phaeophytin, PQ = plastoquinone, PC = plastocyanin, Cyt bf = cytochrome bf.

The electron flow during this process releases sufficient energy to drive the synthesis of ATP. This process of *photophosphorylation* occurs by means of a chemiosmotic mechanism similar to that involved in the electron transport chain of aerobic respiration.

When we look at the way the two photosystems combine to produce ATP and NADPH from light energy and water, it is easy to see why this is some-times known as the 'Z' scheme (Figure 6.30). Only by having the two photo-systems operating in series and at different energy levels can sufficient energy be generated to extract an electron from water on the one hand and generate ATP and NADPH on the other.

As indicated above, there is an alternative pathway by which the electron supply of the chlorophyll P_{700} can be replenished. You will recall that in pho-tosystem I, having reached ferredoxin and reduced it, electrons pass through further carriers before reaching $NADP^+$. Sometimes, however, they may take a different route, joining the electron transport chain at *plastoquinone* (PQ), and generating further ATP (Figure 6.31). The electron ends up back at the P_{700}, so the pathway is termed *cyclic photophosphorylation*. Note that because no splitting of water is involved in the cyclic pathway, no oxygen is

Figure 6.30 Non-cyclic photophosphorylation: the 'Z' scheme. **Photosystems I and II interact to convert water and light energy into ATP and NADPH. P_{680} and P_{700} indicate the wavelength of light maximally absorbed by the two photosystems.**

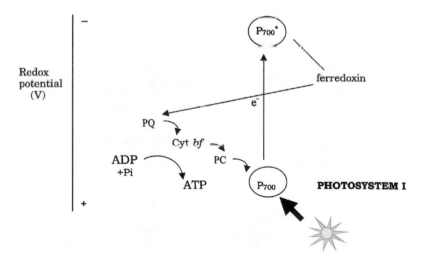

Figure 6.31 Cyclic photophosphorylation. **Photosystem I electrons may enter the electron transport chain at plastoquinone and return to reaction centre chlorophyll P_{700}, generating a proton motive force for use in ATP production.**

generated. Also, because the electrons follow a different pathway, NADPH is not produced.

Non-cyclic photophosphorylation: ATP + NADPH + O_2 produced
Cyclic photophosphorylation: only ATP produced

As we shall see later, the biosynthetic reactions of the Calvin cycle, which follow the light reactions, require more ATP than NADPH. This additional ATP is provided by cyclic photophosphorylation.

6.4.4 'Dark' reactions

The term 'dark reactions' is somewhat misleading; while they *can* take place in the dark, they are dependent upon the ATP and NADPH produced by the light reaction. The most widely used mechanism for the incorporation of carbon dioxide into cellular material is the *Calvin cycle* (Figure 6.32). This is also the means by which many other, non-photosynthetic, autotrophs fix CO_2.

Carbon dioxide entering the Calvin cycle combines with a five-carbon compound called ribulose 1,5-bisphosphate to form two molecules of 3-phosphoglycerate via a transient six-carbon intermediate (not shown). The enzyme responsible for this fixation of CO_2 is called ribulose bisphosphate

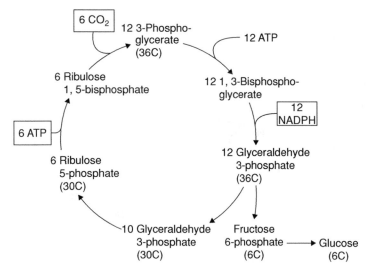

Figure 6.32 The Calvin cycle. Carbon enters the cycle as carbon dioxide and leaves as glyceraldehyde 3-phosphate, from which hexose sugars are formed. Note that only a small part of each hexose molecule produced by this pathway derives directly from the carbon dioxide (only one carbon out of six), thus six complete turns of the cycle would be required to generate a single glucose molecule. (The numbers in the figure relate to six turns.) Although used by most autotrophs, the Calvin cycle is not found in members of the Archaea.

carboxylase (*Rubisco*), and it is the most abundant protein in the natural world. The 3-phosphoglycerate is then reduced to give glyceraldehyde 3-phosphate (G 3-P), in a process that uses both ATP and NADPH from the light reaction. By a series of reactions that are essentially the reverse of the opening steps of glycolysis, some of this G 3-P is converted to glucose. The remaining steps of the Calvin cycle are concerned with regenerating the five-carbon ribulose 5-phosphate. We can summarise the incorporation of CO_2 into glucose as:

$$6CO_2 + 18ATP + 12NADPH + 2H^+ + 12H_2O \rightarrow C_6H_{12}O_6 + 18ADP + 18Pi + 12NADP^+$$

6.4.5 Anoxygenic photosynthesis

Photosynthesis as carried out in the purple and green bacteria is different from the process just described in a number of respects (Table 6.4). Some of the main differences are summarised below:

- No oxygen is generated during this type of photosynthesis, and the bacteria involved grow anaerobically.
- Bacteriochlorophylls absorb light maximally at longer wavelengths than chlorophyll *a* and *b*, allowing them more effectively to utilise the light available in their own particular habitat.
- Purple and green bacteria are not able to utilise water as a donor of electrons, and must instead use a compound that is oxidised more easily, such as hydrogen sulphide or succinate.
- Only a single photosystem is involved in the light reactions of anoxygenic photosynthesis. In the green bacteria this is similar to photosystem I, while in the purples it more closely resembles photosystem II.
- Thylakoid membranes are not found in the green and purple bacteria; light reactions take place in lamellar invaginations of the cytoplasmic membrane in the purples and in vesicles called *chlorosomes* in the greens.

Table 6.4 Comparison of anoxygenic photosynthetic bacteria

	Green sulphur	Green non-sulphur	Purple sulphur	Purple non-sulphur
Bacteriochlorophyll	*c, d, e*	*a, c*	*a, b*	*a, b*
Electron donor	H_2S/reduced S/H_2	Organic compounds	H_2S/reduced S/H_2	Organic compounds
CO_2 fixation	Reverse TCA cycle	Calvin cycle	Calvin cycle	Calvin cycle
Photoheterotrophy	No	Yes	Some	Mainly

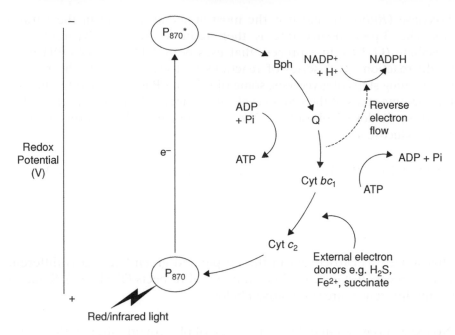

Figure 6.33 Electron flow in the anoxygenic photosynthesis of a purple bacterium. ATP is generated by the passage of electrons down an electron transport chain back to the reaction centre bacteriochlorophyll. Anoxygenic photosynthetic bacteria use molecules such as sulphur and hydrogen sulphide instead of water as external electron donors, hence no oxygen is generated. NADPH for use in CO_2 fixation must be generated against an electrochemical gradient by reverse electron flow. Bph = bacteriophaeophytin.

In the generation of ATP, a form of cyclic photophosporylation is employed; the bacteriochlorophyll acts as both donor and acceptor of electrons (Figure 6.33). In order to generate reducing power in the form of reduced coenzymes, an external electron source is necessary, since this process is non-cyclic. In the green and purple sulphur bacteria, this role is served by sulphur or reduced sulphur compounds such as sulphide or thiosulphate. The non-sulphur bacteria utilise an organic molecule such as succinate as an electron donor. In some cases, the electron donor has a more positive redox potential than the $NADP^+$, so in order to reduce the coenzyme, electrons would have to flow against the electrochemical gradient. An input of energy in the form of ATP is needed to make this possible, in a process known as *reverse electron flow*. This also happens in many chemolithotrophs such as *Acidithiobacillus* and *Nitrobacter*. A special form of anoxygenic photosynthesis occurs in the Heliobacteria (see Chapter 7); these carry out photoheterotrophy using a unique form of bacteriochlorophyll called bacteriochlorophyll *b*. They are unable to use carbon dioxide as a carbon source.

6.5 Anabolic reactions

So far, in describing respiration and photosynthesis, we have considered those reactions whereby a microorganism may generate cellular energy from its environment. As we saw at the beginning of this chapter, one of the uses to which this may be put is the synthesis of new cellular materials. In all of the pathways described in the following section, the conversion of ATP to ADP is required at some point. The term *biosynthesis* is used to describe those reactions by which nutrients are incorporated first into small molecules such as amino acids and sugars and subsequently into biomacromolecules such as proteins and polysaccharides.

6.5.1 Biosynthesis of carbohydrates

We have already seen that autotrophic organisms (not necessarily phototrophic ones) are able to incorporate inorganic carbon as CO_2 or HCO_3^- into hexose sugars, most commonly via the Calvin cycle. Heterotrophic organisms are unable to do this, and must convert a range of organic compounds into glucose by a series of reactions called *gluconeogenesis* (Figure 6.34). Many compounds such as lactate or certain amino acids can be converted to pyruvate either directly or via other members of the TCA cycle, and thence to glucose. To all intents and purposes, gluconeogenesis reverses the steps of glycolysis (see Section 6.3.3), although not all the enzymes involved are exactly the same. This is because three of the reactions are essentially irreversible, so other enzymes must be used to overcome this. These reactions are highlighted in Figure 6.34.

Once glucose or fructose has been produced, it can be converted to other hexose sugars by simple rearrangement reactions. Building up these sugars into bigger carbohydrates (polysaccharides) requires them to be in an energised form: this usually takes the form of either an ADP- or UDP-sugar, and necessitates an input of energy. Pentose sugars such as ribose are important in the synthesis of nucleotides for nucleic acids and coenzymes (see Section 6.5.3).

6.5.2 Biosynthesis of lipids

Fatty acids are synthesised by a stepwise process that involves the addition of two-carbon units to form a chain, most commonly of 16–18 carbons. The starting point of fatty acid metabolism is the two-carbon compound acetyl-CoA.

The basic building blocks in the synthesis of fatty acids are acetyl-CoA (two-carbon) and malonyl-CoA (three-carbon). We have encountered

Figure 6.34 Gluconeogenesis. Non-carbohydrate precursors can feed into a pathway that converts pyruvate to glucose in a series of reactions that are mostly the reverse of glycolysis. Enzymes not found in glycolysis are shown in italics.

acetyl-CoA before, when discussing the TCA cycle; malonyl-CoA is formed by the carboxylation of acetyl-CoA:

$$\text{Acetyl-CoA} + CO_2 \rightarrow \text{Malonyl-CoA}$$

2-carbon   3-carbon

Carbon dioxide is essential for this step, but is not incorporated into the fatty acid as it is removed in a subsequent decarboxylation step. In order to take part in the biosynthesis of fatty acids, both molecules have their coenzyme A element replaced by an *acyl carrier protein* (ACP). In a condensation reaction, one carbon is lost as CO_2 and one of the ACPs is released. The resulting four-carbon molecule is reduced, with the involvement of two NADPH molecules, to *butyryl-ACP*. This is then extended two carbon atoms at a time by a series of further condensations with malonyl-ACP (Figure 6.35).

Thus, extending a fatty acid chain by two carbons involves the expenditure of one ATP and two NADPH molecules. The overall equation for the synthesis of a 16-carbon fatty acid such as palmitic acid can be represented

Figure 6.35 Fatty acid biosynthesis. Acetyl- and malonyl-ACPs condense with the loss of CO_2 to give a four-carbon molecule butyryl-ACP. The addition of further two-carbon acetyl groups is achieved by re-entering the pathway and reacting with further molecules of malonyl-ACP.

thus:

$$8\text{Acetyl-CoA} + 7\text{ATP} + 14\text{NADPH} + 6\text{H}^+ \rightarrow \text{Palmitate} + 14\text{NADP}^+$$
$$+ 8\text{CoA} + 6\text{H}_2\text{O} + 7\text{ADP} + 7\text{Pi}$$

Once formed, fatty acids may be incorporated into phospholipids, the major form of lipid found in microbial cells. Recall from Chapter 2 that a phospholipid molecule has three parts: fatty acid, glycerol and phosphate. These last two are provided in the form of glycerol phosphate, which derives from the dihydroxyacetone of glycolysis (see Figure 6.16). Glycerol phosphate replaces the ACP of two fatty acid-ACP conjugates to yield phosphatidic acid, an important precursor for a variety of membrane lipids. The energy for this reaction is provided, unusually, not by ATP but by CTP (cytidine triphosphate).

6.5.3   Biosynthesis of nucleic acids

Most microorganisms are able to synthesise the purines and pyrimidines that comprise the nitrogenous bases of DNA and RNA. These compounds are synthesised from a number of sources in reactions that require an input of ATP. The contribution of different compounds towards the purine skeleton of guanine or adenine is shown in Figure 6.36. The purines are not synthesised as free bases but are associated with ribose 5-phosphate as complete nucleotides from the outset. Inosinic acid, which is formed initially, acts as a precursor for the other purine nucleotides.

Pyrimidines have a similarly complex synthesis. The amino acids aspartate and glutamine are involved in the synthesis of the precursor orotic acid. Note that unlike the purines, the skeleton of pyrimidines is fully formed *before* association with the ribose 5-phosphate moiety, which is itself derived from glucose (see Section 6.5.1 'Biosynthesis of carbohydrates').

Figure 6.36 Several precursors contribute towards the formation of the purine base skeleton. The nitrogen atoms are donated by the amino acids glutamine, aspartate and glycine. Note the important role played by folic acid. The antimicrobial agent sulphonamide (see Chapter 17) exerts its effect by inhibiting folic acid synthesis, which in turn affects synthesis of purine nucleotides.

Ribonucleotides (as contained in RNA) are converted to deoxyribonucleotides (as contained in DNA) by a reduction reaction, which may involve vitamin B_{12} acting as a cofactor.

6.5.4 Biosynthesis of amino acids

A very limited number of microorganisms are able to utilise molecular nitrogen from the atmosphere by incorporating it initially into ammonia and subsequently into organic compounds (see Chapter 7). Most organisms, however, need to have their nitrogen supplied as nitrate, nitrite or ammonia itself. Ammonia can be incorporated into organic nitrogen compounds in several ways, including glutamate formation from α-ketoglutarate (see TCA cycle in Section 6.3.5):

$$\alpha\text{-Ketoglutarate} + NH_4^+ + NADPH^* + H^+ \rightarrow \text{Glutamate} + NADP^+ + H_2O$$

(*Some species utilise NADH as their electron donor)

The amino group can subsequently be transferred from the glutamate to make other amino acids by *transamination* reactions involving other keto acids; for example:

$$\text{Glutamate} + \text{Oxaloacetate} \rightarrow \text{Aspartate} + \alpha\text{-Ketoglutarate}$$

Glutamate plays a central role in the biosynthesis of other amino acids, as it usually donates the primary amino group of each; for example:

According to the precursor molecule from which they derive, amino acids can be placed into six 'families' (Figure 6.37). The precursors are intermediates in metabolic pathways we have already encountered in this chapter, such as glycolysis or the TCA cycle. When amino acids are broken down, they are likewise broken down into a handful of metabolic intermediates, which then feed into the TCA cycle.

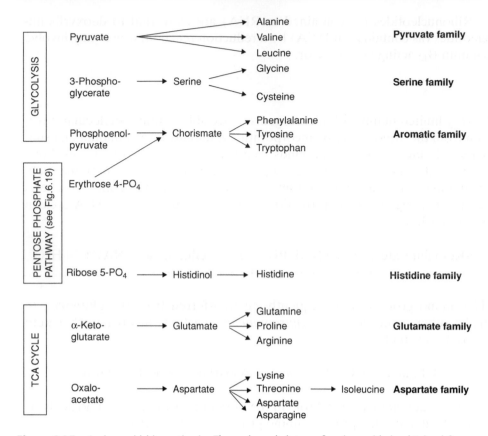

Figure 6.37 Amino acid biosynthesis. **The carbon skeleton of amino acids is obtained from a limited number of precursor molecules, mainly intermediates in glycolysis or the TCA cycle. The amino group originally derives from inorganic sources, but can then be transferred from one organic molecule to another.**

6.6   The regulation of metabolism

Microorganisms, like the rest of us, live in a changing world, and their needs do not always remain the same. It would be highly inefficient and (frequently wasteful) if all their metabolic reactions were going on with equal intensity all the time, regardless of whether they were needed. Over time, regulation systems have evolved, so that metabolism is tailored to the prevailing conditions. Essentially, this regulation involves controlling the activity of enzymes that direct the many biochemical reactions occurring in each cell. This can be done by:

- directly affecting enzyme activity; or
- indirectly, at the genetic level, by controlling the level at which enzymes are synthesised.

Box 6.5 Feedback inhibition

Biosynthetic pathways exist as a series of enzyme-mediated reactions, leading to a final product required by the cell for structural or metabolic purposes. But what happens if for some reason the consumption of the final product slows down, or even stops? Feedback inhibition, also known, perhaps more descriptively, as *'end-product inhibition'*, ensures that excess amounts of the end product are not synthesised. The pathways leading to the synthesis of many amino acids are regulated in this way; for example, isoleucine, which is synthesised from another amino acid, threonine, via a series of intermediates:

Here, the isoleucine itself acts as an inhibitor of threonine deaminase, the enzyme that starts off the pathway. It does this by binding to an *allosteric* site on the enzyme, distorting it and preventing its active site from binding to threonine. Note that by inhibiting the early part of the pathway, we not only prevent further production of isoleucine but also unnecessary breakdown of threonine. When levels of isoleucine starts to run low, less will be available to block the threonine deaminase, and thus the pathway starts to function again.

Direct control of enzymatic activity occurs by the mechanism of *feedback inhibition* (see Box 6.5), whereby the final product of a metabolic pathway acts as an inhibitor to the enzyme that catalyses an early step (usually the first) in the pathway. It thus prevents more of its own formation. When the concentration of the product subsequently falls below a certain level, it is no longer inhibitory, and biosynthesis resumes.

Regulation of metabolic pathways can also be achieved by controlling whether or not an enzyme is synthesised in the first place, and if so, the rate at which it is produced. This is done at the DNA level, by one of two mechanisms, *induction* or *repression*, which respectively 'switch on' and 'switch off' the machinery of protein synthesis discussed earlier in this chapter. These are discussed under the heading 'Regulation of gene expression' in Chapter 11.

III
Microbial Diversity

A few words about classification

In this section, we examine the wide diversity of microbial life. In each of the four chapters – Prokaryotic diversity, The Fungi, The Protista and Viruses – we shall discuss major structural and functional characteristics, and outline the main taxonomic divisions within each group. We shall also consider some specific examples, particularly with respect to their effect on humans. By way of introduction, however, we need to say something on the subject of the classification of microorganisms.

In any discussion on biological classification, it is impossible to avoid mentioning *Carl Linnaeus*, the Swedish botanist who attempted to bring order to the naming of living things by giving each type a Latin name. It was Linnaeus who was responsible for introducing the *binomial* system of nomenclature, by which each organism was assigned to a *genus* and a *species*. To give a few familiar examples, you and I are *Homo sapiens,* the fruit fly that has contributed so much to our understanding of genetics is *Drosophila melanogaster*, and in the microbial world, the bacterium responsible for causing anthrax is *Bacillus anthracis*. Note the following conventions, which apply to the naming of all living things (the naming of viruses is something of a special case, which we'll consider in Chapter 10):

- the *generic (genus) name* is always given an initial *capital* letter;
- the *specific (species) name* is given all in *small* letters;
- the generic and specific names are *italicised*, or, if this isn't possible, underlined.

The science of *taxonomy* involves not just naming organisms, but grouping them with other organisms that share common properties. In the early days, classification appeared relatively straightforward, with all living things apparently fitting into one of two *kingdoms*. To oversimplify the matter, if it ran around, it was

> A *taxon* is a collection of related organisms grouped together for purposes of classification. Thus, genus, family, etc. are taxons.

an animal, if it was green and didn't, it was a plant! As our awareness of the microbial world developed, however, it was clear that such a scheme was not satisfactory to accommodate all life forms, and in the mid-nineteenth century, Ernst Haeckel proposed a third kingdom, the *Protista*, to include the bacteria, fungi, protozoans and algae.

In the twentieth century, an increased focus on the cellular and molecular similarities and dissimilarities between organisms led to proposals for further refinements to the three-kingdom system. One of the most widely accepted of these has been the *five-kingdom system* proposed by Robert Whittaker in 1969 (Figure A.1). Like some of its predecessors, this took into account the fundamental difference in cell structure between prokaryotes and

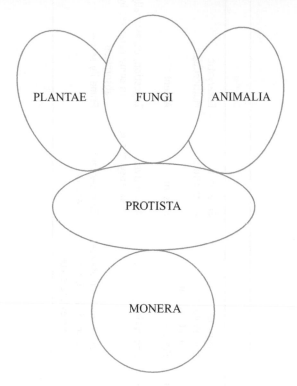

Figure A.1 Whittaker's five kingdom system of classification.

eukaryotes (see Chapter 3), and so placed prokaryotes (bacteria) in their own kingdom, the Monera, separate from single-celled eukaryotes. Another feature of Whittaker's scheme was to assign the Fungi to their own kingdom, largely on account of their distinctive mode of nutrition. Table A.1 shows some of the characteristic features of each kingdom.

Molecular studies in the 1970s revealed that the Archaea differed from all other bacteria in their 16S rRNA sequences, as well as in their cell wall structure, membrane lipids and aspects of protein synthesis. These differences were seen as sufficiently important for the recognition of a third basic cell type to add to the prokaryotes and eukaryotes. This led to the proposal of a *three-domain* scheme of classification, in which prokaryotes are divided into the Archaea and the Bacteria (see Figure 3.1). The third domain, the Eukarya, represents all eukaryotic organisms. The domains thus represent a level of classification that goes even higher than the kingdoms. Although the Archaea (the word means 'ancient') represent a more primitive bacterial form than the Bacteria, they are in certain respects more closely related to the eukaryotes, causing biologists to revise their ideas about the evolution of the eukaryotic state.

Table A.1 Characteristics of Whittaker's five kingdoms

	Monera (procaryotae)	Protista	Fungi	Plantae	Animalia
Cell type	Prokaryotic	Eukaryotic	Eukaryotic	Eukaryotic	Eukaryotic
Cell organization	Unicellular; occasionally grouped	Unicellular; occasionally multicellular	Unicellular or multicellular	Multicellular	Multicellular
Cell wall	Present in most	Present in some, absent in others	Present	Present	Absent
Nutrition	Absorption, some photosynthetic, some chemosynthetic	Ingestion or absorption, some photosynthetic	Absorption	Absorptive, photosynthetic	Ingestion; occasionally in some parasites by absorption
Reproduction	Asexual, usually by binary fission	Mostly asexual, occasionally both sexual and asexual	Both sexual and asexual, often involving a complex life cycle	Both sexual and asexual	Primarily sexual

Table A.2 A modern hierarchical classification for *E. coli*; note that in this classification, there are no kingdoms

Domain	Bacteria
Phylum	Proteobacteria
Class	Zymobacteria
Order	Enterobacteriales
Family	Enterobacteriaceae
Genus	*Escherichia*
Species	*coli*

In hierarchical systems of classification, related species are grouped together in the same genus, genera sharing common features are placed in the same family, and so on. Table A.2 shows a modern classification scheme for the gut bacterium *Escherichia coli*.

The boundaries between longstanding divisions such as algae and protozoa have become considerably blurred in recent years, and alternative classifications based on molecular data have been proposed. This is very much a developing field, and no definitive alternative classification has yet gained universal acceptance.

In the following chapters, we shall broadly follow the five-kingdom scheme, although of course the plant and animal kingdoms, having no microbial members, do not concern us. Thus both the Archaea and Bacteria are considered in Chapter 7, and multicellular fungi form the focus of Chapter 8. Chapter 9 examines the Protists in the modern use of the word, that is, unicellular eukaryotic forms. It retains the traditional distinction between protozoans, algae and other protists (water moulds and slime moulds), but also offers an alternative, 'molecular' scheme, showing the putative phylogenetic relationship between the various groups of organisms. Microbiology has traditionally embraced anomalies such as the giant seaweeds, as it has encompassed all organisms that fall outside of the plant and animal kingdoms. This book, offers only a brief consideration of such macroscopic forms, and for the most part confines itself to the truly microbial world.

The viruses, it ought to be clear by now, are special cases, and are considered in isolation in Chapter 10. Because an understanding of viruses requires an appreciation of the basics of DNA replication and protein synthesis, you may like to jump ahead and read the relevant sections of Chapter 11 before embarking on Chapter 10.

7
Prokaryote Diversity

Ever since bacteria were first identified, microbiologists have attempted to bring order to the way they are named and classified. The range of morphological features useful in the differentiation of bacteria is fairly limited (compared, say, to animals and plants), so other characteristics have also been employed. These include metabolic properties, pathogenicity, nutritional requirements, staining reactions and antigenic properties. The first edition of *Bergey's Manual of Systematic Bacteriology* (henceforth referred to as 'Bergey'), published in the mid-1980s, mainly uses phenotypic characteristics such as these to classify bacteria. The result places bacteria into taxonomic groups that may or may not reflect their evolutionary relationship to one another. In the years since the first edition of *Bergey*, the remarkable advances made in molecular genetics have led to a radical reappraisal of the classification of bacteria. Comparison of nucleic acid sequences, notably those of *16S ribosomal RNA genes*, has led to a new, phylogenetically based scheme of classification, that is, one based on how closely different groups of bacteria are thought to be related, rather than what morphological or physiological features they may share. Ribosomal RNA occurs in all organisms, and serves essentially the same function, thus to a large extent these sequences are *conserved* (remain largely the same) in all organisms. The nature and extent of any differences that have crept in during evolution will, therefore, be an indication of the relatedness of different organisms.

> The *phenotype* of an organism refers to its observable characteristics.
> The *genotype* of an organism refers to its genetic make-up.

The second edition of *Bergey*, issued in five volumes between 2001 and 2012, aims to reflect this change of approach and reassign many bacteria according to their phylogenetic relationship, as deduced from molecular evidence. As an example, the genus *Pseudomonas* previously contained some

Essential Microbiology, Second Edition. Stuart Hogg.
© 2013 John Wiley & Sons, Ltd. Published 2013 by John Wiley & Sons, Ltd.

70 species on the basis of phenotypic similarities, but in the second edition of *Bergey*, taking into account 16S rRNA information, many of these are assigned to newly created genera.

It must be stressed that *Bergey* (second edition) does not represent the definitive final word on the subject, and that the classification of bacteria is very much a developing science, in a constant process of evolution. Indeed, microbiologists are by no means unanimous in their acceptance of the 'molecular' interpretation of bacterial taxonomy. Some point to perceived inadequacies in the collection of data for the scheme, as well as errors in the data arising from the sequencing and amplification techniques utilised. Other critics question the validity of a scheme based on 16S rRNA data when it seems increasingly likely that lateral gene transfer played an important role in bacterial evolution.

> Bacteria may acquire genes from other organisms by a variety of genetic transfer mechanisms (see Chapter 11). This is known as horizontal or *lateral gene transfer*, to distinguish it from vertical inheritance, in which the parental genotype is passed to the offspring.

In the following, the major taxonomic groupings are discussed according to their arrangement in the second edition of *Bergey*. Figure 7.1 shows a phylogenetic tree, reflecting current ideas on the relationship between the major bacterial groups, as determined by 16S rRNA sequencing.

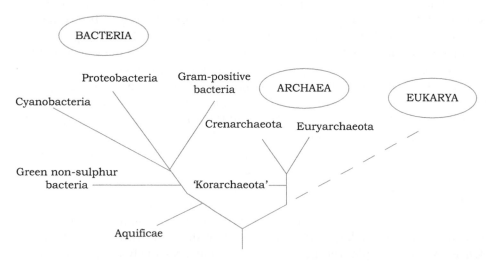

Figure 7.1 Phylogenetic relationships in the Prokaryota. **A phylogenetic tree based on 16S ribosomal RNA data, showing the relationships between selected members of the Archaea and the Bacteria. The relationship to eukaryotic organisms is also indicated. Note that eukaryotes are closer to the Archaea than to the Bacteria.**

7.1 Domain: Archaea

Studies on 16S ribosomal RNA sequences by Carl Woese and colleagues in the 1970s allowed the construction of phylogenetic trees for the prokaryotes, showing their evolutionary relatedness. Figure 7.1 shows how the major prokaryotic groups are thought to be related, based on 16S rRNA data. The work of Woese also revealed that one group of prokaryotes

> *Archaea* are prokaryotes that differ from true bacteria in cell wall and plasma membrane chemistry as well as 16S rRNA sequences.

differed from all the others. As described in Chapter 3, the Archaea are now regarded as being quite distinct from the Bacteria (sometimes called Eubacteria). Together with the Eukarya, these form the three domains of life (see Figure 3.1). As can be seen in Table 7.1, archaeans share some features in common with other bacteria and some with eukaryotes. Extending nucleic

> The *domain* is the highest level of taxonomic grouping.

acid analysis to other genes has shown that members of the Archaea possess many genes not found in any other type of bacteria.

Table 7.1   The three domains of life. Archaea share some features with true bacteria and others with eukaryotes

	Archaea	Bacteria	Eukarya
Main genetic material	Single closed circle of dsDNA	Single closed circle of dsDNA	True nucleus with multiple linear chromosomes
Histones	Present	Absent	Present
Gene structure	Introns absent	Introns absent	Introns present
Plasmids	Common	Common	Rare
Polycistronic mRNA	Present	Present	Absent
Ribosomes	70S	70S	80S
Protein synthesis	Not sensitive to streptomycin, chloramphenicol	Sensitive to streptomycin, chloramphenicol	Not sensitive to streptomycin, chloramphenicol
Initiator tRNA	Methionine	N-formyl methionine	Methionine
Membrane fatty acids	Ether-linked, branched	Ester-linked, straight chain	Ester-linked, straight chain
Internal organelles	Absent	Absent	Present
Site of energy generation	Cytoplasmic membrane	Cytoplasmic membrane	Mitochondria
Cell wall	Muramic acid absent	Muramic acid present	Muramic acid absent

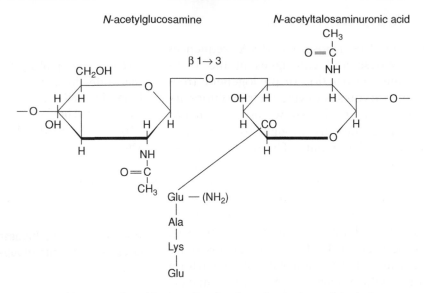

Figure 7.2 Pseudomurein, found in the cell walls of certain members of the Archaea, comprises subunits of *N*-acetylglucosamine and *N*-acetyltalosaminuronic acid. The latter replaces the *N*-acetylmuramic acid in true peptidoglycan (cf. Figure 3.6).

7.1.1 General features of the Archaea

Members of the Archaea show considerable diversity of both morphology and physiology. In view of the fact that the archaeans remained unidentified as a separate group for so many years, it should come as no surprise that they do not display any obvious morphological differences from true bacteria, and all the main cell shapes are represented (see Chapter 2). In addition, however, some archaeal species have flattened square or even triangular cells! The diversity of archaeans extends into their adopted means of nutrition and metabolism, since members may be aerobic or anaerobic, and exhibit autotrophic or heterotrophic nutrition.

Both Gram-positive and Gram-negative archaeans are found, but neither possess true peptidoglycan. Some types have a so-called *pseudomurein*, composed of different substituted polysaccharides and L-amino acids (Figure 7.2). Most archaeans, however, have cell walls composed of a layer of proteinaceous subunits known as an *S-layer*, directly associated with the cell membrane. This difference in cell wall chemistry means

Until recently, no member of the Archaea had been shown to be associated with any disease in humans, but the methanogen *Methanobrevibacter oralis* has been found in a significant number of cases of infected dental root canals. It remains to be seen whether this can be regarded as a true pathogen; if so, it would be a first for the Archaea.

(a)

Glycerol — Ester bond

(b)

— Ether bond

Figure 7.3 Membrane lipids in Archaea and Bacteria. Compositional differences in membrane lipids between (a) Bacteria and (b) Archaea. Note the ether linkages and branched fatty acids in (b).

that members of the Archaea are not susceptible to antibacterial agents such as lysozyme and penicillin, whose action is directed specifically towards peptidoglycan. Differences are also found in the make-up of archaean membranes, where the lipid component of membranes contains branched isoprenes instead of fatty acids, and these are joined to glycerol by ether-linkages, rather than the ester-linkages found in true bacteria (Figure 7.3). The DNA of archaeans, like that of true bacteria, usually comprises a single, circular double-stranded molecule, with no nuclear membrane. Archaeans differ, however, in that histone proteins have been found associated with the DNA. As we saw in Chapter 3, histones are considered to be a eukaryotic characteristic, not found in bacteria.

Many members of the Archaea are found in extreme environments such as deep-sea thermal vents and salt ponds. Some extreme thermophiles are able to grow at temperatures of well over 100°C, while psychrophilic archaeans constitute a substantial proportion of the microbial population of Antarctica. Similarly, examples are to be found of archaeans that are active at extremes of acidity, alkalinity or salinity. Initially it was felt that the Archaea were limited to such environments because there they faced little competition from true bacteria or eukaryotes. However, archaeans are now believed to be more widespread in their distribution, making up a significant proportion of the bacterial biomass found in the world's oceans, and also found in terrestrial and semi-terrestrial niches. The reason why this remained undetected for so long is that these organisms cannot as yet be cultured in the laboratory, and their presence can only be inferred by the use of modern DNA-based analysis.

7.1.2   Classification of the Archaea

In addition to the two phyla of Archaea listed in the new *Bergey* (the Euryarchaeota and the Crenarchaeota), three minor ones have been proposed over the last decade. These are the Korarchaeota, Thaumarchaeota and Nanoarchaeota, the last of which at present comprises just a single species (see Box 7.1). Countless more species of archaeans are thought to exist, which like the Korarchaeota, have not been successfully cultured in the laboratory.

The phylum Euryarchaeota is a bigger and more diverse group than the Crenarchaeota, and includes methanogenic and extreme halophilic forms. The latter are aerobic heterotrophs, requiring a chloride concentration of at least 1.5 M (generally 2.0–4.0 M) for growth. One species, *Halobacterium salinarum*, is able to carry out a unique form of photosynthesis using the bacterial pigment bacteriorhodopsin, and uses the ATP so generated for the active transport into the cell of the chloride ions it requires.

Members of the Euryarchaeota such as *Methanococcus* and *Methanobacterium* are unique among all life forms in their ability to generate methane

Box 7.1   The world's smallest organism

In 2002 a novel microorganism, *Nanoarchaeum equitans*, was isolated from a hydrothermal vent deep below the sea off the Icelandic coast. Found at temperatures around boiling point, its tiny spherical cells were attached to the surface of another archaean, *Ignicoccus* sp., with which it seems to live symbiotically. Ribosomal RNA studies showed *N. equitans* to be sufficiently unlike other members of the Archaea to justify the creation of a new archaean phylum, the Nanoarchaeota. Fully sequenced, the genome of *N. equitans* was the smallest of any living thing at that time (490 kb). It appears to belong to a very deeply rooted branch of the Archaea, leading to speculation that it may resemble the first living cells. At 0.4 μm in diameter, *N. equitans* is also the smallest known living organism. Although at present it is the only recognised species of the Nanoarchaeota, it seems likely to be joined by isolates reported from other locations around the world.

Box 7.2 Prochlorophyta – a missing link?

It has long been thought that the chloroplasts of eukaryotic cells arose as a result of incorporating unicellular cyanobacteria into their cytoplasm. The photosynthetic pigments of plants and green algae, however, are not the same as those of the blue greens; both contain chlorophyll *a*, but while the former also have chlorophyll *b*, the blue greens have a unique group of pigments called phycobilins. In the mid-1970s a group of bacteria was discovered which seemed to offer an explanation for this conundrum. Whilst definitely prokaryotic, the *Prochlorophyta* possess chlorophylls *a* and *b* and lack phycobilins, making them a more likely candidate for the origins of eukaryotic chloroplasts. Initially, these marine bacteria were placed by taxonomists in a phylum of their own, but in the second edition of *Bergey*, they are included among the Cyanobacteria.

from simple carbon compounds. They are strict anaerobes found in environments such as hot springs, marshes and the gut of ruminant mammals. The methane is derived from the metabolism of various simple carbon compounds such as carbon dioxide or methanol in reactions linked to the production of ATP, for example:

$$CO_2 + 4H_2 \rightarrow CH_4 + 2H_2O$$
$$CH_3OH + H_2 \rightarrow CH_4 + H_2O$$

In addition, a few species can cleave acetate to produce methane:

$$CH_3COO^- + H_2O \rightarrow CH_4 + HCO_3^-$$

This acetotrophic reaction is responsible for the much of the methane production in sewage sludges. Although sharing the unique facility to generate methane, some of the methanogenic genera are only distantly related to one another.

Other representatives of the Euryarchaeota include the Thermoplasmatales and the Thermococci. Members of the Thermoplasmatales are highly acidophilic and moderately thermophilic; they completely lack a cell wall, and are pleomorphic. A unique membrane lipid composition allows them to withstand temperatures of well over 50°C. Thermococci are anaerobic extreme thermophiles found in anoxic thermal waters at temperatures as high as 95°C. Enzymes isolated from thermococci have found a variety of applications. A thermostable DNA polymerase from *Pyrococcus furiosus* is used as an alternative to *Taq* polymerase

Pleomorphic means lacking a regular shape.

(see 'Phylum Deinococcus-Thermus' later in this chapter) in the polymerase chain reaction (PCR).

Representative genera: *Methanobacterium, Halobacterium*

Members of the Crenarchaeota are mostly extreme thermophiles, many of them capable of growth at temperatures in excess of 100°C, including *Pyrolobus fumarii*, which has an optimum growth temperature of 106°C, and can survive autoclaving at 121°C. A few apparently mesophilic crenarchaeotes have now been reclassified as a new phylum, Thaumarchaeota.

Many crenarchaeotes utilise inorganic sulphur compounds as either a source or acceptor of electrons (respectively, oxidation to H_2SO_4 or reduction to H_2S). Members of the phylum are mostly anaerobic, and are thought by many to resemble the common ancestors of all bacteria.

Representative genera: *Thermoproteus, Sulfolobus*

7.2   Domain: Bacteria

All the remaining bacterial groups belong to the domain Bacteria. At the time of writing, this is divided into 30 phyla (Table 7.2), the more important of which are discussed in the following pages. As with the Archaea, many other forms are known only through molecular analysis and it is estimated that these represent at least another 20 phyla.

7.2.1   Phylum: Proteobacteria

We start our survey of the Bacteria with the Proteobacteria. This is by far the biggest single phylum, representing around one-third of all known bacterial species, and occupying the whole of Volume 2 in the second edition

Table 7.2 Phyla of domain Bacteria. Those phyla discussed in the text are shown in bold print

Acidobacteria	**Cyanobacteria**	Nitrospira
Actinobacteria	Deferribacteres	**Planctomycetes**
Aquificae	**'Deinococcus-Thermus'***	**Proteobacteria**
Armatimonadetes	Dictyoglomi	**Spirochaetes**
Bacteroidetes	Elusimicrobia	Synergistetes
Caldiserica	Fibrobacteres	**Tenericutes**
Chlamydiae	**Firmicutes**	Thermodesulfobacteria
Chlorobi	Fusobacteria	Thermomicrobia
Chloroflexi	Gemmatimonadetes	**Thermotogae**
Chrysiogenetes	Lentisphaerae	**Verrucomicrobia**

*This phylum has not yet been assigned a formal name.

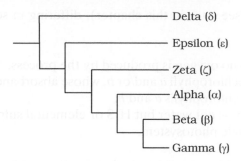

Figure 7.4   The Proteobacteria. Probable phylogenetic relationships of the Proteobacteria, based on 16S rRNA sequences. The Zetaprotobacteria currently only includes a single described species, but many other as yet unculturable representatives are also known.

of *Bergey*. The size of the group is matched by its diversity, both morphological and physiological; most forms of metabolism are represented, and the wide range of morphological forms gives rise to the group's name (Proteus was a mythological Greek god who was able to assume many different forms). The reason such a diverse range of organisms have been assigned to a single taxonomic grouping is that their 16S rRNA indicates that they share a common ancestor (thought to be photosynthetic, though few members of the group now retain this ability). The Proteobacteria includes many of the best known Gram-negative bacteria of medical, industrial and agricultural importance. For taxonomic purposes, they have been divided into six classes reflecting their presumed lines of descent and termed the Alphaproteobacteria, Betaproteobacteria, Gammaproteobacteria, Deltaproteobacteria, Epsilonproteobacteria and Zetaproteobacteria (Figure 7.4). The latter comprises a single iron-oxidising deep-sea species, and in 2010 became the latest class to be created. It should be stressed that because classification is based on molecular relatedness rather than shared phenotypic traits, few if any morphological or physiological properties can be said to characterise all members of each class. Equally, organisms united by a particular feature may be found in more than one of the proteobacterial classes, for example nitrifying bacteria are to be found in the Alpha-, Beta, Gamma- and Deltaproteobacteria. For this reason, in the following paragraphs we describe the Proteobacteria in terms of their *phenotypic* characteristics rather than attempt to group them phylogenetically.

Photosynthetic Proteobacteria   The purple sulphur and purple non-sulphur bacteria are the only members of the Proteobacteria to have retained the photosynthetic ability of their presumed ancestor. The type of photosynthesis they carry out, however, is quite distinct from that carried out by plants, algae and cyanobacteria (see later in this chapter), differing in several important respects:

- it is anoxygenic – no oxygen is produced by the process;
- it utilises bacteriochlorophyll *a* and/or *b*, whose absorbance properties differ from those of chlorophylls *a* and *b*;
- the electron donor is not water but H_2S or elemental sulphur;
- there is only a single photosystem.

Like organisms that carry out green photosynthesis, however, they incorporate CO_2 into carbohydrate by means of the Calvin cycle. All are at least facultatively anaerobic, and are typically found in sediments of stagnant lakes and saltmarsh pools, where they may form extensive coloured blooms. Because the absorption spectrum of bacteriochlorophylls lies mostly in the infrared part of the spectrum, they are able to utilise light energy that penetrates beyond the surface layers of water. The coloration, ranging from orange/brown to purple, is due to the presence of carotenoid pigments such as lycopene and spirillixanthin, which mask the blue/green colour of the bacteriochlorophylls. The photosynthetic pigments are located on highly folded extensions of the plasma membrane. Photosynthetic proteobacteria include rods, cocci and spiral forms.

Under anaerobic conditions, the *purple sulphur bacteria* typically utilise hydrogen sulphide or elemental sulphur as an electron donor for the reduction of CO_2:

$$H_2S + CO_2 \rightarrow (CH_2O)_n + S^0$$
$$S^0 + CO_2 + H_2O \rightarrow (CH_2O)_n + H_2SO_4$$

Many store sulphur in the form of intracellular granules, although there are a few genera that produce sulphur outside of the cell. The purple sulphur bacteria all belong to the Gammaproteobacteria. They are typically found in surface muds and sulphur springs, habitats that provide the right combination of light and anaerobic conditions.

Representative genera: *Thiospirillum, Chromatium*

The *purple non-sulphur bacteria* were distinguished from the purple sulphur bacteria because of their apparent inability to use H_2S as an electron donor. It is now known, however, that the majority *can* do this, but can only tolerate very low concentrations in comparison with the purple sulphur bacteria. The purple non-sulphur bacteria are facultative anaerobes able to grow as photoheterotrophs, that is, with light as an energy source and a wide range of organic molecules such as carbohydrates and organic acids as sources of both carbon and electrons. In addition, many are able to grow aerobically as chemoheterotrophs in the absence of light. Under present classification systems, purple non-sulphur bacteria are divided between the Alpha- and Beta-proteobacteria.

Representative genera: *Rhodospirillum, Rhodopseudomonas*

Nitrifying Proteobacteria This group comprises aerobic Gram-negative chemolithoautotrophs that derive their energy from the oxidation of inorganic nitrogen compounds (either ammonia or nitrite), and their carbon from CO_2. The nitrifying Proteobacteria have representatives in the Alpha-, Beta-, Gamma- and Delta-proteobacteria classes.

The oxidation of ammonia through to nitrate is a two-stage process, with specific bacteria carrying out each stage (ammonia to nitrite and nitrite to nitrate). This is reflected in the generic name of the bacteria, bearing the prefix *Nitroso-* or *Nitro-* according to whether they carry out the first or second reaction. Nitrifying bacteria play an essential role in the cycling of nitrogen in terrestrial, marine and freshwater habitats. Nitrite, which is toxic to many forms of life, rarely accumulates in the environment, due to the activity of nitrifiers.

Representative genera: *Nitrosomonas* $(NH_4^+ \rightarrow NO_2^-)$
Nitrobacter $(NO_2^- \rightarrow NO_3^-)$

Iron- and sulphur-oxidising Proteobacteria Two further groups of environmentally significant chemolithoautotrophs derive their energy through the oxidation of reduced iron and sulphur respectively.

Among the *sulphur oxidisers*, perhaps the best studied are members of the genus *Acidithiobacillus*,[1] which includes extreme acidophiles such as *A. thiooxidans,* which are capable of growth at a pH as low as 1.0! These may utilise sulphur in its elemental form, as H_2S, metal sulphides or other forms of reduced sulphur such as thiosulphate:

$$2S^0 + 3O_2 + 2H_2O \rightarrow 2H_2SO_4$$
$$2S^{2-} + 4O_2 \rightarrow 2SO_4^{2-}$$
$$S_2O_3^{2-} + 2O_2 + H_2O \rightarrow 2SO_4^{2-} + 2H^+$$

The result of all these reactions is the production of sulphuric acid and a lowering of the environmental pH. Bacteria such as these play important roles in the generation of *acid mine drainage* as well as the release of toxic metals such as lead and cadmium. Their environmental impact is discussed in Chapter 14, whilst Chapter 18 describes their use in the extraction of valuable metals from intractable mineral ores. A particularly valuable organism in the

---

[1] *Acidithiobacillus thiooxidans* and *A. ferrooxidans* formerly belonged to the genus *Thiobacillus.* In 2000, several species of *Thiobacillus* were assigned to new genera; however, you may still find them referred to by their old names.

latter context is *A. ferrooxidans*, due to its ability to use not only reduced sulphur compounds as energy sources, but also reduced iron.

A second group of sulphur oxidisers are bacteria that exist not as single cells, but join to form *filaments*, the best known of which is *Beggiotoa*. These are typically found in sulphur springs, marine sediments and hydrothermal vents at the bottom of the sea.

Acidithiobacillus ferrooxidans is also an example of an iron oxidiser. At normal physiological pH values and in the presence of oxygen, reduced iron (iron(II), Fe^{2+}) is spontaneously oxidised to iron(III) (Fe^{3+}). Under very acidic conditions, however, the iron remains in its reduced form, unless acted on by certain bacteria. *A. ferrooxidans* is an obligate aerobe able to use iron(II) as an energy source, converting it to iron(III) at an optimum pH range of around 2:

$$2Fe^{2+} + \frac{1}{2}O_2 + 2H^+ \rightarrow 2Fe^{3+} + H_2O$$

Gallionella ferruginea, on the other hand, grows around neutrality in oxygen-poor environments such as bogs and iron springs. Ferric hydroxide is excreted from the cell and deposited on a stalk-like structure projecting from, and much bigger than, the cell itself. This gives the macroscopic impression of a mass of reddish brown twisted filaments.

Representative genera: *Acidithiobacillus, Beggiotoa* (sulphur oxidisers)
Leptospirillum, Gallionella (iron oxidisers)

Hydrogen-oxidising Proteobacteria   This diverse group of bacteria are united by their ability to derive energy by using hydrogen gas as a donor of electrons, and oxygen as an acceptor:

$$2H_2 + O_2 \rightarrow 2H_2O$$

Nearly all the members of this group are facultative chemolithotrophs, that is, they can also grow as heterotrophs, utilising organic compounds instead of CO_2 as their carbon source, and indeed most grow more efficiently in this way.

Representative genera: *Alcaligenes, Ralstonia*

Nitrogen-fixing Proteobacteria   The Alphaproteobacteria includes certain genera of nitrogen-fixing bacteria. These are able to fix (reduce) atmospheric N_2 as NH_4^+ for subsequent incorporation into cellular materials, a process that requires a considerable input of energy in the form of ATP:

> *Nitrogen fixation* is limited to a few species of bacteria and cyanobacteria. No eukaryotes are known to have this property.

$$N_2 + 8e^- + 8H^+ + 16ATP \rightarrow 2NH_4^+ + 16ADP + 16Pi$$

Nitrogen-fixing bacteria may be free-living in the soil (e.g. *Azotobacter*), or form a symbiotic relationship with cells on the root hairs of leguminous plants such as peas, beans and clover (e.g. *Rhizobium*). The nitrogenase responsible for the reaction (actually a complex of two enzymes) is highly sensitive to oxygen; many nitrogen fixers are anaerobes, while others have devised ways of keeping the cell interior oxygen-free. Nitrogen fixation is discussed further in Chapters 13 and 14.

Closely related to *Rhizobium*, but unable to fix nitrogen, are members of the genus *Agrobacterium*. Like *Rhizobium*, these enter the tissues of plants, but instead of forming a mutually beneficial association, cause cell proliferation and tumour formation. *A. tumefaciens* has proved to be a valuable tool in the genetic engineering of plants, and is discussed further in Chapter 12.

Representative genera: *Rhizobium, Azotobacter*

Methanotrophic Proteobacteria In discussing the Archaea earlier in this chapter, we encountered species capable of producing methane, a gas found widely in such diverse locations as marshes, sewage sludge and animal intestines. Certain proteobacteria are able to utilise this methane as a carbon and energy source and are known as *methanotrophs*.

Methanotrophs are strict aerobes, requiring oxygen for the oxidation of methane. The methane-generating bacteria, however, as we've seen are anaerobes; methanotrophs are consequently to be found at aerobic/anaerobic interfaces such as topsoil, where they can find both the oxygen and the methane they require. The methane is firstly oxidised to methanol, then to formaldehyde, by means of separate enzyme systems. Some of the carbon in formaldehyde is assimilated into organic cellular material, while some is further oxidised to carbon dioxide.

> *Methane* is one of the so-called greenhouse gases, responsible for the phenomenon of global warming. Its effects would be much more pronounced if it were not for the activity of methanotrophic bacteria.

> *Methylophilus methylotrophus* was once produced in huge quantities as a source of *'single cell protein'* for use as animal feed, until the low price of alternatives such as soya and fish meal made it commercially unviable.

Bacteria able to utilise other single-carbon compounds such as methanol (CH_3OH) or methylamine (CH_3NH_2) are termed *methylotrophs*. Depending on whether they possess the enzyme methane monooxygenase (MMO), they may also be methanotrophs.

Representative genera: *Methylomonas, Methylococcus*

Sulphate- and sulphur-reducing Proteobacteria Some 40 or so genera of anaerobic Deltaproteobacteria reduce either elemental sulphur or oxidised forms of sulphur such as sulphate to hydrogen sulphide. Organic compounds such as pyruvate, lactate or certain fatty acids act as electron donors:

$$2CH_3CHOHCOO^- + SO_4^{2-} \xrightarrow{Desulfovibrio} 2CH_3COO^- + H_2S + 2HCO_3^-$$
$$\text{Lactate} \qquad\qquad\qquad\qquad\qquad \text{Acetate}$$

Sulphate- and sulphur-reducers are found in anaerobic muds and play an important role in the global sulphur cycle.

Representative genera: *Desulfovibrio* (sulphate), *Desulfuromonas* (sulphur)

Enteric Proteobacteria   This is a large group of rod-shaped bacteria, mostly motile by means of peritrichous flagella, which all belong to the Gammaproteobacteria. They are facultative aerobes, characterised by their ability in anaerobic conditions to carry out fermentation of glucose and other sugars to give a variety of products. The nature of these products allows division into two principal groups, the mixed acid fermenters and the butanediol fermenters (see Figure 6.23). All the enteric bacteria test negative for cytochrome *c* oxidase (see '*Vibrio* and related genera' later). In view of their similar appearance, members of the group are distinguished from one another largely by means of their biochemical characteristics. An unknown isolate is subjected to a series of tests including its ability to utilise substrates such as lactose and citrate, convert tryptophan to indole, and hydrolyse urea. On the basis of its response to each test, a characteristic profile can be built up for the isolate, and matched against those of known species (see Table 7.3).

The most thoroughly studied of all bacteria, *Escherichia coli* (*E. coli*), is a member of this group, as are a number of important pathogens of humans such as *Salmonella*, *Shigella* and *Yersinia* (the causative agent of plague).

Representative genera: *Escherichia, Enterobacter*

Vibrio and related genera   A few other genera, including *Vibrio* and *Aeromonas,* are also facultative anaerobes able to carry out the fermentative reactions described above, but are differentiated from the enteric bacteria by being oxidase-positive (Table 7.4). *Vibrio* and *Photobacterium* both include examples of marine *bioluminescent* species; these are widely found both in seawater and associated with fish and other marine life. The luminescence, which requires the presence of oxygen, is due to an oxidation reaction carried out by the enzyme *luciferase.*

> *Bioluminescence* is the production of light by living systems.

Vibrio cholerae is the causative agent of cholera, a debilitating and often fatal form of acute diarrhoea transmitted in faecally contaminated water. It

Table 7.3 Identification of enteric bacteria on the basis of their biochemical properties

Some of the tests used to identify isolates of enteric bacteria are listed below. The table on the next page indicates typical results obtained for common genera; note, however, that for many cases, the result of a test may vary for different species within a genus. The symbols + and − indicate that most or all species within a genus give a positive or negative result, whilst +/− denotes that results are more variable within a genus.

Test	Description
Indole	Tests for ability to produce indole from the amino acid tryptophan.
Methyl Red	Acid production causes methyl red indicator to turn red.
Voges-Proskauer	Tests for ability to ferment glucose to acetoin.
Citrate utilisation	Demonstrates ability to utilise citrate as sole carbon source.
Urease	Demonstrates presence of the enzyme urease by detecting rise in pH due to urea being converted to ammonia and CO_2.
Gas from sugars	Production of gas from sugars such as glucose is demonstrated by collection in a Durham tube (a small inverted tube placed in a liquid medium).
H_2S production	Production of H_2S from sulphate reduction or from sulphur-containing amino acids is demonstrated by the formation of black iron sulphide in an iron-rich medium.
Ornithine decarboxylase	Growth on medium enriched in ornithine leads to pH change when enzyme is present.
Motility	Diffusion through soft agar demonstrates cellular movement.
Gelatin liquefaction	Demonstrates presence of proteolytic enzymes capable of liquefying a medium containing gelatin.
% age GC	Nucleotide composition determined by melting point measurements.

(Continued)

Table 7.3   (*Continued*)

	Escherichia	Salmonella	Shigella	Citrobacter	Proteus	Serratia	Klebsiella	Enterobacter	Erwinia
Indole	+	−	+/−	+/−	+/−	−	+/−	−	−
Methyl Red	+	+	+	+	+	+/−	+	+/−	+
Voges-Proskauer	−	−	−	−	+/−	+	+	+	+
Citrate utilisation	−	+/−	−	+	+/−	+	+	+	−
Urease	−	−	−	+	+/−	−	+/−	−	−
Gas from glucose	+	+	−	+	+	+/−	+	+	−
H$_2$S production	−	+	−	+/−	+	−	−	−	−
Ornithine decarboxylase	+	+	+/−	+	+/−	−	+/−	+	+
Motility	+	+	−	+	+	+	−	+	−
Gelatin liquefaction	−	−	−	−	+	+	−	+/−	+
%GC	48–52	50–53	49–53	50–52	38–41	53–59	53–58	52–60	50–58

Table 7.4   Differentiation between enteric bacteria, vibrios and pseudomonads

	Enteric bacteria	Vibrios	Pseudomonads
Oxidase test	−ve	+ve	+ve
Glucose fermentation	+ve	+ve	−ve
Flagella	Peritrichous	Polar*	Polar

*When grown on solid media, some *Vibrio* species also develop lateral flagella, a unique arrangement termed *mixed flagellation.*

remains a major killer in many less developed countries. Several species of *Vibrio*, including *V. cholerae*, have been shown to possess two circular chromosomes instead of the usual one.

Representative genera: *Vibrio, Aeromonas*

The pseudomonads   Members of this group of proteobacteria, the most important genus of which is *Pseudomonas*, are straight or curved rods with polar flagella. They are chemoheterotrophs that generally utilise the Entner–Doudoroff pathway rather than glycolysis for the oxidation of hexoses. They are differentiated from the enteric bacteria (see Table 7.4) by being oxidase-positive and incapable of fermentation. A characteristic of many pseudomonads is the ability to utilise an extremely wide range of organic compounds (maybe over 100!) for carbon and energy, something that makes them very important in the recycling of carbon in the environment.

> The pseudomonad *Burkholderia cepacia* can utilise an exceptionally wide range of organic carbon sources, including sugars, carboxylic acids, alcohols, amino acids, aromatic compounds and amines, to name but a few!

Several species are significant pathogens of animals and plants; *Pseudomonas aeruginosa* is an effective coloniser of wounds and burns in humans, while *P. syringae* causes chlorosis (yellowing of leaves) in a range of plants. Because of their ability to grow at low temperatures, a number of pseudomonads are important in the spoilage of food.

Although most species carry out aerobic respiration with oxygen as the terminal electron acceptor, a few are capable of substituting nitrate (anaerobic respiration, see Chapter 6).

Representative genera: *Pseudomonas, Burkholderia*

Acetic acid bacteria   *Acetobacter* and *Gluconobacter* are two genera of the Gram-negative Alphaproteobacteria that convert ethanol into acetic acid, a highly significant reaction in the food and drink industries (see Chapter 18). Both genera are strict aerobes, but whereas *Acetobacter* can oxidise the acetic acid right through to carbon dioxide and water, *Gluconobacter* lacks all the enzymes of the TCA cycle, and cannot oxidise it further.

Acetobacter also has the ability, rare in bacteria, to synthesise cellulose; the cells become surrounded by a mass of extracellular fibrils, forming a pellicle at the surface of an unshaken liquid culture.

Representative genera: *Acetobacter, Gluconobacter*

Stalked and budding Proteobacteria   The members of this group of aquatic Proteobacteria differ noticeably in their appearance from typical bacteria by their possession of extracellular extensions known as *prosthecae*; these take a variety of forms but are always narrower than the cell itself. They are true extensions of the cell, containing cytoplasm, rather than completely extracellular appendages.

In the *stalked bacteria* such as *Caulobacter* (Figure 7.5), the prostheca serves both as a means of attaching the cell to its substratum, and to enhance nutrient absorption by increasing the surface area-to-volume ratio of the cell. The latter enables such bacteria to live in waters containing extremely low levels of nutrients. Sometimes several cells cluster together, joined by their prosthecae, to form a rosette structure. *Caulobacter* lives part of its life cycle as a free-swimming *swarmer cell* with no prostheca but instead a flagellum for mobility.

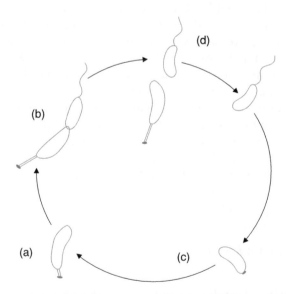

Figure 7.5   The life cycle of *Caulobacter,* a stalked bacterium. The stalked 'mother' cell attaches to a surface by means of a holdfast (a). It grows in length and develops a flagellum (b), before undergoing binary fission. The flagellated swarmer cell swims away (c), and on reaching a suitable substratum, loses its flagellum and develops a stalk, or prostheca (d).

   The iron oxidiser *Gallionella* (see 'Iron- and sulphur-oxidising Proteobacteria' earlier) may be regarded as a stalked bacterium; however, it is not truly
prosthecate, as its stalk does not contain cytoplasm and is thus not strictly
part of the cell.

   In the *budding bacteria,* the prostheca is involved in a distinctive form of
reproduction, in which two cells of unequal size are produced (cf. typical
binary fission, which results in two identical daughter cells). The daughter
cell buds off from the mother cell, either directly, or, as in *Hyphomicrobium*
spp., at the end of a hypha (stalk) (Figure 7.6). Once detached, the daughter
cell grows to full size and eventually produces its own buds. *Hyphomicrobium*
is a methanotroph and a methylotroph, so it also belongs with the 'Methanotrophic Proteobacteria' described earlier.

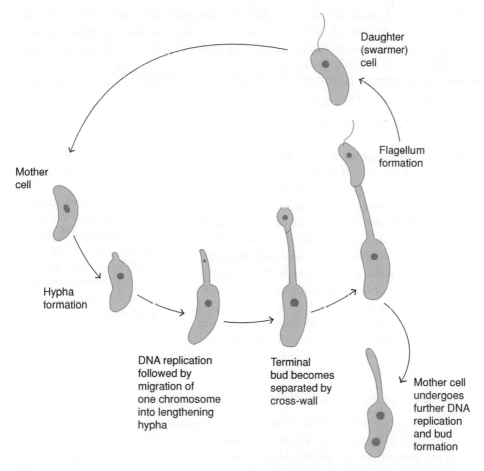

Figure 7.6   The budding bacteria: reproduction in *Hyphomicrobium*. Before reproduction takes
place, the vegetative cell develops a stalk or hypha, at the end of which a bud develops. This
produces a flagellum, and separates to form a motile swarmer cell.

In some bacteria, more than one prostheca is found per cell; these polyprosthecate forms include the genus *Stella*, whose name ('a star') derives from its six symmetrically arranged buds.

Representative genera: *Caulobacter, Hyphomicrobium*

Sheathed Proteobacteria   Some genera of Betaproteobacteria exist as chains of cells surrounded by a tube-like sheath, made up of a carbohydrate–protein–lipid complex. In some cases, the sheath contains deposits of manganese oxide or ferric hydroxide, which may be the product of chemical or biological oxidation. Empty sheaths encrusted with oxides may remain long after the bacterial cells have died off or been released. As with the stalked bacteria discussed above, the sheath helps in the absorption of nutrients, and may also offer protection against predators.

The sheathed bacteria have a relatively complex life cycle. They live in flowing water, and attach with one end of the chain to, for example, a plant or rock. Free-swimming single flagellated cells are released from the distal end and settle at another location, where a new chain and sheath are formed (Figure 7.7). *Sphaerotilus* forms thick 'streamers' in polluted water, and is a familiar sight around sewage outlets.

Representative genera: *Sphaerotilus, Leptothrix*

Predatory Proteobacteria   *Bdellovibrio* is a unique genus belonging to the Deltaproteobacteria. It is a very small (0.25 μ) comma-shaped bacterium, which actually attacks and lives inside other Gram-negative bacteria (Figure 7.8). Powered by its flagellum, it collides with its prey at high speed and penetrates even thick cell walls by a combination of enzyme secretion and mechanical boring. It takes up residence in the periplasmic space, between the plasma membrane and cell wall. Synthesis of its host's nucleic acid and protein ceases, and its macromolecules are degraded, providing nutrients for the invader,

> *Bdellovibrio bacterivorans* has been shown to encode a huge number of lytic enzymes. Its ability to metabolise amino acids, however, is limited, necessitating its unusual mode of existence.

which grows into a long helical cell. This eventually divides into several motile progeny cells, which are then released.

Representative genus: *Bdellovibrio*

Another group of bacteria that may be regarded as predatory are the *Myxobacteria* (Figure 7.9). These are rod-shaped bacteria lacking flagella, which yet are motile by gliding along a solid surface, aided by the excretion of extracellular polysaccharides. For this reason they are sometimes referred to as the *gliding bacteria*. They are heterotrophs, typically requiring complex organic nutrients, which they obtain by the lysis of other types of bacteria.

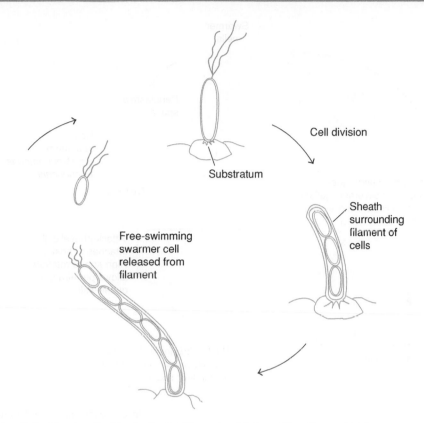

Figure 7.7 The sheathed bacteria: the life cycle of *Sphaerotilus*. Free-swimming swarmer cells settle on an appropriate substratum and give rise to long filaments contained within a sheath. New locations become colonised when flagellated cells are released into the water to complete the cycle.

Thus, unlike *Bdellovibrio*, they digest their prey before they ingest it. When a rich supply of nutrients is not available, they aggregate to form fruiting bodies, inside which *myxospores* develop. These are able to resist drought and lack of nutrients for many years. Myxobacteria exhibit the most complex life cycles of any prokaryote so far studied.

Representative genera: *Myxococcus, Chondromyces*

Spirilla Collected together under this heading are several genera of aerobic (mostly microaerophilic) spiral-shaped bacteria with polar flagella. They include free-living, symbiotic and parasitic types.

Spirilla such as *Aquaspirillum* and *Magnetospirillum* contain magnetosomes, intracellular particles of iron oxide (magnetite, Fe_3O_4). Such magnetotactic bacteria have the remarkable ability to orientate themselves with respect to the Earth's magnetic field.

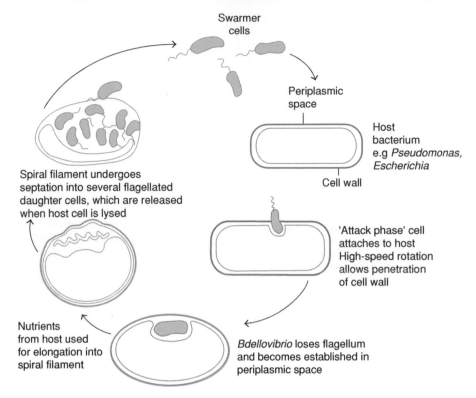

Figure 7.8 The life cycle of *Bdellovibrio*, a bacterial predator. Once *Bdellovibrio* has taken up residence in the periplasmic space of its host, it loses its flagellum and becomes non-motile. In nutrient-rich environments, *Bdellovibrio* is also capable of independent growth.

Two important pathogens of humans are included in the spirilla: *Campylobacter jejuni* is responsible for food-borne gastroenteritis, while *Helicobacter pylori* has in recent times been identified as the cause of many cases of stomach ulcers.

Representative genera: *Magnetospirillum, Campylobacter*

Rickettsiae This group comprises arthropod-borne intracellular parasites of vertebrates, and includes the causative agents of human diseases such as typhus and Rocky Mountain spotted fever. The bacteria are taken up by host phagocytic cells, where they multiply and eventually cause lysis.

The rickettsiae are aerobic organotrophs, but some possess an unusual mode of energy metabolism, only being able to oxidise intermediate metabolites such as glutamate and succinate, which they obtain from their host. *Rickettsia* and *Coxiella*, the two main genera, are not closely related phylogenetically and are placed in the Alpha- and Gamma-proteobacteria respectively.

Representative genera: *Rickettsia, Coxiella*

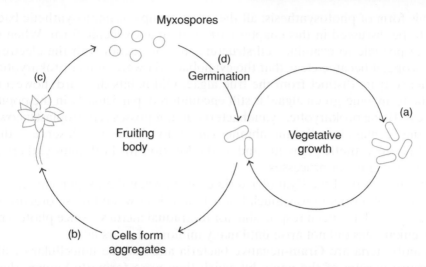

Myxospores

(d)
Germination

(c)

(a)

Fruiting
body

Vegetative
growth

(b)   Cells form
aggregates

Figure 7.9 The Myxobacteria: a complex bacterial life cycle. When nutrients are in plentiful supply, myxobacteria divide by binary fission (a). On depletion of nutrients, they form aggregates of cells, which leads to the formation of a fruiting body (b). Within the fruiting body, some cells form myxospores, enclosed within a sporangium (c). Myxospores remain dormant until environmental conditions are favourable, then germinate into vegetative cells (d).

Neisseria and related proteobacteria All members of this loose collection of bacteria are aerobic non-motile cocci, typically seen as pairs, with flattened sides where they join. Some, however, only assume this morphology during stationary growth phase. Many are found in warm-blooded animals, and some species are pathogenic. The genus *Neisseria* includes species responsible for gonorrhoea and meningitis in humans.

7.2.2 Other Gram-negative phyla

The following section considers those Gram-negative bacteria not included in the Proteobacteria. These phyla are not closely related in the phylogenetic sense, either to each other or to the Proteobacteria.

Phylum Cyanobacteria: the blue-green bacteria The Cyanobacteria are placed in volume 1 of the second edition of *Bergey*, along with the Archaea, the deeply branching bacteria, the 'Deinococcus-Thermus' group, and the green sulphur and green non-sulphur bacteria – see 'Phylum Chlorobi (green sulphur bacteria) and phylum Chloroflexi (green non-sulphur bacteria)' below.

Members of the Cyanobacteria were once known as *blue-green algae* because they carry out the same kind of oxygenic photosynthesis as algae and green plants. They are the only group of prokaryotes capable of carrying out this form of photosynthesis; all the other groups of photosynthetic bacteria to be discussed in this chapter carry out an anoxygenic form. When it became possible to examine cell structure in more detail with the electron microscope, it became clear that the cyanobacteria were in fact prokaryotic, and hence quite distinct from the true algae. Old habits die hard, however, and the term 'blue-green algae' is still encountered, particularly in the popular press. Being prokaryotic, cyanobacteria do not possess chloroplasts; however, they contain lamellar membranes called thylakoids, which serve as the site of photosynthetic pigments and as the location for both light-gathering and electron transfer processes.

Early members of the Cyanobacteria evolved when the oxygen content of the Earth's atmosphere was much lower than it is now, and these organisms are thought to have been responsible for its gradual increase, since photosynthetic eukaryotes did not arise until many millions of years later.

Cyanobacteria are Gram-negative bacteria and may be unicellular or filamentous; in spite of the name by which they were formerly known, they may also appear variously as red, black or purple, according to the pigments they possess. A characteristic of many cyanobacteria is the ability to fix atmospheric nitrogen, that is, to reduce it to ammonium ions (NH_4^+) for incorporation into cellular constituents (see earlier). In filamentous forms, this activity is associated with specialised, enlarged cells called heterocysts (Figure 7.10). The tiny unicellular cyanobacterium *Prochlorococcus* is found in oceans throughout tropical and temperate regions and is thought to be the most abundant photosynthetic organism on our planet. It has several strains adapted to different light conditions. Some cyanobacteria are responsible for the production of unsightly (and smelly!) algal blooms in waters rich in nutrients such as phosphate. When they die, their decomposition by other bacteria leads to oxygen depletion and the death of other aquatic life forms. Bloom-forming species contain gas vacuoles to aid their buoyancy.

Representative genera: *Oscillatoria, Anabaena*

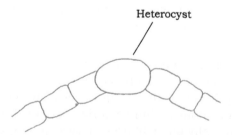

Heterocyst

Figure 7.10   Cyanobacteria. Nitrogen fixation takes place in specialised cells called heterocysts, which develop from ordinary cells when supplies of available nitrogen (e.g. ammonia) are depleted. The heterocyst loses its ability to photosynthesise and therefore to produce oxygen. This is essential because oxygen is highly inhibitory to the nitrogenase enzyme complex.

Phylum Chlorobi (green sulphur bacteria) and phylum Chloroflexi (green non-sulphur bacteria) We have already come across three distinct groups of photosynthetic bacteria in this chapter, the purple sulphur and purple non-sulphur bacteria and the Cyanobacteria; in the following paragraphs we consider the remaining two groups, the green sulphur and green non-sulphur bacteria.

The *green sulphur bacteria* (phylum Chlorobi), like their purple counterparts (see earlier), are anaerobic photolithotrophs that utilise reduced sulphur compounds instead of water as electron donors, and generate elemental sulphur. They differ, however, in a number of respects. The sulphur is deposited *outside* the cell, and CO_2 is assimilated not by the Calvin cycle, but by a reversal of the steps of the TCA cycle. The photosynthetic pigments in the green sulphur bacteria are contained in sac-like structures called *chlorosomes* that are associated with the inside of the plasma membrane.

Most members of the *green non-sulphur bacteria* (phylum Chloroflexi) are filamentous thermophiles, living in non-acid hot springs, where they form thick bacterial mats. Like the purple non-sulphur bacteria, they are photoheterotrophs, but can also grow in the dark as chemoheterotrophs.

Representative genera: *Chlorobium* (green sulphur), *Chloroflexus* (green non-sulphur)

Phylum Aquificae and phylum Thermotogae: the deeply branching bacteria These two phyla are regarded as the two deepest (i.e. oldest) branches in the evolution of the Bacteria and both comprise highly thermophilic Gram-negative rods. They are the only members of the Bacteria that can compare with the Archaea in their ability to live at high temperatures (optimal growth >80°C). The two phyla differ in their mode of nutrition: the Aquificae are autotrophs capable of oxidising hydrogen gas or sulphur, while the Thermotogae are anaerobic heterotrophs, fermenting carbohydrates. Members of the Thermotogae are characteristically surrounded, sometimes in a chain, by a proteinaceous sheath (or 'toga').

Representative genera: *Aquifex* (phylum Aquificae), *Thermotoga* (phylum Thermotogae)

Phylum "Deinococcus-Thermus"[2] Two of the genera included in this phylum are of particular interest because of remarkable physiological properties. *Thermus* species are thermophiles whose best known member is *T. aquaticus*; this is the source of the thermostable enzyme *Taq* polymerase, used in the polymerase chain reaction (see Chapter 12). *Deinococcus* species show an extraordinary degree of resistance to radiation, due, it seems, to an unusually

---

[2] This phylum has not yet been assigned a formal name.

$$
\begin{array}{c}
\text{H} \\
\text{H}_2\text{N} - \text{C} - \text{COOH} \\
| \\
\text{CH}_2 \\
| \\
\text{CH}_2 \\
| \\
\text{CH}_2 \\
| \\
\text{H}_2\text{N} - \text{CH} \\
| \\
\text{COOH}
\end{array}
\qquad\qquad
\begin{array}{c}
\text{H} \\
\text{H}_2\text{N} - \text{C} - \text{COOH} \\
| \\
\text{CH}_2 \\
| \\
\text{CH}_2 \\
| \\
\text{CH}_2 \\
| \\
\text{NH}_2
\end{array}
$$

(a) Diaminopimelic acid   (b)   Ornithine

Figure 7.11   Members of the Deinococcus-Thermus group have an unusual form of peptidogly-can. The diaminopimelic acid at position 3 on the amino acid chain attached to *N*-acetylmuramic acid (see Figure 3.6) is replaced by ornithine.

powerful DNA repair system. This also enables *Deinococcus* to resist chemical mutagens. Both *Deinococcus* and *Thermus* have thick cell walls that give them a Gram-positive staining reaction, but they also have a second membrane and so more closely resemble Gram-negative bacteria in their structure. Both genera have an unusual refinement to the structure of their peptidoglycan, with ornithine replacing diaminopimelic acid (Figure 7.11).

Representative genera: *Deinococcus*, *Thermus*

The remaining phyla of Gram-negative bacteria are grouped together in Volume 4 of the second edition of *Bergey*. The major phyla are described in the following pages.

Phylum Planctomycetes   This very ancient group of bacteria have a number of unusual properties, including cell division by budding, the lack of any pep-tidoglycan in their cell walls and the presence of a degree of internal compart-mentalisation. (Recall from Chapter 3 that membrane-bound compartments are regarded as a quintessentially *eukaryotic* feature.)

Several members of the Planctomycetes, including *Brocadia anammoxidans*, carry out the reaction known as *anammox* (anaerobic ammonium oxidation), whereby ammonium and nitrite are converted to nitrogen gas:

> Among the many unusual features of the *anammox* bacteria is their extremely slow rate of growth, having doubling times that are measured in weeks rather than minutes or hours.

$$ \text{NH}_4^+ + \text{NO}_2^- = \text{N}_2 + 2\text{H}_2\text{O} $$

It is thought that this reaction takes place in *anammoxisomes*, membrane-bound internal structures, and may be responsible for much of the nitrogen

cycling in the world's oceans. These bacteria are anaerobic chemolithoautotrophs; however, this is not typical of the Planctomycetes, most of which
are aerobic chemoorganoheterotrophs.

Representative genera: *Planctomyces, Pirellula*

Phylum Chlamydiae Formerly grouped with the rickettsiae (see 'Rickettsiae'
earlier), these non-motile obligate parasites of birds and mammals are now
assigned a separate phylum comprising only five genera, of which *Chlamydia*
is the most important. Like the rickettsiae, members of the Chlamydiae have
extremely small cells, and very limited metabolic capacities, and depend on
the host cell for energy generation. Unlike that group, however, they are not
dependent on an arthropod vector for transmission from host to host.

   Chlamydia trachomatis is the causative agent of trachoma, a principal cause
of blindness in humans. Different strains of this same species are responsible
for one of the most common forms of sexually transmitted disease. *C. psittaci*
causes the avian disease psittacosis, and *C. pneumoniae* causes chlamydial
pneumonia in humans as well as being linked to some cases of coronary artery
disease.

Representative genus: *Chlamydia*

Phylum Spirochaetae The spirochaetes are distinguished from all other bacteria by their slender helical morphology and corkscrew-like movement. This
is made possible by endoflagella (axial filaments), so-called because they are
enclosed in the space between the cell and a flexible sheath that surrounds it.

   Spirochaetes comprise both aerobic and anaerobic bacteria that inhabit a
wide range of habitats, including water and soil as well as the gut and oral
cavities of both vertebrate and invertebrate animals. Some species are important pathogens of humans, including *Treponema pallidum* (syphilis) and
Leptospira interrogans (leptospirosis).

Representative genera: *Treponema, Leptospira*

Phylum Bacteroidetes No unifying phenotypic feature characterises this
diverse group, but their phylogenetic closeness causes them to be placed
together. In light of this, we can only consider examples, without claiming
them to be in any way representative.

   The genus *Flavobacterium* takes its name from the yellow carotenoid pigments secreted by its members. These are aerobic, free-living, aquatic forms,
although they are also associated with food spoilage. In contrast, *Bacteroides*
species are obligate anaerobes found in the human gut, where they ferment
undigested food to acetate or lactate. Here they outnumber all other microbial forms, and are responsible for a significant percentage of the weight of

human faeces. Some species can also be pathogenic, and may cause peritonitis in cases where the large intestine or appendix has become perforated.

Representative genera: *Bacteroides, Flavobacterium*

Phylum Verrucomicrobia   Members of the Verrucomicrobia form several prosthecae per cell; these are similar to those described for certain proteobacteria (see earlier). Although widespread in terrestrial, freshwater and marine environments, only a handful of representatives have been isolated in pure culture.

Representative genera: *Verrucomicrobium, Prosthecobacter*

7.2.3   The Gram-positive bacteria: phyla Actinobacteria, Firmicutes and Tenericutes

The Gram-positive bacteria are divided into two large phyla, the Firmicutes and the Actinobacteria, and one small one, the Tenericutes. Gram-positive bacteria mostly have a chemoheterotrophic mode of nutrition and include among their number several important human pathogens, as well as industrially significant forms.

The base composition of an organism's DNA can be expressed as the percentage of cytosine and guanine residues (%GC content); this property is used widely in microbial taxonomy, and the Gram-positive bacteria are divided into those whose GC content is significantly over or under 50%. It is convenient to consider groupings within the high GC and low GC forms as follows:

Phylum Actinobacteria: the high GC Gram-positive bacteria   The high GC Gram-positive bacteria make up the fifth and final volume of the second edition of *Bergey*.

Many genera belonging to the Actinobacteria, such as *Streptomyces*, *Actinomyces* and *Frankia*, are often referred to as actinomycetes. They are aerobic filamentous bacteria that form branching mycelia superficially similar to those of the Fungi (see Chapter 8). In some cases, the mycelium extends clear of the substratum and bears asexual conidiospores at the hyphal tips. These are produced by the formation of cross-walls and pinching off of spores, which are often coloured. The best-known actinomycete genus is *Streptomyces*, which contains nearly 600 species, all with a characteristically high GC content (69–73%). *Streptomyces* are very prevalent in soil, where they saprobically degrade a wide range of complex organic substrates by means of extracellular enzymes. Indeed, the characteristic musty smell of many soils is due to the production by *Streptomyces* species of a volatile organic compound called *geosmin*. A high proportion of therapeutically useful antibiotics derive from *Streptomyces* species, including well-known examples such as streptomycin, erythromycin and tetracycline.

Most actinomycetes, including *Streptomyces*, are aerobic; however, members of the genus *Actinomyces* are facultative anaerobes.

Representative genera: *Streptomyces, Actinomyces*

The *coryneform bacteria* are morphologically half way between single-celled bacilli and the branching filamentous actinomycetes. They are rods that show rudimentary branching, giving rise to characteristic 'V' and 'Y' shapes. Among the genera in this group are *Corynebacterium, Mycobacterium, Propionibacterium* and *Nocardia*.

Corynebacterium species are common in soil, and are also found in the mouths of a variety of animals. *C. diphtheriae* is the causative agent of diphtheria; it only becomes pathogenic when it has been infected by a bacteriophage that carries the gene for the diphtheria exotoxin.

Members of the genus *Mycobacterium* are characterised by their unusual cell wall structure; they include unusual complex lipids called mycolic acids. This causes the cells to be positive for the acid-fast staining technique, a useful way of identifying the presence of these bacteria. Mycobacteria are rod-shaped, sometimes becoming filamentous; when filaments are formed, propagation is by means of fragmentation. *M. leprae* and *M. tuberculosis* cause, respectively, leprosy and tuberculosis in humans.

> The *acid–fast test* assesses the ability of an organism to retain hot carbolfuchsin stain when rinsed with acidic alcohol.

Propionibacterium species ferment lactic acid to propionic acid. Some species are important in the production of Swiss cheeses, whilst *P. acnes* is the main cause of acne in humans.

Representative genera: *Corynebacterium, Propionibacterium*

Phylum Firmicutes and Phylum Tenericutes: the low GC Gram-positive bacteria
Volume 3 of the second edition of *Bergey* is devoted to a single phylum, the Firmicutes.

The *spore-forming Gram-positive bacteria* include two large genera, *Clostridium* and *Bacillus*. Although not particularly close in phylogenetic terms, they are both capable of propagation by endospores.

Clostridium species are obligate anaerobes, and common inhabitants of soil. Sugars are fermented to various end-products such as butyric acid, acetone or butanol. Lacking an electron transport system, they obtain all their ATP from substrate-level phosphorylation.

Several species of *Clostridium* are serious human pathogens including *C. botulinum* (botulism) and *C. tetani* (tetanus). *C. perfringens* causes gas gangrene, and if ingested, can also result in gastroenteritis. *C. difficile* can cause severe diarrhoea in individuals whose normal gut flora have been eradicated by the use of broad-spectrum antibiotics, and is a significant cause of nosocomial infections. All these conditions are due to the production of bacterial exotoxins. The resistance of spores to heating is thus highly relevant both in medicine and in the food industry. Related to *Clostridium* are the heliobacteria, two genera of anaerobic photoheterotrophic rods, some of which produce endospores. They are the only known photosynthetic Gram-positive bacteria.

Bacillus species are aerobes or facultative anaerobes. They are chemoheterotrophs and usually motile by means of peritrichous flagella. Only a few species of *Bacillus* are pathogenic in humans, notably *B. anthracis*, the causative agent of anthrax. This is seen by many as a potential agent of bioterrorism, and here again the relative indestructibility of its spores is a crucial factor. Other species, conversely, are positively beneficial to humans; antibiotics such as bacitracin and polymixin are produced by *Bacillus* species, whilst the toxin from *B. thuringiensis* has been used as a natural insecticide (see Chapter 12).

Representative genera: *Bacillus*, *Clostridium*

The *non-spore-forming low GC Gram-positive bacteria* include a number of medically and industrially significant genera, a few of which are discussed below.

The *lactic acid bacteria* are a taxonomically diverse group containing both rods (*Lactobacillus*) and cocci (*Streptococcus*, *Lactococcus*), all characterised by their fermentative metabolism with lactic acid as end-product. Although they are able to tolerate oxygen, these bacteria do not use it in respiration. They are said to be

> *Probiotics* are living organisms that are deliberately ingested by humans with the aim of promoting health.

aerotolerant. Like the clostridia, they lack cytochromes, and are therefore unable to carry out electron transport phosphorylation. The lactic acid bacteria have limited synthetic capabilities, so they are dependent on a supply of nutrients such as amino acids, purines/pyrimidines and vitamins. There has been growing interest in recent years in the use of certain lactic acid bacteria as probiotics.

The genus *Streptococcus* remains a large one, although some members have been assigned to new genera in recent years, for example *Enterococcus* and *Lactococcus*. Streptococci are classified in a number of ways on the basis of phenotypic characteristics, but these do not correspond to phylogenetic relationships.

> *Haemolysis* is the lysis (bursting) of red blood cells. It can be brought about by bacterial toxins called *haemolysins*.

Many species produce haemolysis when grown on blood agar, due to the production of toxins called haemolysins. In α-haemolysis, haemoglobin is reduced to methaemoglobin, resulting in a partial clearing of the medium and a characteristic green colour. β-Haemolysis causes a complete lysis of the red blood cells, leaving an area of clearing in the agar. A few species are non-haemolytic. Streptococci are also classified on the basis of carbohydrate antigens found in the cell wall; this system, which assigns each organism to a lettered group, is named after its devisor, Rebecca Lancefield.

Pathogenic species of *Streptococcus* include *S. pyogenes* ('strep' sore throat, as well as the more serious rheumatic fever), *S. pneumoniae* (pneumococcal pneumonia) and *S. mutans* (tooth decay). Cells of *Streptococcus* exist mostly in chains, but in *S. pneumoniae* they are characteristically paired.

Lactobacillus is used very widely in the food and drink industry in the production of such diverse foodstuffs as yoghurt, cheeses, pickled foods (e.g. sauerkraut) and certain beers. This is discussed further in Chapter 18.

The cells of staphylococci occur in irregular bunches rather than ordered chains. They also produce lactic acid but can additionally carry out aerobic respiration involving cytochromes, and lack the complex nutritional requirements of the lactic acid bacteria. They are resistant to drying and can tolerate relatively high concentrations of salt. These properties allow *Staphylococcus aureus* to be a normal inhabitant of the human skin, where it can sometimes give rise to dermatological conditions such as acne, boils and impetigo. It is also found in the respiratory tract of many healthy individuals, to whom it poses no threat; however, in people whose immune system has been in some way compromised, it can cause serious respiratory infections. *S. aureus* can also cause a type of food poisoning, and is the causative agent of toxic shock syndrome. Widespread antibiotic use has been largely responsible for the development of resistant forms of *S. aureus*, which have become ubiquitous inhabitants of hospitals (meticillin-resistant *Staphylococcus aureus*: MRSA). The problem of antibiotic resistance is discussed at greater length in Chapter 17.

Representative genera: *Streptococcus, Staphylococcus*

Phylum Tenericutes The Tenericutes is a phylum comprising the class Mollicutes, extremely small bacteria that lack a cell wall and hence have a fluid shape (pleomorphic). The Mollicutes are a very difficult group to classify. For many years they were placed in the phylum Firmicutes (see above); however, in the new *Bergey,* they are repositioned in a separate phylum, the Tenericutes. It should be noted that the term Mycoplasma is often used interchangeably with Mollicutes, but this name should now be reserved specifically for members of the best known genus, *Mycoplasma*, rather than as a general term for any member of the Mollicutes.

Having no cell wall, the Mollicutes do not give a positive Gram test; however, they are clearly related at the genetic level to other members of the low GC Gram-positive bacteria. The membranes of mycoplasmas contain sterols; these help in resisting osmotic lysis, and are often essential as a growth requirement. Saprophytic, commensal and parasitic forms are known, and some species are associated with respiratory diseases in animals. Mollicutes frequently occur as contaminants in the culture of animal cells, because their small size allows them to pass through filters, and they are resistant to antibiotics directed at cell wall synthesis. Members of the Mollicutes are among the smallest of all known cells (around 0.1 μ) and have some of the smallest genomes (just over half a million base pairs).

Representative genera: *Mycoplasma, Ureoplasma*

8
The Fungi

As we saw in the introduction to this section on microbial diversity, fungi were for many years classified along with bacteria, algae and the slime moulds (see Chapter 9) as members of the kingdom Plantae. As recently as the 1960s it was possible to find fungi being discussed under this heading, but in more recent times there has been universal agreement that they should be assigned their own kingdom. This is because fungi differ from plants in two quite fundamental respects:

- plants obtain energy from the sun, fungi do not;
- plants utilise CO_2 as a carbon source, fungi do not.

To use the terminology we introduced in Chapter 4, plants are photoautotrophs, whereas fungi are chemoheterotrophs. In fact it now seems on the basis of molecular evidence that fungi are more closely related to animals than they are to plants!

We may define true fungi as primarily terrestrial, spore-bearing eukaryotes, lacking chlorophyll and having a heterotrophic, absorptive mode of nutrition. Around 100 000 species are known and it is thought possible that at least a million more remain to be described; according to one recent article, the figure could be as high as five million! True fungi are a monophyletic group; that is, they are all thought to descend from a common ancestor, some 550 million years ago.

Fungi are of great importance economically and socially, and may have beneficial or detrimental effects. Many fungi, particularly yeasts, are involved in industrial fermentation processes (see Chapter 18). These include, for example, the production of bread and alcoholic drinks, while other fungi are essential to the cheese-making process. Many antibiotics, including penicillin, derive from fungi, as does the immunosuppressive drug cyclosporin

Essential Microbiology, Second Edition. Stuart Hogg.
© 2013 John Wiley & Sons, Ltd. Published 2013 by John Wiley & Sons, Ltd.

(ciclosporin). Along with bacteria, fungi are responsible for the decomposition and reprocessing of vast amounts of complex organic matter; some of this is recycled to the atmosphere as CO_2, while much is rendered into a form that can be utilised by other organisms (see Chapter 14). The other side of this coin is seen in the activity of fungi that degrade and destroy materials of economic importance such as wood, paper and leather, employing essentially the same biochemical processes. Additionally, some fungi may cause disease in plants, and can have a devastating effect on crops of economic importance, either on the living plant or in storage subsequent to harvesting. Rusts, smuts and mildews are all examples of common plant diseases caused by fungi. A number of human diseases, particularly of the skin and scalp, are also caused by fungi (see Chapter 15).

8.1   General biology of the fungi
8.1.1   Morphology

Fungi range in form and size from unicellular yeasts to large mushrooms and puffballs. Yeasts are unicellular, do not have flagella, and reproduce asexually by budding or transverse fission, or sexually by spore formation. Multicellular forms such as moulds have long, branched, threadlike filaments called *hyphae*, which aggregate together to form a tangled *mycelium* (Figure 8.1a). In some fungi the hyphae have cross-walls or septa (sing. septum) separating cells, which may nevertheless be joined by one or more

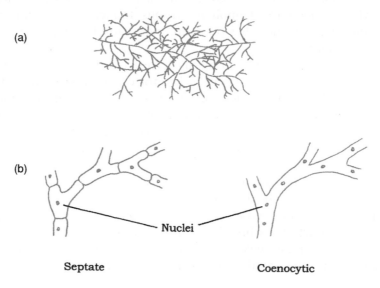

(a)

(b)

Nuclei

Septate   Coenocytic

Figure 8.1 Hyphae and mycelia. (a) Individual hyphae branch and aggregate to form a mycelium. (b) Hyphae may or may not contain cross-walls (septa).

pores, which permit cytoplasmic streaming, a form of internal transport. Such hyphae are said to be *septate*; others have no cross-walls and are therefore *coenocytic* (i.e. many nuclei within a single plasma membrane; Figure 8.1b).

Many fungi are *dimorphic*, that is, they exist in two distinct forms. Some fungi that cause human infections can change from the yeast form in the human to a mycelial form in the environment in response to changes in nutrients, and environmental factors such as CO_2 concentration and temperature. This change in body form is known as the *YM shift*; in fungi associated with plants, the shift often occurs the other way round, that is, the mycelial form exists in the plant and the yeast form in the environment.

One of the features that caused taxonomists finally to remove fungi from the plant kingdom was the distinctive chemical nature of the fungal cell wall. Whereas plant and algal cells have walls composed of cellulose, the cell wall of fungi is made up principally of chitin(see Figure 3.17b), a strong but flexible polysaccharide that is also found in the exoskeleton of insects.

8.1.2 Nutrition

Most fungi are saprobic (although some have other modes of nutrition), that is, they obtain their nutrients from decaying matter, which they grow over and through, frequently secreting enzymes extracellularly to break down complex molecules to simpler forms, which can then be absorbed by the hyphae. Most fungi are able to synthesise their own amino acids and proteins from carbohydrates and simple nitrogenous compounds. Although fungi are unable to move, they can swiftly colonise new territory as a result of the rapid rate at which their hyphae grow. All energy is concentrated on adding length rather than thickness; this growth pattern leads to an increase in surface area and is an adaptation to an absorptive way of life. Carbohydrates are stored mainly in the form of glycogen (cf. starch in higher plants, green algae). Metabolism is generally aerobic, but some yeasts can function as facultative anaerobes.

> The term *saprobe* describes an organism that feeds on dead and decaying organic materials. The older term *saprophyte* is no longer used, since the name perpetuates the idea of fungi being plants (*phyton* = a plant). Saprobes contribute greatly to the recycling of carbon and other elements.

8.1.3 Reproduction

Although there is a good deal of variety among the patterns of reproduction among the fungi, all share in common the feature of reproducing by *spores*; these are non-motile reproductive cells that rely on being carried by animals or the wind for their dispersal. The hyphae that bear the spores usually project up into the air, aiding their dispersal. One of the main reasons that we have to practise aseptic techniques in the laboratory is that fungal spores are pretty well ubiquitous, and will germinate and grow if they find a suitable growth medium. Spores of the common black bread mould, *Rhizopus* (see later) have been found in the air over the North Pole, as well as hundreds of miles out to sea. In some fungi these aerial spore-bearing hyphae are developed into large complex structures called fruiting bodies, the most familiar example of which is the mushroom. Many people think of the mushroom itself as the whole fungus, but it only represents a part of it. The main part of the organism is a network of almost invisible hyphae, buried out of sight below the surface of the soil or other organic matter in which the organism grows.

8.2   Classification of the Fungi

Fungi have traditionally been arranged into four major phyla on the basis of differences in their sexual reproduction. Their classification is constantly being reviewed, however, and there is no single current universally accepted system for arranging the different types. This is because the use of molecular techniques has resulted in a radical change to the way we view the relationship between different fungal groups. Matters are not helped by the fact that different schemes of classification sometimes use a different name to describe the same group of organisms! As one writer put it: 'such inconsistencies create confusion, especially for students and non-specialists, and they hamper efforts to develop taxonomic databases.'

The system adopted here is based on a study published in 2007, which relies on molecular phylogenetic relationships, rather than reproductive strategies. It expands the number of phyla to seven – the Microsporidia, Chytridiomycota, Blastocladiomycota, Neocallimastigomycota, Glomeromycota, Ascomycota and Basidiomycota (Figure 8.2) – and in addition there are four subphyla *incertae sedis* ('of uncertain place'), groups whose relationship to the other phyla is as yet unclear. To illustrate just how fluid the whole business of fungal classification is, in 2011, a widespread group of hitherto unknown fungi was described, identified on the basis of their DNA sequences. They were sufficiently different from other fungi for the authors of the paper to suggest they may represent an entirely new phylum, tentatively named the Cryptomycota ('hidden fungi').

The largest phyla are the Ascomycota and the Basidiomycota, both of which remain unchanged from previous classification schemes. Together they form the subkingdom Dikarya; they are also sometimes known as the Higher Fungi.

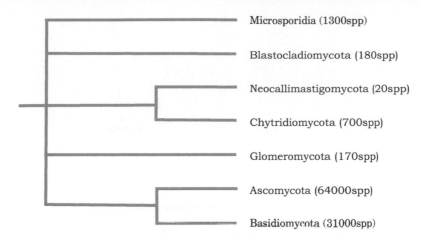

Microsporidia (1300spp)

Blastocladiomycota (180spp)

Neocallimastigomycota (20spp)

Chytridiomycota (700spp)

Glomeromycota (170spp)

Ascomycota (64000spp)

Basidiomycota (31000spp)

Figure 8.2 A proposed scheme for the seven-phylum classification of the Fungi. The subphyla *incertae sedis*, comprising at least another 1000 species, are not shown.

8.2.1 Phylum Ascomycota

The Ascomycota are characterised by the production of haploid *ascospores* through the meiosis of a diploid nucleus in a small sac called an ascus. For this reason they are sometimes called the sac fungi or cap fungi. They include over 60 000 species, among them yeasts, food spoilage moulds, brown fruit-rotting fungi and truffles. Note that the latter, often regarded as the most prized type of mushrooms by gourmets, are assigned to a completely different group to the true mushrooms, which belong to the Basidiomycota. Many of the fungi that cause serious plant diseases such as Dutch elm disease and powdery mildew belong to the Ascomycota, as does *Hymenoscyphus pseudoalbidus*,[1] the cause of ash dieback, which finally reached the UK during 2012, and is likely to have a devastating effect on British ash trees in the coming years. Around half of ascomycete species exist in associations with algae to form *lichens*; these will be discussed further in Chapter 13.

> A *lichen* is formed by the symbiotic association of a fungus (usually an ascomycete) and an alga or cyanophyte.

Asexual reproduction in most ascomycetes involves the production of airborne spores called *conidia*. These are carried on the ends of specialised hyphae called conidiophores, where they may be pinched off as chains or clusters (Figure 8.3). The conidia may be naked or protected by a flask-like structure called the pycnidium. Asexual reproduction by conidia formation is a means of rapid propagation for the fungus in favourable conditions. The characteristic green, pink or brown colour of many moulds is due to the

---

[1] Frequently still referred to by its previous name, *Chalara fraxinea*.

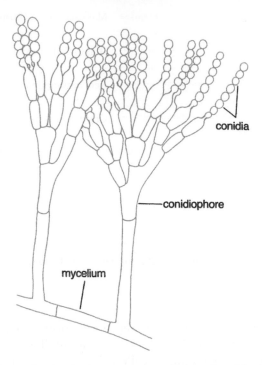

Figure 8.3 Asexual reproduction in the Ascomycota. Chains of conidia develop at the end of specialised hyphae called conidiophores.

pigmentation of the conidia, which are produced in huge numbers and dispersed by air or water currents. The conidia germinate to form another haploid mycelium.

In the case of the unicellular yeasts, asexual reproduction occurs as the result of *budding*, a pinching off of a protuberance from the cell, which eventually grows to full size (Figure 8.4).

Although some ascomycetes are self-fertile, sexual reproduction often involves separate 'plus' and 'minus' mating strains. Reproductively distinct, these two types are, however, morphologically identical, so it is not appropriate to refer to them as 'male' and 'female'. The hyphae involved in reproduction are termed the *antheridium* (+ strain) and *ascogonium* (− strain). Hyphae from the different strains grow together and there is a fusion of their cytoplasm (Figure 8.5). Within this fused structure, nuclei pair, but do not fuse; the resulting structure is a dikaryon. Following cytoplasmic fusion (plasmogamy), branching hyphae develop. These hyphae are septate – i.e. partitioned off into

> *Plasmogamy* is the fusion of the cytoplasmic content of two cells. *Karyogamy* is the fusion of nuclei from two different cells.

> A *dikaryon* is a structure formed by two cells whose contents, but not nuclei, have fused.

Figure 8.4 Budding in yeasts. Yeast cells in various stages of budding. A protuberance or bud develops on the parent yeast; the nucleus undergoes division and one copy passes into the bud. Eventually the bud is walled off and separated to form a new cell.

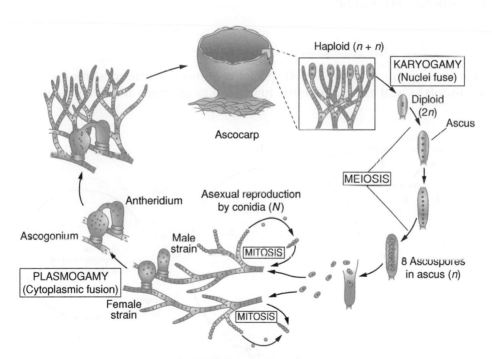

Figure 8.5 Sexual reproduction in the Ascomycota. Nuclei in the cells at the tips of the ascogenous hyphae fuse to give a diploid zygote. Meiotic and mitotic divisions result in the formation of eight haploid ascospores inside a tubular ascus. On germination, each ascospore is capable of giving rise to a new mycelium. Reproduced from Black, JG (1999) Microbiology: Principles and Explorations, 4th edn, with permission from John Wiley & Sons.

separate cells – but each cell is dikaryotic, having a nucleus from each parental type.

As we have seen, a defining feature characteristic of ascomycetes is the production of sexual spores in saclike structures called asci. These develop in distinct macroscopic fruiting bodies called ascocarps, which arise from the aggregation of dikaryotic hyphae with sterile haploid hyphae. At the tip of each dikaryotic hypha, pairs of nuclei fuse to give a diploid zygote; this is followed by one meiotic and one mitotic division, giving rise to eight haploid ascospores. An ascocarp may contain thousands of asci, each with eight ascospores.

When the ascus is mature, it splits open at its tip and the ascospores are released. They are dispersed, often over long distances, by air currents. If a mature ascocarp is disturbed, it may release smokelike puffs containing thousands of ascospores. The germinating ascospore forms a new mycelium. Thus the diploid state plays only a very brief part in the life cycle. Frequently there are many rounds of asexual reproduction between successive rounds of sexual ascospore production.

8.2.2 Phylum Basidiomycota

This large group of some 30 000 species contains the true mushrooms and toadstools as well as other familiar fungi such as puffballs and bracket fungi.

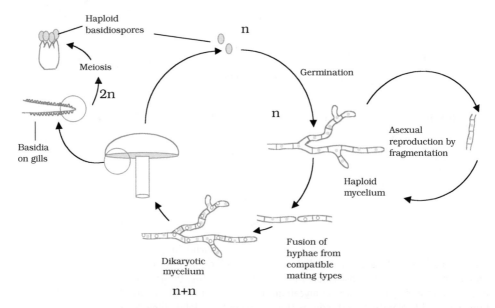

Figure 8.6 Life cycle of the Basidiomycota. Most of the fungus exists as a mycelial mass underground. The mushroom is an aerial fruiting body that facilitates the dispersal of spores.

In fact the great majority of the fungi that we see in fields and woodlands belong to the Basidiomycota. They are of great economic importance in the breakdown of wood and other plant material (see Chapter 18). The group derives its common name of 'club fungi' from the way that the spore-bearing hyphae involved in reproduction are swollen at the tips, resembling clubs (the *basidia*). Asexual reproduction occurs much less frequently in basidiomycetes than in the Ascomycota. When it does occur, it is generally by means of conidia, although some types are capable of fragmenting their hyphae into individual cells, each of which then acts like a spore and germinates to form a new mycelium.

Box 8.1 *Clamp connections*

Imagine the situation in a dikaryotic cell when mitosis took place. The most likely outcome would be that as the two nuclei divided, both of one type would end up in one daughter cell and both of the other type in the other. As growth and cell division proceeds in the secondary mycelium, it is therefore necessary to ensure that the dikaryotic state is maintained, i.e. that each new cell inherits one nucleus of each type from its parent. This is achieved by means of *clamp connections*.

As the terminal hyphal cell elongates (b), a tubelike clamp connection grows out back towards (c). As this happens, one nucleus enters the clamp and mitosis occurs simultaneously in both parental nuclei. A septum is formed to separate the first pair of nuclei (d), then, as the loop of the clamp connection is completed, a second septum separates the second pair (e). The result is two new daughter cells, each with one copy of each nuclear type.

Sexual reproduction in a typical mushroom involves the fusion of haploid hyphae belonging to two compatible mating types to produce a dikaryotic mycelium in which each cell has two haploid nuclei (Figure 8.6). The most striking feature of this secondary mycelium is the *clamp connection*; this is unique to the Basidiomycota and is a device for ensuring that as growth continues, each new cell has one nucleus from each of the parent mating strains (see Box 8.1). This dikaryotic secondary mycelium continues to grow, overwhelming any remaining haploid hyphae from the parent fungi.

When the secondary mycelium has been developing for some time, it forms a dense compact ball or button, which pushes up just above the surface and expands into a *basidiocarp*; this is the mushroom itself. Stalk formation and upward growth are extremely rapid; a stalk or stipe of 10 cm can be formed in only about 6–9 hours. The growth is initially towards light (positive phototropism) and then upward (negative geotropism). As the cap expands, fleshy flaps radiating from the centre of its underside open up. These are the gills, made up of compacted hyphae with numerous basidia arranged at right angles. As each basidium matures, its two nuclei finally fuse, and then undergo meiosis to produce four haploid basidiospores. A single large mushroom can produce millions of basidiospores in the space of a few days. They are discharged from the end of the basidia and then fall by gravity from the gills. Air currents then carry them away for dispersal. Upon finding a suitable substratum, the spores germinate into a haploid mycelium just below the surface of the soil, thus completing the life cycle.

Box 8.2   Ergot

Members of the genus *Claviceps* may infect a variety of grains, particularly rye, when the plants come into flower, giving rise to the condition called *ergot*. No great damage is caused to the crop, but as the fungus develops in the maturing grain, powerful hallucinatory compounds are produced, which cause *ergotism* in those who consume bread made from the affected grain. This was relatively common in the Middle Ages, when it was known as St Anthony's Fire. The hallucinatory effects of ergotism have been put forward by some as an explanation for outbreaks of mass hysteria such as witch hunts and also as the cause of the abandonment of the Mary Celeste. The effects can go beyond the psychological, causing convulsions and even death. In small controlled amounts, the drugs derived from ergot can be medically useful in certain situations such as the induction of childbirth and the relief of migraine headaches.

8.2.3 Phylum Microsporidia

The 1000+ members of this phylum are all unicellular, spore-forming, obligate intracellular parasites of animals, mostly invertebrates (especially insects), but also some vertebrates, including humans. In recent times, infections of humans by microsporidians have become more common, particularly in immunocompromised individuals. Some species are used in the biological control of certain insects, including the *Anopheles* mosquito that spreads malaria.

In terms of their cellular and genomic structure, the microsporidians have undergone considerable reduction. Their cells lack mitochondria, and it is thought that they obtain their energy by transporting ATP from their host's cells using special carrier proteins. Microsporidians contain some of the smallest genomes known in eukaryotes; however, this does not mean that they are primitive, rather that their genome has become highly adapted to their parasitic way of life.

> When they were first discovered in the 1850s, microsporidia were classified as fungi. In the 1980s they were reclassified as protists, but have now been returned to the fungi!

Microsporidians are characterised by a polar tube or filament, which may be many times the length of the cell itself. As a result of a build-up of internal osmotic pressure, it is propelled from the germinating spore and the spore's contents are transferred along it (Figure 8.7). Should it encounter a host cell, it will penetrate it, internalising the spore's contents. Once inside the host cell, the microsporidian acts as an intracellular parasite and undergoes growth

Anchoring disc
Polarplast
Polar tube
Nucleus
Posterior vacuole

Figure 8.7 Eversion and ejection of polar filament in microsporidians. In response to an environmental stimulus, the spore begins to swell, leading to the rupture of the anchoring disc and the release and eversion of the polar tube. Continuing pressure within the spore causes its contents to be forced through the tube, emerging surrounded by a new membrane at the tip. Reproduced from Keeling and Fast (2002), with permission from Annual Reviews Inc.

and division. Eventually mature spores are released, which may quickly infect nearby cells of the same host, or be released into the environment.

The Microsporidia have none of the characteristics of a 'typical' fungus, and there is still some debate as to whether they should be included in the Fungi, with some systems excluding them.

8.2.4   Phylum Chytridiomycota

Along with the Blastocladiomycota and Neocallimastigomycota, the chytrids are characterised by the possession of flagellated zoospores. At one time, lack of flagella was regarded as one of the defining features of the Fungi, so these organisms were assigned to the Protista. Molecular evidence, however, including their possession of a chitinous cell wall, suggests that they would be more appropriately placed among the Fungi. They are believed to have been the first of the fungal groups to diverge from a common ancestor many millions of years ago, and constitute a separate evolutionary branch from the other fungi. Some forms are unicellular, whilst multicellular forms are coenocytic.

Most members of the chytrids live saprobically on decaying matter in freshwater environments, while others are parasites of plants (e.g. potato wart) and algae. In recent years a dramatic decline in frog populations in certain parts of the world has resulted due to chytridiomycosis, caused by a parasitic species of chytrid, *Batrachochytrium dendrobatidis*.

Some chytrids are unicellular, while others form mycelia of coenocytic hyphae. Reproduction may be asexual by means of motile zoospores, or sexual. The latter may involve fusion of gametes to produce a diploid zygote, but there is no dikaryotic stage in the life cycle.

8.2.5   Phylum Blastocladiomycota and phylum Neocallimastigomycota

According to classification schemes in existence at the time of writing the first edition of this book, both these groups of flagellated unicellular fungi were included in the Chytridiomycota, but on the basis of molecular data they have now been assigned phylum status in their own right.

The Blastocladiomycota are mostly saprobic; however, a significant number cause disease in invertebrates. They show an alternation of generations between haploid and diploid forms, differentiating them from the chytrids, in which the zygote is the only diploid stage. The Neocallimastigomycota are a small group of anaerobic fungi that lack true mitochondria and are found in the rumen of herbivores. They contain the enzyme xylanase, essential for the breakdown of hemicellulose in plant cell walls.

8.2.6   Phylum Glomeromycota

This small phylum contains some of the fungi previously placed in the now discontinued phylum Zygomycota. The remainder have been reassigned to subphyla whose members have yet to be accurately placed in relation to other groups (subphyla *incertae sedis*).

Members of the Glomeromycota have coenocytic hyphae, and are of great ecological importance in forming an internal symbiotic association known as a *mycorrhiza* with plant roots. This will be discussed further in Chapter 13.

8.2.7   Subphyla *incertae sedis*

The four currently non-aligned subphyla are the Mucoromycotina, Zoopagomycotina, Kickxellomycotina and Entomophthoromycotina. The

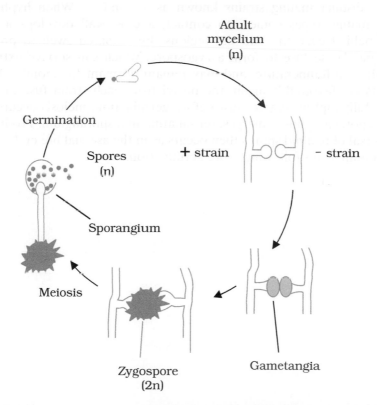

Figure 8.8   The life cycle of *Rhizopus*. Both sexual and asexual cycles involve the production of sporangiospores. In sexual reproduction, hyphae from different mating strains fuse to form a diploid zygospore, via a short-lived dikaryotic intermediate. Germination of the zygospore gives rise to an aerial sporangium; this contains many haploid sporangiospores, which give rise to another vegetative mycelium.

Mucoromycotina contains familiar fungi such as *Mucor* and the black bread mould *Rhizopus*, previously representative members of the Zygomycota.

The life cycle of *Rhizopus* is shown in Figure 8.8. Hyphae spread rapidly over the surface of the substrate (bread, fruit, etc.) and penetrate it, absorbing soluble nutrients such as sugars. Upright hyphae develop, carrying at their tip *sporangia*, full of black haploid spores. The spores give the characteristic colour to the

> A *sporangium* is a structure inside which spores develop. It is held aloft on an aerial hypha called a *sporangophore*.

mould; they are the asexual reproductive structures, and are released when the thin wall of the sporangium ruptures. When conditions are favourable, *Rhizopus* reproduces in this way; each spore, upon finding a suitable substratum for growth, is capable of germinating and initiating a new mycelium.

Sexual reproduction occurs when environmental conditions are unfavourable. Most species of *Rhizopus* are *heterothallic*; that is, there exist two distinct mating strains known as + and −. When hyphae of opposite mating types come into contact, a cross-wall develops a short distance behind each tip, and the regions thus isolated swell to produce gametangia. These fuse to form a zygospore, which can survive extremes of drought and temperature, and may remain dormant for months. When conditions are favourable again, the nuclei from each strain fuse in pairs, to give a fully diploid zygote. Just before germination, meiosis occurs, and an aerial sporangiophore emerges, terminating in a sporangium. Production and dispersal of haploid spores then occurs as in the asexual life cycle, and a new mycelium forms following spore germination.

9
The Protista

Although not such an all-embracing a term as originally envisaged by Haeckel, the Protista represents a very diverse group of organisms, united by their possession of eukaryotic characteristics, and failure to fit satisfactorily into the animal, plant or fungal kingdoms. Some scientists limit use of the name to unicellular organisms, while others also include organisms such as the macroscopic algae, which are not accommodated conveniently elsewhere.

It has become clear from molecular studies that some members of the Protista bear only a very distant relationship to each other, making it an unsatisfactory grouping in many respects. However, in an introductory text such as this it is a convenient grouping, and hopefully it is not confusing to retain the name Protista as a chapter heading, as long as the student understands that the term does not represent a coherent taxonomic grouping of phylogenetically related organisms. At the end of the chapter we shall look at how members of the Protista are placed in modern, phylogenetic taxonomic schemes.

> The *Protista* is a grouping of convenience, containing organisms not easily accommodated elsewhere. It includes all unicellular and colonial eukaryotic organisms, but is often expanded to include multicellular algae.

It has been found helpful in the past to think of protists as being divided into those with characteristics that are plant-like (the Algae), animal-like (the Protozoa) and fungus-like (the water moulds and slime moulds), and we shall discuss each of these groups in turn. It should be borne in mind, however, that molecular evidence suggests such a division to be artificial; on the basis of molecular and cytological comparison, the 'animal-like' protozoan *Trypanosoma*, for example, is closely related to the photosynthetic (and therefore 'plant-like') *Euglena*.

Essential Microbiology, Second Edition. Stuart Hogg.
© 2013 John Wiley & Sons, Ltd. Published 2013 by John Wiley & Sons, Ltd.

9.1   The 'algae'

The term 'algae' is traditionally given to several phyla of primitive, and mostly aquatic plants, making up a highly diverse group of over 30 000 species. They do not form a coherent taxonomic group. Algae display a wide variety of structure, habitat and life cycle, ranging from single-celled forms to massive seaweeds tens of metres in length. Most algae share a number of common features that caused them to be grouped together. Among these are:

- possession of the pigment chlorophyll;
- deriving energy from the sun by means of oxygenic photosynthesis;
- fixing carbon from CO_2 or dissolved bicarbonate.

Modern taxonomy attempts to reflect more accurately the relationship between organisms with an assumed common ancestor. Thus, in the following pages, the unicellular 'algae' are discussed in relation to other unicellular eukaryotes. Multicellular forms, including the Phaeophyta (brown algae) and Rhodophyta (red algae), are not discussed at great length, but are included for the sake of completeness.

9.1.1   Structural characteristics of algal protists

All algal types are eukaryotic, and therefore contain the internal organelles we encountered in Chapter 3, such as nuclei, mitochondria, endoplasmic reticulum, ribosomes, Golgi body and, in most instances, chloroplasts. With the exception of one group (the Euglenophyta), all have a cellulose cell wall, which is frequently modified with other polysaccharides, including pectin and alginic acids. In some cases, the cell wall may be fortified with deposits of calcium carbonate or silica. This is permeable to small molecules and ions, but impermeable to macromolecules. To the exterior of the cell may be one or two flagella, with the typical eukaryotic $9+2$ microstructure (see Figure 3.18), which may allow unicellular types to move through the water; cilia are not found in any algae.

The characteristics used to place algal protists into different taxa include the type of chlorophyll present, the form in which carbohydrate is stored, and the structure of the cell wall (Table 9.1). A group not considered here are the cyanophytes, previously known as the blue-green algae; although they carry out oxygenic photosynthesis, they are prokaryotes, and as such are more closely related to certain bacteria. They are therefore discussed in Chapter 7.

9.1.2   Euglenophyta

This is a group of unicellular flagellated organisms, which probably represent the most ancient group of algal protists. Individuals range in size from

Table 9.1 Characteristics of major algal groups

	Common name	Morphology	Pigments	Storage compound	Cell wall
Euglenophyta	Euglenids	Unicellular	Chlorophylls *a* and *b*	Paramylon	None
Pyrrophyta	Dinoflagellates	Unicellular	Chlorophylls *a* and *c*, xanthophylls	Starch	Cellulose/none
Chrysophyta	Golden-brown algae, diatoms	Unicellular	Chlorophylls *a* and *c*	Lipids	Cellulose, silica, CaCO₃, etc.
Chlorophyta	Green algae	Unicellular to multicellular	Chlorophylls *a* and *b*	Starch	Cellulose
Phaeophyta	Brown algae	Multicellular	Chlorophylls *a* and *c*, xanthophylls	Laminarin	Cellulose
Rhodophyta	Red algae	Multicellular	Chlorophylls *a* and *d*, phycocyanin, phycoerythrin	Starch	Cellulose

10 to 500 μm. Euglenophytes are commonly found in fresh water, particularly that with a high organic content, and to a lesser extent, in soil, brackish water and salt water. Members of this group have a well-defined nucleus, and chloroplasts containing chlorophylls *a* and *b* (Figure 9.1). The storage product of photosynthesis is a β-1,3-linked glucan called *paramylon*, found almost exclusively in this group. Euglenophytes lack a cellulose cell wall but have instead, situated within the plasma membrane, a flexible *pellicle* made up of interlocking protein strips, a characteristic that links them to certain protozoan species. A further similarity is the way in which locomotion is achieved by the undulation of a terminal flagellum. Movement towards a light source is facilitated in many euglenids by two structures situated near the base of the flagellum; these are the *paraflagellar body* and the *stigma*, or *eyespot*. The latter is particularly conspicuous, as it is typically an orange-red colour, and relatively large.

> A *pellicle* is a semi-rigid structure composed of protein strips found surrounding the cell of many unicellular protozoans and algae.

Reproduction is by binary fission (i.e. by asexual means only). Division starts at the anterior end, and proceeds longitudinally down the length of the cell, giving the cell a characteristic 'two-headed' appearance. During mitosis, the chromosomes within the nucleus replicate, forming pairs that split longitudinally. Since the euglenophyte is usually haploid, it thus becomes diploid

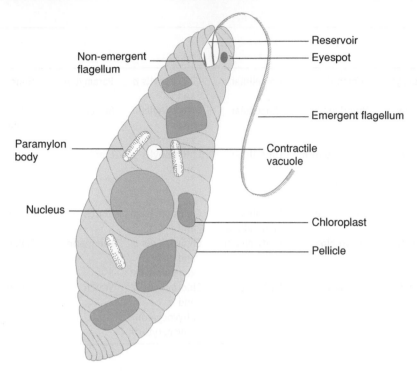

Figure 9.1 *Euglena. Euglena* shares a number of features in common with the zooflagellates (see Figure 9.11), but its possession of chloroplasts has meant that it has traditionally been classified among the algae.

for a short period. As fission proceeds, one daughter cell retains the old flagellum, while the other later generates a new one. As in the binary fission of bacteria, the progeny are genetically identical, that is, clones. When conditions are unfavourable for survival due to failing nutrient supplies, the cells round up to form cysts surrounded by a gelatinous covering; these have an increased complement of paramylon granules, but no flagella. An important respect in which euglenids may be at variance with the notion of 'plant-like protists' is their ability to exist as heterotrophs under certain conditions. When this happens, they lose their photosynthetic pigments and feed saprobically on dead organic material in the water.

9.1.3   Dinoflagellata

The dinoflagellates (also known variously as Pyrrophyta, or 'fire algae') are chiefly marine planktonic types, comprising over 2000 species. This is another unicellular group, but one whose cells are often covered with armoured cellulose plates known as *thecae* (sing. *theca*). They are generally biflagellate, with the two dissimilar flagella lying in part within the longitudinal and lateral

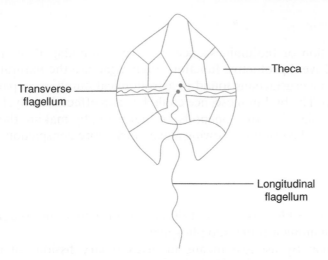

Figure 9.2 *Peridinium*, a dinoflagellate. Note the two flagella in perpendicular grooves. Each plays its part in the organism's locomotion.

grooves that run around the cell (Figure 9.2). The beating of the flagella causes the cell to spin like a top as it moves through the water (the group takes its name from the Greek word 'to whirl'). Although many non-photosynthetic (chemoheterotrophic) types exist, most dinoflagellates are photosynthetic, containing chlorophylls *a* and *c* plus certain carotenoids and xanthophylls, which give them a red to golden appearance. As a group, they are second only to the diatoms (see below) as the primary photosynthetic producers in the marine environment. Some dinoflagellates form endosymbiotic relationships with marine animals such as corals and sea anemones; these are termed *zooxanthellae*. An unusual feature of dinoflagellate ultrastructure is that the chromosomes contain little, if any, histone protein, and exist almost permanently in the condensed form.

Some tropical species of dinoflagellate emit light – the only algae to do so (see Box 9.1). Due to an enzyme–substrate (luciferin–luciferase) interaction, this can cause a spectacular glow in the water at night, especially when the water is disturbed, for example by a ship. *Bioluminescence* of this kind has proved to a useful molecular 'tagging' system for cells in biological research. Other marine dinoflagellates can produce metabolites that act as nerve toxins to higher animals. Shellfish such as mussels and oysters can concentrate these with no harm to themselves, but they can be fatal to humans who consume them. Sometimes, when conditions are highly favourable, an explosion of growth results in the development of huge 'red tides' of dinoflagellates in coastal waters. This produces a build-up of toxins, and may lead to the death of massive numbers of fish and other marine life. The greatly increased

---

Box 9.1   Why would an alga want to glow in the dark?

The production of bioluminescence by several dinoflagellate species is thought to have a protective function. Such algae are the natural prey of copepods, tiny crustaceans found in astronomical numbers as part of the zooplankton. The bioluminescence could have an effect directly, by acting as a warning signal to the copepods, or indirectly, by making those crustaceans that had consumed glowing algae much more conspicuous to their own predators.

---

incidence of these blooms in recent decades is probably due to pollution by fertilisers containing nitrates and phosphates.

Reproduction by asexual means involves binary fission. In armoured forms, the theca may be shed before cell division, or split along suture lines; in either case, daughter cells must regenerate the missing sections. Sexual reproduction is known to occur in some dinoflagellates, and is probably more widespread. Gametes produced by mitosis fuse to produce a diploid zygote; this undergoes meiosis to reinstate the haploid condition in the offspring. In some species we see *isogamy*, the fusion of identical, motile gametes, while in others, *anisogamy* occurs, in which gametes of dissimilar size fuse. Fusion may occur between genetically identical gametes, or only when the gametes come from genetically distinct populations.

9.1.4   Diatoms

The diatoms, which belong to the division Chrysophyta (the golden-brown algae), make up the majority of phytoplankton in marine food chains, and as such are the most important group of algal protists in terms of photosynthetic production. As many as 100 000 species of diatom are recognised. As with the dinoflagellates, chlorophylls *a* and *c* are present in diatoms, but not chlorophyll *b*. Their colour is due to carotenoids and xanthophylls (particularly fucoxanthin) masking the chlorophyll.

Diatoms have their cells surrounded by a silica-based shell known as a *frustule*, comprising two overlapping halves (the epitheca and the hypotheca). Writers are rarely able to resist the temptation to liken this structure to that of a Petri dish, and this one is no exception. With the electron microscope it can be seen that the frustule is perforated with numerous tiny pores, which connect the protoplast of the cell with the outside environment. Diatom classification is based almost entirely on the shape and pattern of these shells, which are uniform for a particular species, and often have a very striking appearance (Figure 9.3). When diatoms die, their shells fall to the bottom of the sea, and can accumulate in thick layers where they represent a valuable

Figure 9.3 Diatoms are covered by an intricate two-part siliceous shell called a frustule, whose often very striking appearance makes them among the most beautiful of microorganisms.

mineral resource. This fine, light material (diatomaceous earth) has a number of applications, for example in filtration systems, and also as a light abrasive in products such as silver polish or toothpaste.

Reproduction is usually asexual by binary fission, but a sexual phase with the production of haploid gametes can occur. Chrysophytes are unusual among the three primitive groups of algae in that they are diploid. In diatoms, asexual reproduction involves mitotic cell division, with each daughter cell receiving one half of the parental frustule, and synthesising a new one to complement it. The newly formed half, however, always acts as the hypotheca (lower half) of the new cell; consequently, one in two daughter cells will be slightly smaller than the parent, an effect that is heightened over a number of generations (Figure 9.4). This process continues until a critical size is reached, and the diatoms undergo a phase of sexual reproduction, which re-establishes the normal frustule size. In species whose frustules have a degree of elasticity, the daughter cells are able to expand, and the problem of cell diminution does not arise. In bilaterally symmetrical (long, thin) diatoms, meiosis in parental cells produces identical, non-motile gametes, which fuse to form a zygote. The radially symmetrical (round) forms provide an example of the third pattern of gamete fusion found in the algae: *oogamy*. Here, there is a clear distinction between the small, motile sperm cell and the larger, immobile egg cell. Both

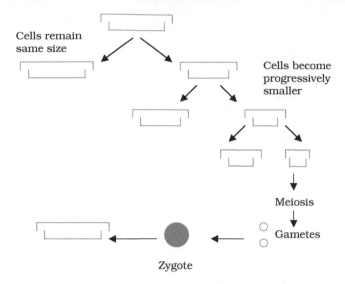

Cells remain
same size

Cells become
progressively
smaller

Meiosis

Gametes

Zygote

Figure 9.4 Asexual reproduction in diatoms. Two halves of the parent cell separate following mitosis, and a new half develops to fit inside it. Thus the parental shell always forms the upper (larger half) of the new individual. For half of the population, this means a gradual diminution of size over a number of generations. Eventually, gametes are produced by meiosis and a sexual cycle is entered.

are produced by meiosis in the parental cell, followed, in the case of the male, by several rounds of mitosis, to give a large number of sperm cells.

9.1.5 Chlorophyta

The green algae have always attracted a lot of interest because, as a group, they share a good deal in common with the higher plants in terms of ultra-structure, metabolism and photosynthetic pigments, pointing to the likelihood of a common ancestor. They possess both chlorophylls *a* and *b* and certain carotenoids, store carbohydrate in the form of starch, and generally have a rigid cell wall containing cellulose. The starch is stored in structures called *pyrenoids*, which are found within chloroplasts. There are two phylogenetically distinct lines of green algae, the Charophyta and the Chlorophyta; the latter are much the bigger group, but the charophytes seem to be more closely related to green plants (see Figure 9.17).

Chlorophytes demonstrate a wide variety of morphology, ranging from unicellular types to colonial, filamentous, membranous and tubular forms. The vast majority of species are freshwater aquatic, but a few marine and pseudoterrestrial representatives exist.

The genus usually chosen to illustrate the unicellular condition in chlorophytes is *Chlamydomonas* (Figure 9.5). This has a single chloroplast, similar

Figure 9.5 *Chlamydomonas*, a unicellular green alga. Sexual reproduction only occurs under adverse conditions, resulting in the production of a resistant zygote. When conditions are favourable, asexual reproduction by means of zoospores predominates.

in structure and shape to that of a higher plant, and containing a pyrenoid. Situated together at the anterior end are a pair of smooth or whiplash flagella, whose regular, ordered contractions propel the organism through the water. A further structural feature found in *Chlamydomonas* and other motile forms of green algae is the stigma, or eyespot; this is made up of granules of a carotenoid pigment and is at least partially responsible for orienting the cell with respect to light.

Reproduction in *Chlamydomonas* and other unicellular types under favourable conditions of light, temperature and nutrients, occurs asexually by the production of *zoospores*. A single haploid adult loses its flagella and undergoes mitosis to produce several daughter cells, which then secrete cell walls and flagella and take up an independent existence of their own. This can result in a tremendous increase in numbers; a single cell can divide as many as eight times in one day. Sexual reproduction in *Chlamydomonas*, which occurs when conditions are less favourable, differs in detail according to the species (see Figure 9.5). Any one of the three variants of gamete production seen in the algae may be seen (isogamy, anisogamy and oogamy – see discussion earlier). In all cases, two haploid gametes undergo a fusion of both cytoplasm

Daughter colony

Figure 9.6 *Volvox*, a colonial green alga. The colony comprises thousands of biflagellated cells embedded in mucilage. Note the presence of daughter colonies, produced by asexual reproduction; these are eventually liberated and assume an independent existence.

and nuclei to give a diploid zygote. The gametes may simply be unmodified haploid adult cells, or they may arise through mitotic cleavage of the adult, depending on the species. The process of isogamy, where the two gametes are morphologically alike and can't be differentiated visually, only occurs in relatively lowly organisms such as *Chlamydomonas*. In some species we see the beginnings of sexual differentiation (there are two mating strains, designated + and −) and fusion will only take place between individuals of opposite strains. The diploid zygote, once formed, often develops into a tough-walled protective spore called a *zygospore*, which tides the organism over conditions of cold or drought. At an appropriate time the zygospore is stimulated to recommence the life cycle, and meiosis occurs, to produce haploid cells, which then mature into adult individuals.

In *C. braunii*, sexual reproduction is anisogamous; a 'plus' strain produces eight microgametes and a 'minus' strain produces four macrogametes. In *C. coccifera*, simple oogamy occurs, in which a vegetative cell loses its flagella, rounds off and enlarges; this acts as the female gamete or ovum, and is fertilised by male gametes formed by other cells.

The next level of organisation in the green algae is seen in the *colonial* types, typified by *Volvox*. These, like the unicellular types, are motile by means of flagella, and exist as a number of cells embedded in a jelly-like matrix (Figure 9.6). Both the number of cells and the way they are arranged is fixed and characteristic of a particular species. During growth, the number of cells does not increase. In simpler types, all cells seem to be identical but in more complex forms there are distinct anterior and posterior ends, with the stigma more prominent at the anterior, and the posterior cells becoming larger. Reproduction can occur sexually or asexually.

The diversity of body forms in multicellular chlorophytes referred to earlier is matched by that of their life cycles. Two examples are described here.

Oedogonium is a filamentous type. When young it attaches to its substratum by a basal holdfast, but unless it lives in flowing water the adult form is free-floating. Asexual reproduction occurs by means of motile zoospores, which swim free for around an hour before becoming fixed to a substratum

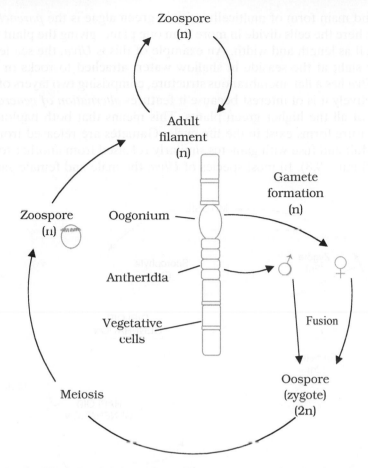

Figure 9.7 The life cycle of *Oedogonium*, a filamentous green alga. In the example shown, gametangia of both sexes are present on the same filament. The oospore (zygote) that results from the fusion of gametes is the only diploid part of the life cycle. Some species of *Oedogonium* have male and female gametangia on separate filaments.

and developing into a new filament. In sexual reproduction, the process of sexual differentiation is carried a step further than we've seen so far, with two separate filaments producing gametes from specialised cells called *gametangia* (Figure 9.7). These are morphologically distinct, with the male being termed an *antheridium* and the female an *oogonium*. Gametes (morphologically distinct: anisogamy) fuse to form a resistant zygote or *oospore*, which, when conditions are favourable, undergoes meiosis to produce four haploid zoospores, each of which can germinate into a young haploid filament. The oospore is thus the only diploid phase in the life cycle. In *Oedogonium* there are species with separate male and female filaments (*dioecious*) as well as ones with both sexes on the same filament (*monoecious*).

A second main form of multicellularity in green algae is the *parenchymatous* state: here the cells divide in more than one plane, giving the plant thickness as well as length and width. An example of this is *Ulva*, the sea lettuce, a familiar sight at the seaside in shallow water, attached to rocks or other objects. *Ulva* has a flat, membranous structure, comprising two layers of cells. Reproductively it is of interest because it features *alternation of generations*, a feature of all the higher green plants. This means that both haploid and diploid mature forms exist in the life cycle. Gametes are released from one haploid adult and fuse with gametes similarly released from another to form a zygote (Figure 9.8). In most species of *Ulva*, the male and female gametes

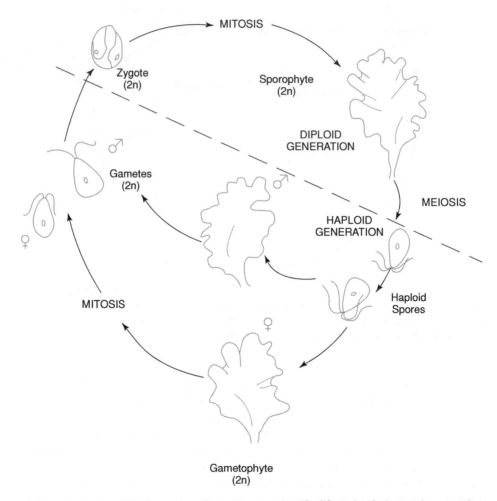

Figure 9.8   Isomorphic alternation of generations: *Ulva*. The life cycle of *Ulva* involves morphologically identical haploid and diploid plants. Fusion of gametes forms a zygote, which grows into the mature diploid plant. Meiosis produces haploid zoospores, which give rise to separate male and female haploid plants.

are morphologically identical (isogamy). The zygote germinates to form a diploid plant, indistinguishable from the plant that produced the gametes, except for its complement of chromosomes. When the diploid plant is mature, it undergoes meiosis to produce haploid zoospores, which settle on an appropriate substratum and develop into haploid *Ulva* plants. This form of alternation of generations is called *isomorphic*, because both haploid and diploid forms look alike and each assumes an equal dominance in the life cycle. It is, however, more usual for alternation of generations to be *heteromorphic*, with the *sporophyte* and the *gametophyte* being physically dissimilar, and with one form or other dominating.

> The *sporophyte* is the diploid, spore-forming stage in a life cycle with alternation of generations.
> The haploid, gamete-forming stage is called the *gametophyte*

9.1.6 Phaeophyta

The brown algae are multicellular, large and complex seaweeds, which dominate rocky shores in temperate and polar regions. Apart from one or two freshwater types, they are all marine. The presence of fucoxanthin masks the presence of chlorophylls *a* and *c*. (In this context it must be stated here that not all 'brown' seaweeds look brown, nor indeed do all the 'red' ones look red.) Unlike the higher plants and green algae, which use starch as a food reserve, the phaeophytes use an unusual polysaccharide called laminarin (a β-1,3-glucan).

The level of tissue organisation in the brown algae is greatly in advance of any of the types we've discussed so far. The simplest thallus of a brown alga resembles the most complex found in the greens.

> *Thallus* is the term used to describe a simple vegetative plant body showing no differentiation into root, stem and leaf.

The phaeophytes also represent an advance in terms of sexual reproduction; here oogamy is the usual state of affairs and alternation of generations has developed to such an extent that diploid and haploid stages frequently assume separate morphological forms. Again, we shall use two examples to illustrate life cycle diversity in the brown algae.

Laminaria is one of the kelps, the largest group of brown algae. They grow attached to underwater rocks or other objects by means of holdfasts, rootlike structures that anchor the plant. The thallus is further subdivided into a stalklike stipe and a broader, bladelike lamina. Reproduction in *Laminaria* involves sporophyte and gametophyte plants that are morphologically quite distinct (heteromorphic alternation of generations). Reproductive areas called *sori* (sing. *sorus*) develop on the blade of the diploid sporophyte at certain times of year (Figure 9.9). These consist of many unilocular sporangia, interspersed with thick protective hairs called paraphyses. As the sori develop, meiosis occurs, leading to the production of haploid zoospores.

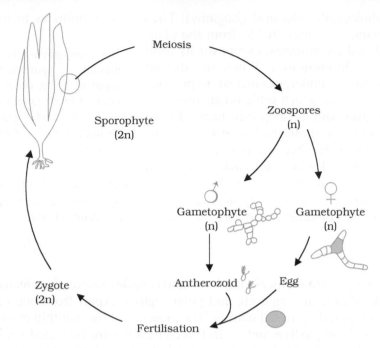

Figure 9.9 Heteromorphic alternation of generations: *Laminaria*. The haploid and diploid generations are morphologically quite distinct: in contrast to the large robust sporophyte, the gametophyte is a tiny filamentous structure.

These in turn develop into haploid filamentous gametophyte plants, much smaller and quite different in morphology from the more highly organised sporophyte. Indeed, in contrast to the large sporophyte the gametophyte is a microscopic structure. The gametophytes are dioecious, that is the male and female reproductive structures are borne on separate individuals. The female plant bears a number of *oogonia*, each of which produces a single egg, which escapes through a pore at the apex of the oogonium, but remains attached in a sort of cup, formed by the surrounds of the pore. In similar fashion the male plant bears several antheridia, each liberating a single antherozoid; this is motile by means of flagella and fertilises the egg. The diploid zygote so produced grows immediately into a new sporophyte plant.

In our second example of a phaeophyte life cycle, there is no alternation of generations at all, the gametophyte generation having been completely lost. The wracks are familiar seaside seaweeds found in the intertidal zone, and *Fucus vesiculosus*, known commonly as the bladder wrack, is one of the best known (Figure 9.10). It gets its name from the air bladders distributed on its surface, which assist buoyancy.

The adult has reproductive structures called receptacles, slight swellings situated at the tip of the thallus; within these are flask-like invaginations

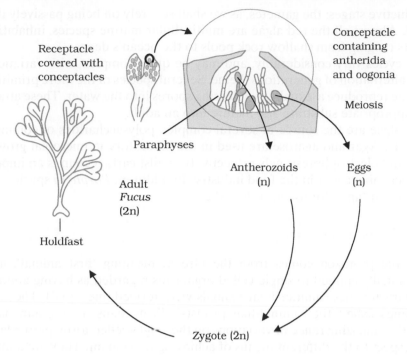

Figure 9.10 *Fucus*: a life cycle with no haploid plant. The only representatives of the haploid state in the life cycle of *Fucus* are the gametes produced by meiosis within the male and female gametangia.

called *conceptacles*, which contain the male or female gametangia, again interspersed with sterile paraphyses. *F. vesiculosus* is monoecious but some other *Fucus* species are dioecious. Each antheridium undergoes meiosis, followed by mitosis to produce 64 antherozoids or sperm, while by the same means eight eggs are produced in the oogonium. At high tide these gametes are released into the open water. Fertilisation results in a diploid zygote, which continues to drift quite free, while secreting a mucilaginous covering. It eventually settles, becoming anchored by the mucilage, and germinates into an adult individual. Here then, we have a life cycle in which there is no gametophyte generation, and no specialised asexual reproduction (although in certain conditions fragments may regenerate to form adults).

9.1.7 Rhodophyta

The red colouration of the rhodophytes is due to the pigments phycoery-thrin and phycocyanin, which mask the chlorophylls present, in this case *a* and *d*. The biggest single difference between the red algae and the other groups we've looked at so far is that they lack flagella at any stage of their life cycle. Thus they are completely lacking in any motile forms, even in the reproductive stages; the gametes, as we shall see, rely on being passively dispersed. Almost all the red algae are multicellular marine species, inhabiting habitats ranging from shallow rock pools to the ocean's deeps.

Life cycles vary considerably, and may be quite complex, with variations on the alternation of generations theme. Several species of the more primitive red algae reproduce asexually by releasing spores into the water. These attach to an appropriate substrate and mature into an adult.

Red algae are the source of several complex polysaccharides of commercial value. Agar and agarose are used in the laboratory in microbial growth media and electrophoresis gels respectively, whilst carrageenan is an important thickening agent in the food industry. In addition, *Porphyra* species are cultivated in Japan for use in sushi dishes.

9.2   The 'protozoa'

The name protozoa comes from the Greek, meaning 'first animal', and was originally applied to single-celled organisms regarded as having animal-like characteristics (multicellular animals were termed 'metazoa'). They are sometimes called the 'animal-like protists'. Protozoans as a group have evolved an amazing range of variations on the single-celled form, particularly with respect to the different means of achieving movement. They are a morphologically diverse group of well over 60 000 species; although the majority are free-living, the group also includes *commensal* forms and some extremely important parasites of animals and humans.

Most protozoans are found in freshwater or marine habitats, where they form a significant component of plankton, and represent an important link in the food chain. Although water is essential for the survival of protozoans, many are terrestrial, living saprobically in moist soil.

> *Plankton* are the floating microscopic organisms of aquatic systems.

Remember that a protozoan needs to pack all the functions of an entire eukaryotic organism into a single cell; consequently a protozoan cell may be much more complex than a single animal cell, which is dedicated to a single function. Thus, protozoans display most of the typical features of a eukaryotic cell discussed in Chapter 3, but they may also have evolved certain specialised features. The single cell is bounded by the typical bilayer membrane discussed earlier, but depending on the type in question, this may in turn be covered by a variety of organic or inorganic substances to form an envelope or shell.

One of the most characteristic structural features of protozoans is the *contractile vacuole*, whose role is to pump out excess amounts of water that enter the cell by osmosis. The activity of the contractile vacuole is directly related to the difference in osmotic potential between

> The *contractile vacuole* is a fluid-filled vacuole involved in the osmoregulation of certain protists.

the cell and its surroundings. This is vitally important for freshwater protozoans, since the hypotonic nature of their environment means that water is continually entering the cell. The contractile vacuole often has a star-shaped appearance, the radiating arms being canals that drain water from the cytoplasm into the vacuole.

Most protozoans have a heterotrophic mode of nutrition, typically ingesting particulate food such as bacteria, and digesting them in phagocytic vacuoles. Since they actively 'hunt' their food rather than simply absorbing it across the cell surface, it is not surprising that the majority of protozoans are capable of movement. The structural features used to achieve locomotion (e.g. cilia, flagella) are among the characteristics used to classify the protozoans.

We shall now examine the characteristics of the principal groupings into which the protozoans have traditionally been divided, largely on the basis of their mode of locomotion. It should be repeated, however, that the protozoa do not represent a coherent taxonomic grouping with a common ancestor, but rather a phylogenetically diverse collection of species with certain features in common. Indeed, each of the four groups is now regarded as having a closer evolutionary relationship with certain 'algal' groups than with each other. See Figure 9.17 for a modern view of how the various taxonomic groupings of protozoans are related.

9.2.1   The zooflagellates (Mastigophora)

Members of this, the biggest and most primitive group of protozoans, are characterised by the long flagellum (*mastigos*, 'a whip'), by which they propel themselves around. Although typical zooflagellates have a single flagellum, some types possess several. The prefix 'zoo-' distinguishes them from plant-like flagellates such as *Euglena*, but as we have already mentioned, such a distinction is not necessarily warranted on molecular and structural grounds (see Section 9.4).

Zooflagellates may be free-living, symbiotic or parasitic. An example of the latter is *Trypanosoma*, the causative agent of African sleeping sickness and Chagas disease in humans (see Chapter 15). *Trypanosoma* belongs to the kinetoplastids, a group characterised by the possession of a unique organelle called the *kinetoplast*, found within the cell's single, large, tubular mitochondrion, and containing its own DNA. The flagellum extends back to form the edge of a long *undulating membrane*, which gives *Trypanosoma* its characteristic locomotion (Figure 9.11).

The *choanoflagellates* (Figure 9.12) are a group of zooflagellates of particular interest, as it is thought that they represent the closest single-celled relatives of animals. They are characterised by a 'collar' of microvilli that surrounds the base of a single flagellum, an arrangement that is also seen in the simplest multicellular animals, the sponges. This connection is made even

Figure 9.11   *Trypanosoma*, a zooflagellate. Note the kinetoplast contained within the single large mitochondrion. The flagellum is continuous with the undulating membrane. Reproduced from Baron, EJ, et al. (1994) Medical Microbiology: A Short Course, with permission from John Wiley & Sons.

more apparent in colonial forms of choanoflagellates, and is supported by molecular evidence. Also, both choanoflagellates and animals share the flat, lamellar type of cristae in their mitochondria.

A third group of zooflagellates worthy of note are the *diplomonads*. These have two nuclei per cell and multiple flagella, but their most remarkable feature is that they do not apparently possess any mitochondria (but see Chapter 3). *Giardia lamblia*, the causative agent of the intestinal disease giardiasis, is a member of this group. Unlike *Trypanosoma*, *Giardia* does not have a

Figure 9.12   Choanoflagellates are free-living zooflagellates characterised by a collar of tentacles or microvilli that surround the single flagellum. They are often colonial, as in the example shown.

secondary host, but survives outside the body as a resistant cyst, before it is taken up again in infected water. The diplomonads occupy a branch of suggested phylogenetic trees very distant from the kinetoplastids and choanoflagellates (see Figure 9.17). The *parabasalians* are another group of amitochondriate flagellates, whose best-known member is *Trichomonas vaginalis*, a cause of infections of the female urogenital tract.

9.2.2 The amoebas (Sarcodina)

The amoebas are characterised by the possession of *pseudopodia* ('false feet'), temporary projections from the cell into which cytoplasm flows until the organism has moved forwards (Figure 9.13). This means that amoebas are continually changing their body shape and the position of their internal organelles. Pseudopodia are also used to capture and engulf food, forming a vacuole around it. Once again, digestive enzymes are released from lysosomes and the food particle dissolved. Once absorption of soluble nutrients has taken place, undigested waste is ejected by the vacuole moving back to the cell surface.

Reproduction in the amoebas is by simple binary fission. Most amoebas are free-living, in aquatic environments; their mode of movement and feeding makes them well adapted to life on the bottom of ponds and lakes, where there is a good supply of prey organisms and suspended organic matter. Also included in the group, however, are some important parasites, including *Entamoeba histolytica*, which causes amoebic dysentery in humans (see Chapter 15).

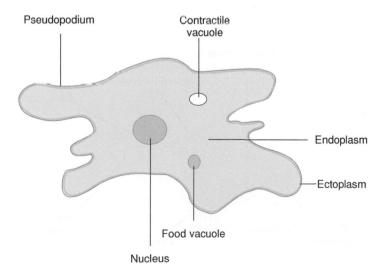

Figure 9.13 Amoeboid structure. The internal features of an amoeba change position as the cell changes its shape by cytoplasmic streaming.

9.2.3   Amoebas with external shells: Foraminifera and Radiolaria

Some types of amoeba have an external shell covering the cell. The *Foraminifera* secrete a shell of protein coated with calcium carbonate; their shells are covered with pores through which their long, filamentous pseudopodia project. Some foraminiferans are zooplankton, microscopic organisms living at the surface of the sea, while others are bottom-dwellers. It is the discarded shells of countless long-dead creatures such as these that make up the deposits of limestone many hundreds of metres in depth. Many thousands of shells are needed to form just one gram of such a deposit! The white cliffs of Dover are an example of a limestone deposit made up largely of foraminiferan shells.

The outer surface of *radiolarians* is composed of silica, which again is perforated to allow the passage of many very fine pseudopodia.

9.2.4   The ciliates (Ciliophora)

The ciliates are not only the largest group of protozoans, but also the most complex, showing the highest level of internal organisation in any single-celled organism. Most are free-living types such as *Paramecium* (Figure 9.14), and as the name suggests, they are characterised by the possession of cilia, which may be present all over the cell surface or arranged in rows or bands. They beat in a coordinated fashion to propel the organism, or assist in the ingestion of food particles.

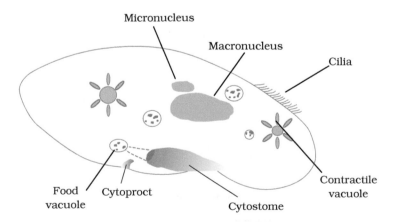

Figure 9.14   Ciliates such as *Paramecium* have specialised structures for the ingestion of food particles and elimination of waste. Note the rows of cilia covering the surface.

A unique feature of the ciliates is that they possess two distinct types of nuclei:

*   *Macronuclei* are concerned with encoding the enzymes and other proteins required for the cell's essential metabolic processes. They are polyploid, containing many copies of the genome.
*   *Micronuclei*, of which there may be as many as 80 per cell, are involved solely in sexual reproduction by conjugation.

As might be expected, removal of the macronucleus leads quickly to the death of the cell; however, cells lacking micronuclei can continue to live, and reproduce asexually by binary fission.

Most ciliates possess a specialised 'mouth' structure, the *cytostome*, through which food particles are ingested (see Figure 9.14). The beating of cilia directs the particles to a *cytopharynx*, a membrane-covered passage or tube, which enlarges and detaches to form a food vacuole. Fusion with lysosomes and digestion by enzymes occurs as described earlier. Undigested particles are ejected from a region on the surface (the anal pore or *cytoproct*).

As well as cilia, some members of the group have *trichocysts* projecting from the cell surface, harpoon-like structures that can be used for attachment or defence.

Some ciliated protozoans are anaerobic, such as those found in the rumen of cattle. The only ciliate known to cause disease in humans is *Balantidium coli*, which causes a form of dysentery.

9.2.5 The Sporozoans (Apicomplexa)

Members of this group are all parasitic, infecting a range of vertebrates and invertebrates. They have complex life cycles involving both haploid and diploid phases and infecting more than one host. Probably the best known is *Plasmodium*, the causative agent of malaria, which is discussed in further detail in Chapter 15. Sporozoans are characterised by a spore-like stage called a *sporozoite*, which is involved in the transmission of the parasite to a new host. The tip of the sporozoite contains a complex of structures (the *apicoplast*) that assists in the penetration of the host's tissues. Unlike the protozoans discussed above, sporozoans are generally non-motile, and absorb soluble nutrients across the cell surface rather than ingesting particulate matter.

> A *sporozoite* is a motile infective stage of members of the Sporozoa that gives rise to an asexual stage within the new host.

9.3   The slime moulds and water moulds (the fungus-like protists)

The final group to consider in this chapter are the so-called 'fungus-like' protists. Its members are phylogenetically diverse, and as we'll see in Section 9.4, its two principal groupings, the slime moulds and the water moulds, are placed far apart from each other in modern classification systems.

9.3.1   Oomycota (water moulds)

Water moulds resemble true fungi in their gross structure, comprising a mass of branched hyphae. At the cellular and molecular level, however, they bear little resemblance to fungi; they have a cellulose cell wall and DNA analysis shows their nearest living relatives to be the diatoms and brown algae (see Figure 9.17). The name Oomycota refers to the single large egg cell that is fertilised to produce a diploid zygote as part of the sexual reproduction cycle.

Many water moulds play an important role in the decomposition of dead plants and animals in freshwater ecosystems, while others are parasitic on the gills of fish. Terrestrial members of the Oomycota include a number of important plant pathogens, such as rusts and mildews, which can have a devastating effect on crops such as tobacco and potatoes.

9.3.2   Myxogastrida (Myxomycota, the plasmodial slime moulds)

At one stage in their life cycle, the plasmodial or acellular slime moulds exist as a single-celled amoeboid form. Two of these haploid amoebas fuse to give a diploid cell, which then undergoes repeated divisions of the nucleus, without any accompanying cell division; the result is a *plasmodium*, a mass of cytoplasm that contains numerous nuclei surrounded by a single membrane (Figure 9.15). This retains the amoeboid property of cytoplasmic streaming, so the whole multinucleate structure is able to move in a creeping fashion. This 'feeding plasmodium', which may be several centimetres in length and often brightly coloured, feeds phagocytically on rotting vegetation. Fruiting bodies develop from the plasmodium when it is mature or when conditions are unfavourable, and a cycle of sexual reproduction is entered. When favourable conditions return, meiosis

> A *plasmodium* is a mass of protoplasm containing several nuclei and bounded by a cytoplasmic membrane.

> Some slime moulds have colourful common names. My own favourite is *Fuligo septica*, 'the dog vomit slime mould'!

Figure 9.15 The plasmodial slime moulds. Acellular slime moulds such as *Physarum* produce amorphous coenocytic plasmodia, which move by amoeboid movement and phagocytically engulf particles of food. Fruiting bodies bearing sporangia release haploid spores, which germinate to form new amoebas.

gives rise to haploid spores, which germinate to produce the amoeboid form once more.

9.3.3 Dictyostelida (cellular slime moulds)

A unicellular amoeboid form also figures in the life cycle of the other group of slime moulds, the Dictyostelida (Figure 9.16). This haploid amoeba is the main vegetative form, but when food supplies are scarce, large numbers aggregate to form a sluglike blob, superficially not unlike the plasmodium described above. Unlike the plasmodium, however, this aggregate is fully cellular, so each component cell retains its plasma membrane.

Compare the life cycle of cellular slime moulds (Figure 9.16) with that of the plasmodial kind (Figure 9.15). Fruiting bodies again develop, giving rise to spores that germinate into new amoebas. No meiosis step is required,

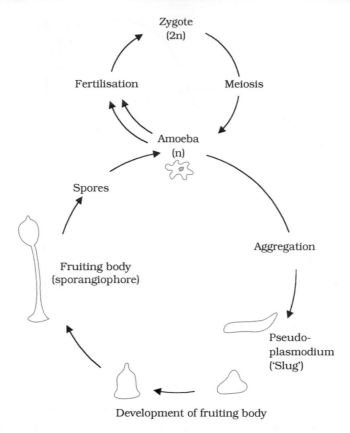

Figure 9.16   The cellular slime moulds. Fruiting bodies develop from the pseudoplasmodium, or 'slug', and release haploid spores that develop into individual amoebas. Only haploid forms participate in this cycle, which is therefore asexual. Sexual reproduction can also occur, involving the production of dormant diploid spores called macrocysts. Note that the pseudoplasmodium of cellular slime moulds is entirely cellular.

however, because the whole cycle comprises haploid forms, and this is therefore a form of asexual reproduction. A simple sexual cycle may also occur, when haploid amoebas fuse to give a diploid zygote.

9.4 Protistan taxonomy: a modern view

Traditional classification of the protists was made on the basis of physical features such as the possession of flagella, chloroplasts and other structures. This had the effect of placing into separate groups organisms that at the molecular level may be closely related. The kinetoplastids and the euglenophytes, for example, although both are flagellate forms, were distantly separated because the latter contained chloroplasts. They are now known to be closely related

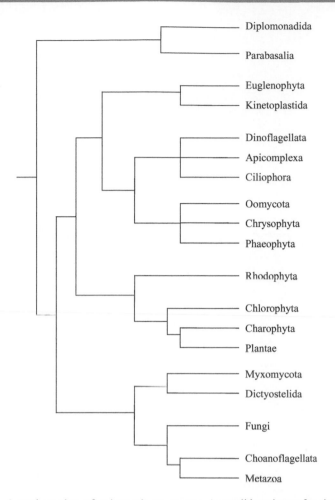

Figure 9.17 A modern view of eukaryotic taxonomy. A possible scheme for the relationship between protistan groups based on 18S RNA data. The positions of the fungi, plants and multi-cellular animals (Metazoa) are also shown. Note that some protistan groups placed together in traditional schemes (e.g. kinetoplastids and choanoflagellates) are very distant in phylogenetic terms. The diplomonads and parabasilians are shown as diverging before the acquisition of mitochondria through endosymbiosis with bacteria. This hypothesis may need to be revised in light of recent evidence that these organisms did once possess mitochondria but have since lost them.

in phylogenetic terms (Figure 9.17). On the other hand, the kinetoplastids were previously placed together with choanoflagellates as zooflagellates, but molecular analysis now shows the two groups to have little in common in evolutionary terms.

10

Viruses

In Chapter 1, we saw how, in the late nineteenth century, one disease after another, in plants as well as in animals, was shown to have a bacterial cause. In 1892, however, the Russian Dimitri Iwanowsky made a surprising discovery concerning the condition known as tobacco mosaic disease. He showed that an extract from an infected leaf retained the ability to transmit the disease to another plant even after being passed through a porcelain filter. This recently developed device was believed to remove even the smallest bacteria, and it was therefore proposed that perhaps the cause of the disease was not an organism, but a filterable toxin. The work of the Dutch botanist Martinus Beijerinck and others around the turn of the century, however (see Table 10.1), led to the discovery of *viruses*, filterable entities that were much smaller than bacteria, and responsible for a wide range of diseases in plants, animals and members of the microbial world.

10.1   What are viruses?

All viruses are obligate intracellular parasites; they inhabit a no-man's land between the living and the non-living worlds, and possess characteristics of both. We now know them to differ radically from the simplest cellular organisms, the bacteria, in a number of respects:

- they cannot be observed using a light microscope;
- they have no internal cellular structure;
- they contain either DNA or RNA, but not both;[1]

---

[1] Some viruses have DNA and RNA at different phases of their growth cycle. See Section 10.4.4.

Essential Microbiology, Second Edition. Stuart Hogg.
© 2013 John Wiley & Sons, Ltd. Published 2013 by John Wiley & Sons, Ltd.

Table 10.1　Some milestones in the history of virology

Date	Event	Key individual(s)
1892	Tobacco mosaic disease (TMD) shown to be caused by a filterable agent	Iwanowsky
1898	Proposal that TMD is due to a novel type of infectious agent	Beijerinck
	Demonstration of first viral disease in animals (foot and mouth)	Loeffler and Frosch
1901	Demonstration of first human viral disease (yellow fever)	Reed
1915/1917	Discovery of bacterial viruses (bacteriophages)	Twort, d'Herelle
1918	Spanish influenza pandemic	
1935	TMV is first virus to be crystallised	Stanley
1937	Separation of TMV into protein and nucleic acid fractions	Bawden and Pirie
1939	Viruses visible under electron microscope	Kausch, Pfankuch and Ruska
1955	Spontaneous reassembly of TMV from protein and RNA components	Fraenkel-Courat and Williams
1971	Discovery of viroids	Diener
1980	Sequencing of first complete viral genome (cauliflower mosaic virus)	Frank
1982	Sequencing of first RNA genome (TMV)	
	Recombinant hepatitis B vaccine	
	Discovery of prions	Prusiner
1983	Discovery of HIV, thought to be causative agent of AIDS	Montaigner and Gallo
1990	Retrovirus used as vector in first human gene therapy trial	Anderson
2001	Bovine spongiform encephalitis (BSE) outbreak in the UK	
2003	Outbreak of new human viral disease –severe acute respiratory syndrome (SARS) – in SE Asia	
2008	WHO Global Polio Eradication Initiative launched	
2009	Global swine flu pandemic	

- they are incapable of replication unless occupying an appropriate living host cell;
- they are incapable of metabolism;
- individuals show no increase in size.

Although most viruses fall within the size range 20–300 nm, in 2011 French scientists announced the discovery of one that was 700 nm in diameter, big enough to be seen with a light microscope. *Megavirus*, as they christened it, was isolated from the Pacific Ocean off Chile, and its host is thought to be a type of marine amoeba.

When inside a host cell, viruses show some of the features of a living organism, such as the ability to replicate themselves, but outside the cell they are just inert chemical structures, thus fuelling the debate as to whether they can be considered to be life forms. A particular virus has a limited host range, that is, it is only able to infect certain cell types.

10.2 Viral structure

The demonstration by Wendel Stanley in 1935 that a preparation of tobacco mosaic virus could be crystallised was an indication of the relative chemical homogeneity of viruses, and meant that they could not be thought of in the same terms as other living things. Compared to even the most primitive cellular organism, viruses have a very simple structure (Figure 10.1). An intact viral particle, or *virion*, has in essence just two components: a core of nucleic acid, surrounded and protected by a protein coat or *capsid*, the combination of the two being known as the *nucleocapsid*. In certain virus types, the nucleocapsid is further surrounded by a membranous *envelope*, partly derived from host cell material. Most viruses are smaller than even the smallest bacterial cells; Figure 10.2 shows the size of some viruses compared to that of typical bacterial and eukaryotic cells.

Figure 10.1 Viral structure. Viruses comprise a nucleic acid genome surrounded by a protein coat (capsid).

10.2.1 The viral genome

The genetic material of a virus may be either RNA or DNA, and either of these may be single-stranded or double-stranded (Figure 10.3). As shown in Figure 10.4, the genome may, furthermore, be circular or linear. An additional variation in the viral genome is seen in certain RNA viruses, such as the influenza virus; here, instead of existing as a single molecule, it is *segmented*, existing as several pieces, each of which may encode a separate protein. In some plant viruses, the segments may be present in separate particles, so in order for replication to occur, a number of virions need to co-infect a cell,

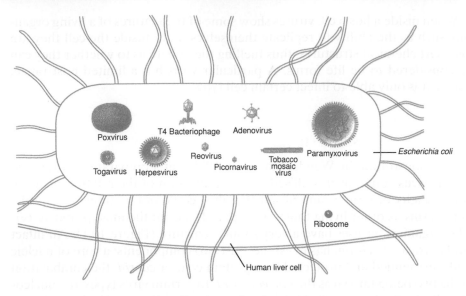

Figure 10.2 Viruses are much smaller than cells. The viruses shown are drawn to scale and compared to an *E. coli* cell and a human liver cell. As a guide, *E. coli* cells are 1–2 μm in length. Reproduced from Black, JG (1999) Microbiology: Principles and Explorations, 4th edn, with permission from John Wiley & Sons.

thereby complementing each other (multipartite genomes). Double-stranded RNA is always present in the segmented form.

The size of the genome varies greatly; it may contain as few as four genes or as many as over two hundred (see Box 10.2). These genes may code for both structural and non-structural proteins; the latter include enzymes such as RNA polymerases and DNA polymerases required for viral replication.

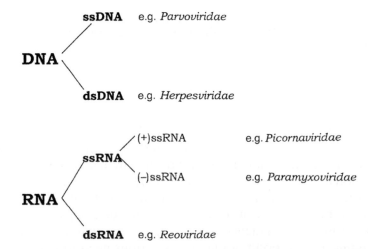

Figure 10.3 The diversity of viral genomes. ss, single-stranded; ds, double-stranded.

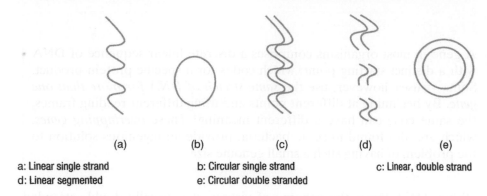

(a) (b) (c) (d) (e)

a: Linear single strand b: Circular single strand c: Linear, double strand
d: Linear segmented e: Circular double stranded

Figure 10.4 Viral genomes may be circular or linear. Some RNA viruses have their genome broken up into segments, each encoding a separate protein. Reproduced from Hardy, SP (2002) Human Microbiology, with permission from Taylor & Francis Group.

Box 10.1 Where do viruses come from?

Nobody is really sure how viruses evolved. Three major mechanisms have been proposed:

- *'Escaped gene' theory:* Viruses derive from normal cellular nucleic acids and 'gain independence' from the cell. DNA viruses could come from plasmids or transposable elements (see Chapter 12), while RNA viruses could derive from mRNA.
- *Regressive theory:* Gradual degeneration of prokaryotes living parasitically in eukaryotic cells. Enveloped forms such as poxviruses are most likely to have been formed in this way.
- *Coevolution theory:* Independent evolution alongside cellular forms from primordial soup.

Some scientists consider it unlikely that the same mechanism could account for the diversity of viruses we see today, and therefore propose that viruses must have evolved many times over. A study published in 2004 conversely proposes that all viruses share a common ancestor and may even have developed before cellular life forms.

Single-stranded RNA viral genomes can be divided into two types, known as *(+) sense* and *(−) sense RNA*. The former is able to act as messenger RNA (mRNA), attach to ribosomes and become translated into the relevant proteins within the host cell. As such, it is infectious in its own right. Minus (−) sense RNA, on the other hand, is only infectious in the presence of a capsid protein possessing RNA polymerase activity. This is needed to convert the (−) RNA into its complementary (+) strand, which then acts as a template for protein production as described above.

Box 10.2   *The mother of invention*

A gene in most organisms comprises a discrete linear sequence of DNA with a distinct starting point, which codes for a specific protein product. Some viruses, however, use the *same stretch of DNA for more than one gene*. By beginning at different points and using different reading frames, the same code can have a different meaning! These *overlapping genes*, which are also found in some bacteria, provide an ingenious solution to the problem of having such a small genome size.

When DNA forms the genome of viruses, it is usually double-stranded DNA (dsDNA), although some of the smaller ones such as the parvoviruses have single-stranded DNA (ssDNA) (see Figure 10.3).

10.2.2   Capsid structure

The characteristic shape of a virus particle is determined by its protein coat, or capsid. In the non-enveloped viruses, the capsid represents the outermost layer, and plays a role in attaching the virus to the surface of a host cell. It also acts to protect the nucleic acid against harmful environmental factors such as ultraviolet (UV) light and desiccation, as well as the acid and degradative enzymes encountered in the gastrointestinal tract.

The capsid is made up of a number of subunits called *capsomers* (Figure 10.5), and may comprise a few different protein types or just one. The number of capsomers is constant for a particular viral type. This repetitive

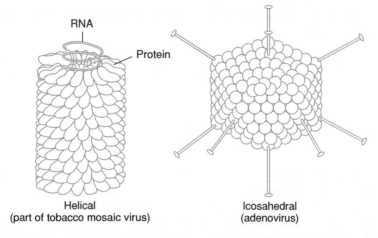

Figure 10.5 Viral capsids have two basic forms, helical and icosahedral. Complex viruses represent a fusion of both forms. Reproduced from Harper, D (1998) Molecular Virology, 2nd edn, with permission from Taylor & Francis Group.

subunit construction is necessitated by the small amount of protein-encoding RNA/DNA in the viral genome. The capsomers have the ability to interact with each other spontaneously to form the completed capsid by a process of self-assembly. This would be less easily achieved if there were large numbers of different protein types.

Capsomers are arranged symmetrically, giving rise to two principal capsid shapes, *icosahedral* and *helical* (see Figure 10.5). Both shapes can be found in either enveloped or non-enveloped viruses. *Complex* viruses, such as certain bacteriophages, contain elements of both helical and icosahedral symmetry.

Helical capsids A number of plant viruses, including the well-studied tobacco mosaic virus, have a rodlike structure when viewed under the electron microscope (Figure 10.5a). This is due to a helical arrangement of capsomers, resulting in a tube or cylinder, with room in the centre for the nucleic acid element, which fits into a groove on the inside. The diameter of the helix is determined by the nature of the protein(s) making up the capsomers; its length depends on the size of the nucleic acid core.

Icosahedral capsids An icosahedron is a regular three-dimensional shape with 20 triangular faces, and 12 points or corners (Figure 10.5a). The overall effect is of a roughly spherical structure.

> The *icosahedron* has a low surface-area to volume ratio, allowing for the maximum amount of nucleic acid to be packaged.

10.2.3 The viral envelope

Envelopes are much more common in animal viruses than in plant viruses. The lipid bilayer covering an enveloped virus is derived from the nuclear or cytoplasmic membrane of a previous host. Embedded in this, however, are proteins (usually glycoproteins) encoded by the virus's own genome. These may project from the surface of the virion as *spikes*, which may be instrumental in allowing the virus to bind to or penetrate its host cell (Figure 10.6). The envelope is more susceptible than the capsid to environmental pressures, and the virus needs to remain moist in order to survive. Consequently such viruses are transmitted by means of body fluids such as blood (e.g. hepatitis B virus) or respiratory secretions (e.g. influenza virus).

10.3 Classification of viruses

As we saw at the beginning of this chapter, viruses are not considered to be strictly living, and their classification is a complex issue. As with true organisms we have species, genera, families and orders of viruses, but none of the higher groupings (class, phylum, kingdom). Latin binomials (e.g. *Homo*

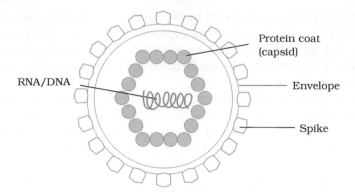

Figure 10.6 An enveloped virus. The envelope derives from the host cell's membrane, and includes virus-encoded proteins. Some viral envelopes contain projecting spikes, which may assist in attachment to the host.

sapiens, Escherichia coli), familiar from conventional biological taxonomy, are not used for viruses; however, a proposal for non-Latinised viral binomials is under discussion at the time of writing. Originally, no attempt was made to draw up any sort of phylogenetic relationship between the viruses, but more recent developments in sequencing of viral genomes has meant that insights are being gained in this area.

Factors taken into account in the classification of viruses include:

- host range (vertebrate/invertebrate, plant, algae/fungi, bacteria);
- morphology (capsid symmetry, enveloped/non-enveloped, capsomer number);
- genome type/mode of replication (see Figure 10.3).

> Orders of viruses always end in *-ales*, families in *-viridae*, subfamilies in *-virinae* and genera in *-virus*. Such names are italicised and capitalised, whereas this is not done for species. For example: order *Mononegavirales*, family *Paramyxoviridae*, subfamily *Paramyxovirinae*, genus *Morbillivirus* and species measles virus. For informal usage, we would talk about, e.g., 'the picornavirus family', or the 'enterovirus genus'.

The ninth report of the International Commission on Taxonomy of Viruses (ICTV) (2011 update) recognises six orders, 94 families, 395 genera and 2475 species of virus (Table 10.2). Countless others, undiscovered or insufficiently characterised, remain unclassified.

An alternative approach to classifying viruses was proposed by David Baltimore in 1971. This orders them with respect to the strategies they use for mRNA production, and results in seven major groupings (Table 10.3).

Table 10.2 Virus taxonomy. The International Commission on Taxonomy of Viruses (ICTV) currently recognises six orders and 94 families of viruses. Most families (72) have not yet been assigned to an order

Order	Family
Caudovirales	Myoviridae
	Podoviridae
Herpesvirales	Alloherpesviridae
	Herpesviridae
Mononegavirales	Paramyxoviridae
	Rhabdoviridae
Nidovirales	Arteriviridae
	Coronaviridae
Picornavirales	Dicistroviridae
	Picornaviridae
	Secoviridae
Tymovirales	Alphaflexiviridae
	Betaflexiviridae
	Tymoviridae

Families assigned to an order but comprising only a handful of genera are not shown.

Table 10.3 Major groupings of viruses based on the Baltimore system

Group I	dsDNA viruses
Group II	ssDNA viruses
Group III	dsRNA viruses
Group IV	($+$) sense ssRNA viruses
Group V	($-$) sense ssRNA viruses
Group VI	Single-stranded ($+$) sense RNA with DNA intermediate
Group VII	Double-stranded DNA with RNA intermediate

ds, double-stranded; ss, single-stranded.

10.4   Viral replication cycles

One characteristic that viruses share with true living organisms is the need to reproduce themselves.[2] As we have seen, all viruses are obligate intracellular

---

[2] Since the processes involved proceed at the molecular rather than the organismal level, it is more appropriate to speak of viral replication than of reproduction.

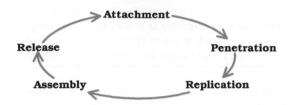

Figure 10.7   Main stages in a viral replication cycle. The replication cycle of all viruses is based on this generalised pattern.

parasites, and so in order to replicate, a host cell must be successfully entered. It is the host cell that provides much of the 'machinery' necessary for viral replication.

All viral growth cycles follow the same general sequence of events (Figure 10.7), with some differences from one type to another, determined by viral structure and the nature of the host cell.

10.4.1   Replication cycles in bacteriophages

Viruses that infect bacterial cells are called *bacteriophages* ('phages' for short), which means, literally, 'bacteria eaters'. Perhaps the best understood of all viral replication cycles are those of a class of bacteriophages that infect *E. coli*, known as the *T-even phages*. These are large, complex viruses with a characteristic head and tail structure (Figure 10.8). The double-stranded

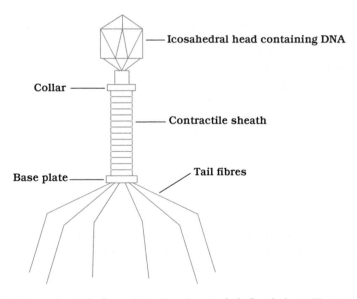

Figure 10.8   T-even bacteriophage. Note the characteristic 'head plus tail' structure. The tail fibres and base plate are involved in the attachment of the phage to its host cell's surface.

linear DNA genome contains over 100 genes, and is contained within the icosahedral head. The growth cycle is said to be *lytic*, because it culminates in the lysis (i.e. bursting) of the host cell. Figure 10.9 shows the lytic cycle of phage T4, and the main stages are described below.

1. *Adsorption (attachment):* T4 attaches by means of specific tail fibre proteins to complementary receptors on the host cell's surface. The nature of these receptors and the virus's ability to attach to them is the principal factor in determining a virus's host specificity.

2. *Penetration:* The enzyme lysozyme, present in the tail of the phage, weakens the cell wall at the point of attachment, and a contraction of the tail sheath of the phage causes the core to be pushed down into the cell, releasing the viral DNA into the interior of the bacterium. A viral *pilot protein* assists in the passage of the DNA across the host plasma membrane. The capsid remains entirely outside the cell, as elegantly demonstrated in the famous experiment by Hershey and Chase (see Chapter 11).

> You might reasonably ask yourself why cells would evolve receptor molecules for viruses, when the outcome is clearly not in their interests. The answer is, of course, that they haven't; the receptors have other biological properties, but the viruses have 'taken them over'. HIV attaches to its host cell through an interaction between its gp120 surface protein and CD4, a glycoprotein on the surface of T lymphocytes involved in normal immune function.

3. *Replication:* Phage genes cause host protein and nucleic acid synthesis to be switched off, so that all of the host's metabolic machinery becomes dedicated to the synthesis of phage DNA and proteins. Host nucleic acids are degraded by phage-encoded enzymes, thereby providing a supply of nucleotide building blocks. Host enzymes are employed to replicate phage DNA, which is then transcribed into mRNA and translated into protein.

4. *Assembly:* Once synthesised in sufficient quantities, capsid and DNA components assemble spontaneously into viral particles. The head and tail regions are synthesised separately, then the head is filled with the DNA genome, and joined onto the tail.

5. *Release:* Phage-encoded lysozyme weakens the cell wall, and leads to lysis of the cell and release of viral particles. These are able to infect new host cells, and in so doing recommence the cycle. During the early phase of infection, the host cell

> 'Early' proteins are viral enzymes including DNA polymerase, used to synthesise more phage DNA, and others which disrupt normal host processes. Later, production switches to 'late', structural proteins, required for the construction of capsids; these are produced in much greater quantities.

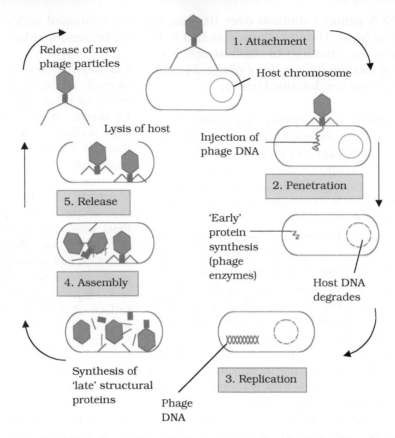

Figure 10.9   The lytic cycle of phage T4. The cycle comprises the five main stages described in the text; from injection of phage of DNA to cell lysis takes 22 minutes. The number of phage particles released per cell is called the burst size, and for T4 it ranges from 50 to 200.

contains components of phage, but no complete particles. This period is known as the *eclipse period*. The time that elapses between the attachment of a phage particle to the cell surface and the release of newly synthesised phages is the *latent period* (sometimes known as the *burst time*); for T4 under optimal conditions, this is around 22 minutes. This can be seen in a one-step growth curve, as shown in Figure 10.10.

10.4.2   Lysogenic replication cycle

Phages such as T4, which cause the lysis of their cells, are termed *virulent* phages. *Temperate* phages, in addition to following a lytic cycle as outlined above, are able to undergo an alternative form of replication cycle. Here, the phage DNA actually becomes incorporated into the host's genome as a *prophage* (Figure 10.11). In this condition of *lysogeny*, the host cell suffers no

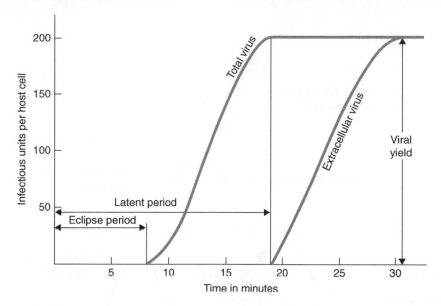

Figure 10.10 The one-step growth curve. During the eclipse period early phase the host cell does not contain complete phage particles. Following synthesis of new particles, they are released, signalling the end of the latent period. The left-hand curve represents the number of phage particles, while the number of free (extracellular) particles is shown by the right-hand curve.

harm. This is because the action of repressor proteins, encoded by the phage, prevents most of the other phage genes being transcribed. These genes are, however, replicated along with the bacterial chromosome, so all the bacterial offspring contain the incorporated prophage. The lysogenic state is ended when the survival of the host cell is threatened, usually by an environmental factor such as UV light or a chemical mutagen. Inactivation of the repressor protein allows the phage DNA to be excised, and adopt a circular form in the cytoplasm. In this form, it initiates a lytic cycle, resulting in destruction of the host cell. An example of a temperate phage is bacteriophage λ (lambda), which infects certain strains of *E. coli*. Bacterial strains that can incorporate phage DNA in this way are termed *lysogens*.

10.4.3 Replication cycles in animal viruses

Viruses that infect multicellular organisms such as animals may be specific not only to a particular organism, but also to a particular cell or tissue type. This is known as the *tissue tropism* of the virus, and is due to the fact that attachment occurs via specific receptors on the host cell surface.

The growth cycles of animal viruses have the same main stages as described for bacteriophages (see Figure 10.7), but may differ a good deal in some of

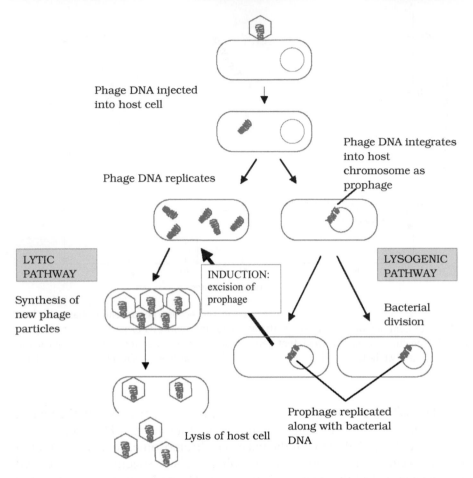

Phage DNA injected
into host cell

Phage DNA replicates

Phage DNA integrates
into host
chromosome as
prophage

LYTIC
PATHWAY

INDUCTION:
excision of
prophage

LYSOGENIC
PATHWAY

Synthesis of
new phage
particles

Bacterial
division

Lysis of host cell

Prophage replicated
along with bacterial
DNA

Figure 10.11 Replication cycle of phage lambda, a temperate phage. In the lysogenic pathway, the phage DNA is integrated as a prophage into the host genome, and replicated along with it. Upon induction by an appropriate stimulus, the phage DNA is removed and enters a lytic cycle.

the details. Most of these variations are a reflection of differences in structure between bacterial and animal host cells.

- *Adsorption and Penetration:* Animal viruses do not have the head and tail structure of phages, so it follows that their method of attachment is different. The specific interaction with a host receptor is made via some component of the capsid, or, in the case of enveloped viruses, by special structures such as spikes (*peplomers*). Viral attachment sites can frequently be blocked by host antibody molecules; however, some viruses (e.g. the rhinoviruses) have overcome this by having their sites situated in deep depressions, inaccessible to the antibodies.

Whereas bacteriophages inject their nucleic acid component from the outside, the process in animal viruses is more complex, a fact reflected in the time taken for completion of the process. Animal viruses do not have to cope with a thick cell wall, and in many such cases the entire virion is internalised. This necessitates the extra step of uncoating, a process carried out by host enzymes. Many animal viruses possess an envelope, and such viruses are taken into the cell either by fusion with the cell membrane or by endocytosis (Figure 10.12). While some non-enveloped types release only their nucleic acid component into the cytoplasm, others require additionally that virus-encoded enzymes be introduced to ensure successful replication.

- *Replication (DNA viruses):* The DNA of animal cells, unlike that of bacteria, is compartmentalised within a nucleus, and it is here that replication and transcription of viral DNA generally occur.[3] Messenger RNA then passes to ribosomes in the cytoplasm for translation (Figure 10.13). In the case of viruses with a ssDNA genome, a double-stranded intermediate is formed, which serves as a template for mRNA synthesis.

- *Assembly:* Translation products are finally returned to the nucleus for assembly into new virus particles.

- *Release:* Naked (non-enveloped) viruses are generally released by lysis of the host cell. In the case of enveloped forms, release is more gradual. The host's plasma membrane is modified by the insertion of virus-encoded proteins, before engulfing the virus particle and releasing it by a process of *budding*. This can be seen as essentially the reverse of the process of internalisation by fusion (Figure 10.13a).

> *Herpesviruses* are unusual in deriving their envelope from the nuclear, rather than the cytoplasmic membrane.

10.4.4 Replication of RNA viruses

The phage and animal virus growth cycles we have described so far have all involved double-stranded DNA genomes. As you will remember from the start of this chapter, however, many viruses contain RNA instead of DNA as their genetic material, and we now need to consider briefly how these viruses complete their replication cycles.

Replication of RNA viruses occurs in the cytoplasm of the host; but the details differ, depending on whether the RNA is single- or double-stranded,

---

[3] Poxviruses are an exception. Both replication and assembly occur in the cytoplasm.

Figure 10.12 Enveloped viruses enter the host cell by fusion or endocytosis. (a) Fusion between viral envelope and host membrane results in release of nucleocapsid into the cell. Fusion depends on the interaction between spikes in the envelope and specific surface receptors. (b) Viral particles bound to the plasma membrane are internalised by endocytosis. Acidification within the endosome allows the release of the nucleocapsid into the cell.

and (+) or (−) sense. The genome of a *(+) sense single-stranded RNA virus* functions directly as an mRNA molecule, producing a giant polyprotein, which is then cleaved into the various structural and functional proteins of the virus. In order for the (+) sense RNA to be replicated, a complementary (−) sense strand must be made, which acts as a template for the

dsDNA virus

Figure 10.13   Replication in double-stranded DNA (dsDNA) viruses. Replication of viral DNA and transcription to mRNA take place in the nucleus of the host cell. The mRNA then passes out into the cytoplasm, where protein synthesis occurs on ribosomes. The capsid protein so produced returns to the nucleus for assembly into new viral particles. Newly synthesised DNA polymerase also returns to the nucleus, for further DNA replication. Reproduced from Hardy, SP (2002) Human Microbiology, with permission from Taylor & Francis Group.

production of more (+) sense RNA (Figure 10.14). The RNA of a *(−) sense RNA virus* must first act as a template for the formation of its complementary sequence by a virally encoded RNA polymerase. The (+) sense RNA so formed has two functions: (i) to act as mRNA and undergo translation into the virus's various proteins, and (ii) to act as template for the production of more genomic (−) sense RNA (Figure 10.15).

Double-stranded RNA viruses are all segmented. They form separate mRNAs for each of their proteins by transcription of the (−) strand of their

Positive strand RNA virus

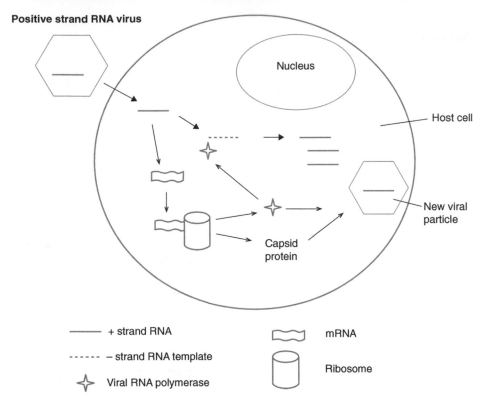

Figure 10.14 Replication in (+) sense single-stranded RNA (ssRNA) viruses. On entering the cell, the (+) sense ssRNA genome is able to act directly as mRNA, directing the synthesis of capsid protein and RNA polymerase. In addition, it replicates itself, being converted firstly into a (−) sense ssRNA intermediate. All steps take place outside the nucleus. Reproduced from Hardy, SP (2002) Human Microbiology, with permission from Taylor & Francis Group.

genome. These are each translated, and later form an aggregate (*subviral particle*) with specific proteins, in this form they act as templates for the synthesis of a double-stranded RNA genome, ready for incorporation into a new viral particle.

Two final, rather complicated variations on the viral replication cycles involve the enzyme *reverse transcriptase*, first discovered in 1970 (see Box 10.3), and found in retroviruses and hepadnaviruses.

Retroviruses These viruses, which include some important human pathogens, have a genome that exists as RNA and DNA at different part of their replication cycle. Retroviruses have a (+) sense ssRNA genome, which is unique among viruses in being *diploid*. The two copies of the genome serve as templates for the enzyme *reverse transcriptase* to produce a complementary

Negative strand RNA virus

——— + strand RNA	⌒⌒⌒ mRNA
------ − strand RNA	
✦ RNA-dependent RNA polymerase	▯ Ribosome

Figure 10.15 Replication in (−) sense single-stranded RNA (ssRNA) viruses. Before it can function as mRNA, the (−) sense ssRNA must be converted to its complementary (+) sense sequence. This serves both as mRNA and as template for the production of new (−) sense ssRNA genomes. Reproduced from Hardy, SP (2002) Human Microbiology, with permission from Taylor & Francis Group.

Box 10.3 *The enzyme that breaks the rules*

The discovery in 1970 of an enzyme that could convert an RNA template into DNA caused great surprise in the scientific world. The action of this *reverse transcriptase*, or RNA-dependent DNA polymerase, is a startling exception to the 'central dogma' of molecular biology, that the flow of genetic information is unidirectional, from DNA to RNA to protein (see Chapter 11). This in its revised form can be represented:

$$DNA \longrightarrow RNA \longrightarrow Protein$$

(The circular arrow to the left denotes DNA's ability to be replicated; the dashed arrow represents the action of reverse transcriptase)

Retrovirus

Nucleus

Host cell

Capsid proteins &

New viral particle

——— Single stranded RNA

- - - - - - - + strand RNA

mRNA

Ribosome

Host chromosome with integrated provirus

Reverse transcriptase

Figure 10.16 Replication in retroviruses. Reverse transcriptase makes a DNA copy of the single-stranded RNA retroviral genome. This is integrated into the host genome and is transcribed by the cellular machinery. Messenger RNA passes out to the ribosomes, where translation into viral coat proteins and more reverse transcriptase occurs. Retrovirus packaging takes place outside the nucleus. Reproduced from Hardy, SP (2002) Human Microbiology, with permission from Taylor & Francis Group.

strand of DNA. The RNA component of this hybrid is then degraded, allowing the synthesis of a second strand of DNA. This *proviral DNA* passes to the nucleus, where it is incorporated into the host's genome (Figure 10.16). Transcription by means of a host RNA polymerase results in mRNA, which is translated into viral proteins and also serves as genomic material for the new retrovirus particles. The human immunodeficiency virus (HIV), the causative agent of AIDS, is an important example of a retrovirus.

> Incorporation of retroviral DNA into the host genome parallels the integration seen in lysogenic phage growth cycles. Unlike the prophage, however, the provirus is not capable of a separate existence away from the host chromosome.

Hepadnaviruses In two families of viruses (hepadnaviruses and caulimoviruses), DNA and RNA phases again alternate, but their order of appearance is reversed, so that a dsDNA genome is produced. This is made possible by reverse transcriptase occurring at a later stage, during the maturation of the virus particle.

10.4.5 Replication cycles in plant viruses

Plant viruses face a formidable barrier to penetration in the form of the plant's cellulose cell wall, and in some cases a waxy cuticle. Entry depends on mechanical damage to this barrier, often caused by insects that use their mouthparts to penetrate the cell wall. These insects are also responsible for the spread of a virus from one plant to another. This form of transmission is called *horizontal transmission* and involves the introduction of a virus from the outside. This form of transmission can also occur by means of inanimate objects such as garden tools. Compare this with *vertical transmission*, in which the virus is passed from a plant to its offspring, either by asexual propagation or through infected seeds.

The majority of plant viruses discovered so far have an RNA genome, although DNA forms such as the caulimoviruses (see above) are also known. Replication is similar to that of animal viruses, depending on the nature of the viral genome. An infection only becomes significant if it spreads throughout the plant (a *systemic* infection). Viral particles do this by moving through the plasmodesmata, naturally occurring cytoplasmic strands linking adjacent plant cells.

10.5 Viroids

In 1971, Theodore Diener proposed the name *viroid* to describe a newly discovered pathogen of potatoes. Viroids are many times smaller than the smallest virus, and consist solely of a small circle of ssRNA containing some 300–400 nucleotide bases and no protein coat. Enzymes in the host's nucleus are used to replicate the RNA, which does not appear to be translated into protein. Appreciable sequence homology suggests that viroids arose from *transposable elements* (see Chapter 11), segments of DNA capable of movement within or between DNA molecules. To date, viroids have only been found in plants, where they cause a variety of diseases.

> A *viroid* is a plant pathogen that comprises only ssRNA and does not code for protein product.

10.6 Prions

A decade after the discovery of viroids, Stanley Prusiner made the startling claim that *scrapie*, a neurodegenerative disease of sheep, was caused by a self-replicating agent composed *solely of pro-tein*. He called this type of entity a *prion*, and in the years that followed, several related diseases of humans and animals were shown to have a similar cause. These include *bovine spongi-form encephalopathy* (BSE, 'mad cow disease') and its human equivalent, *Creutzfeldt–Jakob disease*.

> A *prion* (from 'proteina-ceous *in*fectious particle') is a self-replicating protein responsible for a range of neurodegenerative disor-ders in humans and other mammals.

How could something that contains no nucleic acid be capable of replicat-ing itself? Prusiner's idea seemed to go against the basic rules of biology. It appears that prions may be altered versions of normal animal proteins that somehow have the ability to cause the normal version to refold itself into the mutant form. Thus the prion propagates itself. All prion diseases described thus far are similar conditions, involving a degeneration of brain tissue.

10.7   Cultivating viruses

Whilst the growth of bacteria in the laboratory generally demands only a supply of the relevant nutrients and appropriate environmental conditions, maintaining viruses presents special challenges. Think back to the start of this chapter, and you will realise why this is so; all viruses are obligate intracellular parasites, and therefore need an appropriate host cell if they are to replicate.

Bacteriophages, for example, are grown in culture with their bacterial hosts. Stock cultures of phages are prepared by allowing them to infect a broth culture of the appropriate bacterium. Successful propagation of phages results in a clearing of the culture's turbidity; centrifugation removes any remaining bacteria, leaving the phage particles in the supernatant. A quantitative measure of phages, known as the *titre*, can be obtained by mixing them with a much greater number of bacteria and immobilising them in agar. Due to their numbers, the bacteria grow as a confluent *lawn* following incubation. Some become infected by phage, and when new viral particles are released following lysis of their host, they infect more host cells. Because they are immobilised in agar, the phages are only able to infect cells in the immediate vicinity. As more and more cells in the same area are lysed, an area of clearing called a *plaque* appears in the lawn of bacteria (Figure 10.17). Quantification is based on the assumption that each visible plaque arises from infection by a single phage particle. Thus, we speak of *plaque-forming units* (pfu: see Box 10.4).

Animal viruses used to be propagated in the host animal; clearly there are limitations to this, not least when the host is human! One of the major breakthroughs in the field of virus cultivation was made in 1931 when it was shown by Alice Woodruff and Ernest Goodpasture that fertilised chicken's eggs could serve as a host for a number of animal and human viruses, such as

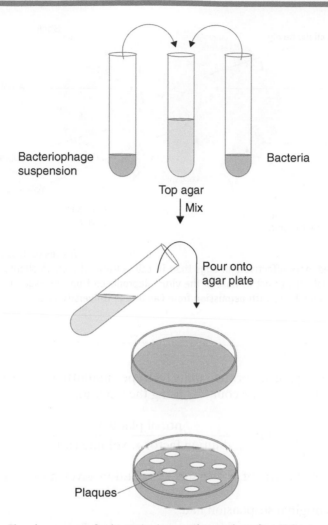

Figure 10.17 The plaque assay for bacteriophages. The number of particles in a phage preparation can be estimated by means of a plaque assay. Phages and bacteria are mixed in a soft agar then poured onto the surface of an agar plate. Bacteria grow to develop a confluent lawn, and the presence of phage is indicated by areas of clearing (plaques) where bacteria have been lysed. Reproduced from Reece, RJ (2003) Analysis of Genes and Genomes, with permission from John Wiley & Sons.

those that cause rabies and influenza. It has been said that the chicken embryo did for virus culture what agar did for the growth of bacteria. Depending on the virus in question, inoculation can be made into the developing embryo itself or into one of the various membranes and cavities such as the chorioallantoic membrane or the allantoic cavity (Figure 10.18). Viral propagation is demonstrated by death of the embryo, or the appearance of lesions on the membranes.

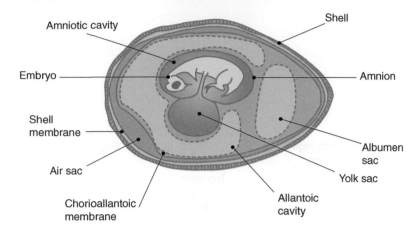

Figure 10.18 The culture of animal viruses in embryonated eggs. Viruses such as the influenza virus are cultured more effectively in eggs than in cell culture. The chorioallantoic membrane provides epithelial cells to act as host to the virus. Reproduced from Heritage, J, et al. (1996) Introductory Microbiology, with permission from Cambridge University Press.

Box 10.4   *Plaque counts*

The number of plaque-forming units (pfu) per millilitre of suspension can be calculated by a plaque count by using the equation:

$$\text{pfu/mL} = \frac{\text{no. of plaques}}{\text{dilution factor} \times \text{volume (mL)}}$$

For example: 100 μL of a 1 in 10 000 dilution gives 53 colonies when plated out.
   Thus the original suspension had:

$$\frac{53}{10^{-4} \times 0.1} = 5.3 \times 10^{6} \text{ pfu/mL}$$

   In the 1950s, cell culture techniques advanced, thanks in part to the widespread availability of antibiotics, making the control of bacterial contamination much more readily achieved. Cells are usually grown as *monolayers* in tissue culture flasks containing a suitable liquid growth medium. Treatment with the protease trypsin dissolves the connective tissue matrix between the cells, allowing them to be harvested, and used to seed new cultures. Changes in cell morphology, known generically as *cytopathic effects* (CPE), are indicators of viral infection, and may be used diagnostically in the identification of specific viral types.

Plant viruses need to overcome the barrier presented by the cellulose cell wall of the plant; in nature this is often achieved by the piercing mouthparts of an insect vector or by entering areas of damaged tissue. Experimentally, viruses can be introduced into an appropriate host by rubbing the surface of a leaf with the virus together with a mild abrasive to create a minor wound.

IV
Microbial Genetics

11
Microbial Genetics

When any living organism reproduces, it passes on genetic information to its offspring. This information takes the form of *genes*, linear sequences of DNA that can be thought of as the basic units of heredity. The total complement of an organism's genetic material is called its *genome*.

> A *gene* is a sequence of DNA that usually encodes a polypeptide.

11.1 How do we know genes are made of DNA?

The concept of the gene as an inherited physical entity determining some aspect of an organism's *phenotype* dates back to the earliest days of genetics. The question of what genes are actually made of was a major concern of molecular biologists (not that they would have described themselves as such!) in the first half of the twentieth century. Since it was recognised by this time that genes must be located on chromosomes, and that chromosomes (in eukaryotes) comprised largely protein and DNA, the reasonable assumption was made that genes must be made up of one of these substances. In the early years, protein was regarded as the more likely candidate, since, from what was known of molecular structure at the time, it offered far more scope for the variation that would be essential to account for the thousands of genes that any organism must possess. The road to proving that DNA is in fact the 'stuff of life' was a long and hard one, which can be read about elsewhere; we shall mention below just some of the key experiments that provided crucial evidence.

Essential Microbiology, Second Edition. Stuart Hogg.
© 2013 John Wiley & Sons, Ltd. Published 2013 by John Wiley & Sons, Ltd.

In 1928 the Englishman Fred Griffith carried out a seminal series of experiments that not only demonstrated for the first time the phenomenon of genetic transfer in bacteria (a subject we shall consider in more detail later in this chapter), but also acted as the first step towards proving that DNA was the genetic material. As we shall see, Griffith showed that it was possible for heritable characteristics to be transferred from one type of bacterium to another, but the cellular component responsible for this phenomenon was not known at this time.

Attempts were made throughout the 1930s to isolate and identify the *transforming principle*, as it became known, and in 1944 Avery, MacLeod and McCarty published a paper, which for the first time, proposed DNA as the genetic material. Avery and his colleagues demonstrated that when DNA was rendered inactive by enzymatic treatment, transforming ability was lost from a cell extract, but if proteins, RNA or any other cellular component was similarly inactivated, the ability was retained. In spite of this apparently convincing proof, the pro-protein lobby was not easily persuaded. It was to be several more years before the experimental results of Alfred Hershey and Martha Chase (Box 11.1) coupled with Watson and Crick's model for DNA structure (see Figure 2.23) finally cemented the universal acceptance of DNA's central role in genetics.

Box 11.1   Hershey and Chase: the Waring blender experiment

In 1952 Alfred Hershey and Martha Chase provided further experimental evidence that DNA was the genetic material. In their experiment, the bacteriophage T2 was grown with *E. coli* cells in a medium with radiolabelled ingredients, so that their proteins contained ^{35}S, and their DNA ^{32}P. The phages were harvested, and mixed with a fresh culture of *E. coli*. They were left long enough for the phage particles to infect the bacteria, but not long enough to produce new phage particles and lyse the cells. The culture was then subjected to mechanical agitation in a Waring blender, which, it was hoped, would remove the 'shell' of the phages from the outside of the bacteria, but leave the injected genetic material inside.

The bacteria were sedimented by centrifugation, leaving the much lighter phage 'shells' in the supernatant. When solid and liquid phases were analysed for ^{32}P and ^{35}S, it was found that nearly all the ^{32}P was associated with the bacterial cells, while the great majority of the ^{35}S remained in the supernatant. The conclusion drawn from these results is that it was the ^{32}P-labelled DNA that had been injected into the bacteria, and was therefore the genetic material.

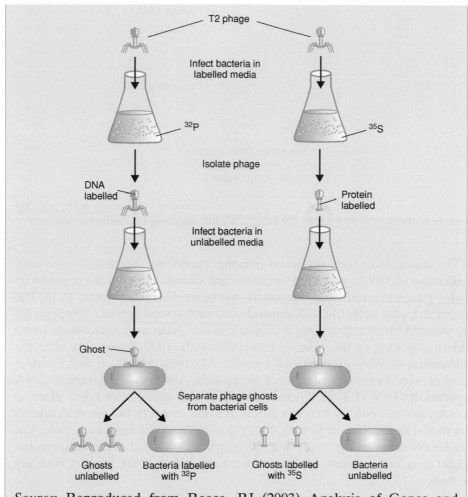

Source: Reproduced from Reece, RJ (2003) Analysis of Genes and Genomes, with permission from John Wiley & Sons.

11.2   DNA replication

The elucidation of the structure of DNA by Watson and Crick in 1953 stands as one of the great scientific breakthroughs of the twentieth century. An important implication of their model was suggested by the complementary nature of the two strands, as they commented themselves in their famous paper in *Nature*:

> It has not escaped our notice that the specific pairing we have postulated immediately suggests a possible copying mechanism for the genetic material.

Parent DNA

Daughter DNA

————— Inherited from parent

————— Newly synthesised

Figure 11.1 DNA replication is semi-conservative. Following replication, each new DNA molecule comprises one strand from the parent DNA and one newly synthesised strand.

The mechanism they proposed became known as the *semi-conservative replication* of DNA, so called because each daughter molecule is made up of one parental strand and one newly synthesised strand (Figure 11.1). The parental double helix of DNA unwinds and each strand acts as a template for the production of a new complementary strand, with new nucleotides being added according to the rules of base-pairing. In 1957, four years after the publication of Watson and Crick's model, Matthew Meselson and Franklin Stahl provided experimental proof of the semi-conservative nature of DNA replication (Box 11.2). The exact way in which replication takes place in prokaryotic and eukaryotic cells differs in some respects, but we shall take as our model replication in *E. coli*, since it has been studied so extensively. For a fuller coverage of this and other mechanisms described in this chapter, the reader is referred to one of the genetics texts listed in the Further Reading section.

11.2.1 DNA replication in prokaryotes

You may recall from Chapter 4 that bacteria multiply by a process of binary fission; before this occurs, each cell must duplicate its genetic information so that each daughter cell has a copy.

DNA replication involves the action of a number of specialised enzymes:

- helicases
- DNA topoisomerases
- DNA polymerase I
- DNA polymerase III
- DNA primase
- DNA ligase

Box 11.2   Experimental proof of semi-conservative replication

The semi-conservative model of DNA replication, as suggested by Watson and Crick, was not the only one in circulation during the 1950s. Meselson and Stahl provided experimental evidence that not only supported the semi-conservative model, but also showed the other models to be unworkable. Their elegantly designed experiment used newly developed techniques to differentiate between parental DNA and newly synthesised material.

First, they grew *E. coli* in a medium with ammonium salts containing the heavy isotope ^{15}N as the only source of nitrogen. This was done for several generations of growth, so that all the cells contained nitrogen exclusively in the heavy form.

The bacteria were then transferred to a medium containing nitrogen in the normal, ^{14}N form. After a single round of replication, DNA was isolated from the culture and subjected to *density gradient centrifugation*. This is able to differentiate between DNA containing the two forms of nitrogen, as ^{15}N has a greater buoyant density and therefore settles at a lower position in the tube.

Meselson and Stahl found just a single band of DNA after centrifugation, with a density intermediate between that of ^{14}N and ^{15}N, indicating a hybrid molecule, as predicted by the semi-conservative model. After a second round of *E. coli* replication in a medium containing ^{14}N,

Bacteria grown
in ^{15}N only

^{15}N

Bacteria
transferred to
^{14}N medium

$^{14}N/^{15}N$

DNA after
one round of
replication

Each DNA molecule
comprises one
strand from the
parent and one
newly synthesised

^{14}N

$^{14}N/^{15}N$

DNA after
two rounds
of replication

two bands of DNA were produced, one hybrid and one containing exclusively ^{14}N, exactly as predicted by the semi-conservative model, but inconsistent with other hypotheses.

Figure 11.2 DNA replication takes place at a replication fork. Strands of DNA are separated and unwound by helicase and DNA gyrase, and prevented from rejoining by the attachment of single-strand binding proteins. Starting from an RNA primer, DNA polymerase III adds the complementary nucleotides to form a second strand. On the leading strand, a single primer is set down (not shown), and replication proceeds uninterrupted in the same direction as the fork. Because DNA polymerase III only works in the 5′ → 3′ direction, on the lagging strand a new primer must be added periodically as the strands open up. Replication here is thus discontinuous, as a series of Okazaki fragments (see the text). Reproduced from Bolsover, SR, et al. (1997) From Genes to Cells, with permission from John Wiley & Sons.

Replication begins at a specific sequence called an *origin of replication*. The two strands of DNA are caused to separate by *helicases* (Figure 11.2), while *single-stranded DNA binding proteins* (SSB) prevent them from rejoining. Opening out part of the double helix causes increased tension (super-coiling) elsewhere in the molecule, which is relieved by the enzyme *DNA topoisomerase* (sometimes known as *DNA gyrase*). As the 'zipper' moves along, and more single-stranded

> DNA replication takes place at a *replication fork*, a Y-shaped structure formed by the separating strands. The fork moves along the DNA as replication proceeds.

DNA is exposed, *DNA polymerase III* adds new nucleotides to form a complementary second strand, according to the rules of base-pairing (A with T, G with C). DNA polymerases are not capable of initiating the synthesis of an entirely new strand, but can only extend an existing one. This is because they require a free 3′-OH group on which to attach new nucleotides (see Figure 2.22). Thus, DNA polymerase III *can only work in the 5′→3′ direction*. A form of RNA polymerase called *DNA primase* synthesises a short

Figure 11.3   DNA replication is bidirectional. Two replication forks form simultaneously, moving away from each other and developing a replication bubble.

single-stranded *primer* of RNA, which supplies the required 3'-OH group for the DNA polymerase III (see Figure 11.2).

When replication occurs, complementary nucleotides are added to one of the strands (the *leading* strand) in a continuous fashion (Figure 11.2). The other strand (the *lagging* strand), however, runs in the opposite polarity, so how is a complementary sequence synthesised here? The answer is that DNA polymerase III allows a little unwinding to take place and then, starting at the fork, works back over, in the 5'→3' direction. Thus the second strand is synthesised discontinuously, in short bursts, about 1000–2000 nucleotides at a time. These short stretches of DNA are called *Okazaki fragments*, after their discoverers, Reiji and Tsuneko Okazaki.

> A *primer* is a short sequence of single-stranded DNA or RNA required by DNA polymerase as a starting point for chain extension.

On the lagging strand, a new RNA primer is needed at the start of every Okazaki fragment. These short sequences of RNA are later removed by *DNA polymerase I*, which then replaces them with DNA nucleotides. Finally, the fragments are joined together by the action of *DNA ligase*. Replication is bidirectional (Figure 11.3), with two forks moving in opposite directions; when they meet, the whole chromosome is copied and replication is complete.[1]

> *DNA ligase* repairs breaks in DNA by re-establishing phosphodiester bonds in the sugar–phosphate backbone.

11.2.2   What happens when replication goes wrong?

It is clearly important that synthesis of a complementary second strand of DNA should occur with complete accuracy, but occasionally a non-complementary nucleotide is inserted; this may happen as frequently as once

---

[1] This rather long-winded description hardly does justice to a process that incorporates new nucleotides at a rate of around 1000 per second!

Figure 11.4   DNA replication in eukaryotes. Many replication bubbles develop simultaneously; they extend towards each other and eventually merge. The arrows denote the direction of the replication forks.

in every 10 000 nucleotides. Cells, however, are able to have a second attempt to incorporate the correct base because of the *proofreading* activity of the enzymes DNA polymerase I and III. These are able to cut out the 'wrong' nucleotide and replace it with the correct one. As a result of this monitoring, mistakes are very rare; they are thought to occur at a frequency of around one in every billion (10^9) nucleotides copied. Mistakes that do slip through the net result in *mutations*, which are discussed later in this chapter.

11.2.3   DNA replication in eukaryotes

In *eukaryotic* organisms, the DNA is linear, not circular, so the process of replication differs in some respects. Genome sizes are generally much bigger in eukaryotes, and replication much slower, so numerous replication forks are active simultaneously on each chromosome. Replication takes place bidirectionally, creating numerous replication bubbles (Figure 11.4), which merge with one another until the whole chromosome has been covered.

11.3   What exactly do genes do?

At the start of the twentieth century Archibald Garrod had proposed that inherited disorders such as alkaptonuria may be due to a defect in certain key metabolic enzymes, thus offering for the first time an explanation of how genetic information is expressed. His ideas were not really developed, however, until the work of George Beadle and Edward Tatum in the 1940s; their experiments with the bread mould *Neurospora* led to the formulation of the *one gene, one enzyme* hypothesis. Although now acknowledged to be somewhat over-generalised (see Box 11.3), this model proved useful in the years when the molecular basis of gene action was being elucidated.

Having established that genes are made of DNA, and having a model for the structure of DNA that explained how it was able to copy itself, the way was open in the 1950s for scientists to work out the mechanism by which

Box 11.3 One gene, one enzyme: not quite true

Beadle and Tatum proposed that each gene was responsible for the production of a specific enzyme. However, all proteins, not just enzymes, are encoded by DNA, and furthermore, some have a quaternary structure (see Chapter 2) with different polypeptide subunits being encoded by different genes. The hypothesis was therefore modified to *one gene, one polypeptide*. Later still, it emerged that even this is not always the case, as some genes do not encode proteins at all, but forms of RNA.

the information encoded in a DNA sequence was converted into a specific protein.

11.3.1 How does a gene direct the synthesis of a protein?

You may recall from Chapter 2 that both DNA and proteins are polymers whose 'building blocks' (nucleotides and amino acids respectively) can be put together in an almost infinite number of sequences. The sequence of amino acids making up the primary structure of a protein is determined by the sequence of nucleotides in the particular gene responsible for its production. It does this not directly, but through an intermediary molecule, now known to be a form of RNA called *messenger RNA* (mRNA). It is this intermediary that carries out the crucial task of passing the information encoded in the DNA sequence to the site of protein synthesis. This unidirectional flow of information can be summarised thus:

$$\text{DNA} \longrightarrow \text{mRNA} \longrightarrow \text{Protein}$$

DNA replication

and is often referred to as the *central dogma* of biology, because of its applicability to all forms of life. Proposed by Crick in the late 1950s, this is still accepted as being a true model of the basic events in protein synthesis. Sometimes the message encoded in DNA is transcribed into either *ribosomal RNA* (rRNA) or *transfer RNA* (tRNA); these types of RNA are not translated into proteins but represent end-products in themselves. (In Chapter 10 we saw that *retroviruses* have proved an exception to one part of the central dogma, since they possess reverse transcriptase enzymes capable of forming DNA from an RNA template.)

Figure 11.5 Transcription and translation. The sequence on one strand of DNA is transcribed as a molecule of mRNA (with uracil replacing thymine). The triplet code on the mRNA is then translated at ribosomes as a series of amino acids.

The conversion of information encoded as DNA into the synthesis of a polypeptide chain occurs in two distinct phases (Figure 11.5). First the 'message' encoded in the DNA sequence of a gene is converted to mRNA by *transcription*; then this directs the assembly of a specific sequence of amino acids during *translation*. We shall discuss how this happens shortly, but first we need to consider the question: how does the sequence of nucleotides in a gene serve as an instruction in the synthesis of proteins?

> Messenger RNA (mRNA) is formed from a DNA template (transcription) and carries its encoded message to the ribosome, where it directs synthesis of a polypeptide (translation).

11.3.2 The genetic code

Messenger RNA carries information (copied from a DNA template) in the form of a *genetic code,* which directs the synthesis of a particular protein. The nature of this code was worked out in the early 1960s by Marshall Nirenberg, Har Gobind Khorana and others. The message encoded in a gene takes the form of a series of triplets (*codons*). Of the 64 possible three-letter com-

> Remember that RNA molecules have uracil (U) instead of thymine (T).

binations of A, C, G and U, 61 correspond to specific amino acids, while the remaining three act as 'stop' messages, indicating that reading of the message should cease at that point (Figure 11.6). It is essential that the reading of the message starts at the correct place, otherwise the *reading frame* (sets of three nucleotides) may become disrupted. This would lead to a completely inappropriate sequence of amino acids being produced. *Frameshift mutations*

UUU	Phe	UCU	Ser	UAU	Tyr	UGU	Cys
UUC	Phe	UCC	Ser	UAC	Tyr	UGC	Cys
UUA	Leu	UCA	Ser	UAA	**STOP**	UGA	**STOP**
UUG	Leu	UCG	Ser	UAG	**STOP**	UGG	Trp
CUU	Leu	CCU	Pro	CAU	His	CGU	Arg
CUC	Leu	CCC	Pro	CAC	His	CGC	Arg
CUA	Leu	CCA	Pro	CAA	Gln	CGA	Arg
CUG	Leu	CCG	Pro	CAG	Gln	CGG	Arg
AUU	Ile	ACU	Thr	AAU	Asn	AGU	Ser
AUC	Ile	ACC	Thr	AAC	Asn	AGC	Ser
AUA	Ile	ACA	Thr	AAA	Lys	AGA	Arg
AUG	**Met**	ACG	Thr	AAG	Lys	AGG	Arg
GUU	Val	GCU	Ala	GAU	Asp	GGU	Gly
GUC	Val	GCC	Ala	GAC	Asp	GGC	Gly
GUA	Val	GCA	Ala	GAA	Glu	GGA	Gly
GUG	Val	GCG	Ala	GAG	Glu	GGG	Gly

Figure 11.6 The genetic code. Apart from methionine and tryptophan, all amino acids can be coded for by more than one triplet codon (for some, e.g. leucine, there may be as many as six). The code is thus said to be degenerate. Three of the triplet sequences are stop codons, and represent the point at which translation of the mRNA message must end. Translation always begins at an AUG codon, meaning that newly synthesised proteins always begin with a methionine residue. See Box 2.4 for full names of amino acids.

(see below) have this effect. Since there are only 20 amino acids to account for, it follows that the genetic code is *degenerate*, that is, a particular amino acid may be coded for by more than one triplet. Amino acids such as serine and leucine are encoded by as many as six triplets each, whilst tryptophan and methionine are the only amino acids to have just a single codon (see Figure 11.6).

Box 11.4   The genetic code is (almost) universal

The genetic code was first worked out using *E. coli*, but soon found to apply to other organisms too. It seemed reasonable to assume that the code was universal, i.e. applicable to all life forms. It has been shown, however, that certain genes such as those found in the mitochondria of some eukaryotes, employ a slight variation of the code. Mitochondria have their own transcription enzymes, ribosomes and tRNAs and so are able to use a modified system.

11.3.3   Transcription in prokaryotes

In the first phase of gene expression, one strand of DNA acts as a template for the production of a complementary strand of RNA. In the outline that follows, we shall describe how mRNA is synthesised, but remember that

Promoter   Terminator

RNA polymerase

mRNA

5'   3'

5'   3'

Figure 11.7   Transcription. (a) The gene to be transcribed is flanked by a promoter and a terminator sequence. (b) Following localised unwinding of the double-stranded DNA, one strand acts as a template for RNA polymerase to make a complementary copy of mRNA. (c) The mRNA is extended; only the most recently copied part remains associated with the DNA. (d) After reaching the termination sequence, both RNA polymerase and newly synthesised mRNA detach from the DNA, which reverts to its fully double-stranded state.

sometimes the product of transcription is rRNA or tRNA. An important point to note is that the coding strand is not the same for all genes; some are encoded on one strand, some on the other. Whereas in DNA replication the whole molecule is copied, an RNA transcript is made only of specific sections of DNA, typically single genes. The enzyme *RNA polymerase*, unlike DNA polymerase, is able to use completely single-stranded material, that is, no primer is required. It is able to synthesise an mRNA chain from scratch, according to the coded sequence on the template, and working in the 5' to 3' direction (Figure 11.7). In order to do this the RNA polymerase needs instructions for when to start and finish. Firstly, it recognises a short sequence of DNA called a *promoter*, which occurs upstream of a gene. A protein cofactor called *sigma* (σ) assists in attachment to this, and is released again shortly after transcription commences. The promoter tells the RNA polymerase where transcription should start, and also on which strand. The efficiency with which a promoter binds the RNA polymerase determines how effectively a particular gene will be transcribed. The promoter comprises two parts, one 10 bases upstream (known as the *Pribnow box*), and the other 35 bases upstream (Figure 11.8). RNA polymerase binds to the promoter, and the double helix of the DNA is caused to unwind a little at a time,

Figure 11.8 The promoter sequence in *E. coli*. RNA polymerase attaches at a point 35 nucleotides upstream of the start of transcription; as the DNA unwinds, it binds to the Pribnow box, situated at −10 nucleotides. The actual sequences may differ from gene to gene; the sequences shown in the figure are consensus sequences. Note: only the non-template DNA strand is shown.

exposing the coding sequence on one strand. Ribonucleotides are added one by one to form a growing RNA chain, according to the sequence on the template; this occurs at a rate of some 30–50 nucleotides per second. Because RNA has uracil rather than thymine, a 'U' is incorporated into the mRNA whenever an 'A' appears on the template. Transcription stops when a *terminator* sequence is recognised by the RNA polymerase; both the enzyme and the newly synthesised mRNA are released. Unlike the promoter sequence, the terminator is not transcribed. Some termination sequences are dependent on the presence of a protein called the rho factor (ρ). Groups of bacterial proteins having related functions may have their genes grouped together. Only the last one has a termination sequence, so a single, contiguous mRNA is produced, encoding several proteins (*polycistronic* mRNA).

Transcription in eukaryotes proceeds along similar lines, but with certain differences. The most important of these is that in eukaryotes, the product of transcription does not act directly as mRNA, but must be modified before it can undergo translation. This is because of the presence within eukaryotic genes of DNA sequences not involved in coding for amino acids. These are called *introns* (cf. coding sequences = *exons*), and are removed to give the final mRNA by a process of *RNA splicing* (Figure 11.9).

> Eukaryotic genes generally contain non-coding sequences (*introns*) in between the coding sequences (*exons*).

11.3.4 Translation

The message encoded in mRNA is translated into a sequence of amino acids at the ribosome. The ribosomes are not protein-specific; they can translate any mRNA to synthesise its protein. Amino acids are brought to the ribosome by a *transfer RNA* (tRNA) molecule. Each tRNA acts as an adaptor, bearing at one end the complementary sequence for a particular triplet codon, and at the other the corresponding amino acid (Figure 11.10). It recognises a specific codon and binds to it by complementary base pairing,

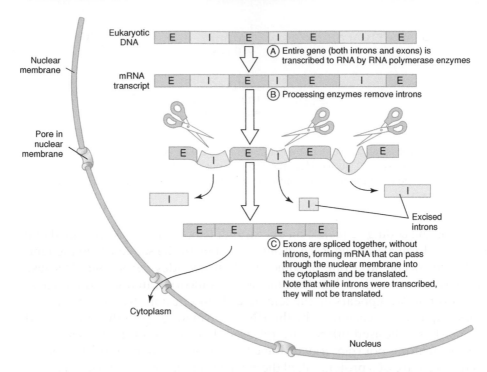

Figure 11.9 Eukaryotic genes contain non-coding sequences. The product of transcription (primary RNA transcript) cannot act as mRNA because it contains sequences that do not code for the final polypeptide. These introns must be removed and the remaining coding sequences (exons) spliced together to give the mature mRNA. Reproduced from Black, JG (1999) Microbiology: Principles and Explorations, 4th edn, with permission from John Wiley & Sons.

thus ensuring that the appropriate amino acid is added to the growing peptide chain at that point. Enzymes called *aminoacyl-tRNA synthetases* ensure that each tRNA is coupled with the correct amino acid in an ATP-dependent process.

There is at least one type of tRNA for each amino acid, each with a three-base *anticodon*, enabling it to bind to the complementary triplet sequence on the mRNA. However, there is not a different anticodon for each of the 61 possible codons, in fact there are fewer than 40. To explain this, the *wobble hypothesis* proposed that certain non-standard pairings are allowed between the third nucleotide of the codon and the first of the anticodon (Table 11.1). This means that a single anticodon may pair with more than one codon (Figure 11.11).

> The initial amino acid in the chain is always a *methionine*, corresponding to the AUG start codon. In prokaryotes there is a modified form called *formylmethionine* (fMet); a special tRNA carries it to the initiation site.

Translation starts when the small ribosomal subunit binds to a specific sequence on the mRNA upstream of where translation is to begin. This is the

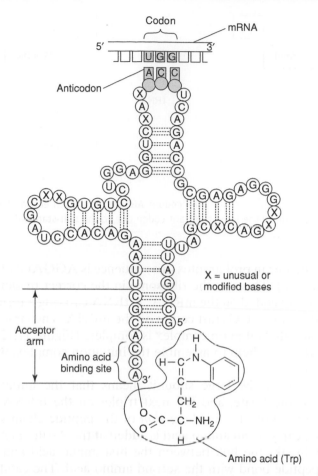

Figure 11.10 Transfer RNA. The single-stranded tRNA adopts its characteristic clover leaf appearance due to partial base pairing between complementary sequences. The molecule contains some modified nucleotides such as inosine and methylguanosine. Transfer RNA acts as an adaptor between the triplet codon on the mRNA and the corresponding amino acid (tryptophan in the example shown). It base pairs with the mRNA via a complementary anticodon, thus bringing the amino acid into position for incorporation into the growing peptide chain. Reproduced from Black, JG (1999) Microbiology: Principles and Explorations, 4th edn, with permission from John Wiley & Sons.

Table 11.1 The wobble hypothesis: permitted pairings and mispairings

First anticodon base	Third codon base
C	G
A	U
U	A or G
G	U or C
I (inosine)	U, C or A

Figure 11.11 The wobble hypothesis. The codons AGC and AGU each encode the amino acid serine. Both can be 'read' by the same tRNA anticodon due to the non-standard pairing allowed at the codon's final base.

ribosome binding site; in prokaryotes this sequence is AGGAGG (the *Shine–Dalgarno* sequence). This sets the ribosome in the correct reading frame to read the message encoded on the mRNA. A tRNA carrying a formylmethionine then binds to the AUG start codon on the mRNA. The large ribosomal subunit joins, and the initiation complex is complete (Figure 11.12). Proteins called *initiation factors* help to assemble the initiation complex, with energy provided by GTP.

The positioning of the large subunit means that the initiation codon (AUG) fits into the P-site, and the next triplet on the mRNA is aligned with the A-site (Figure 11.13). *Elongation* of the peptide chain starts when a second tRNA carrying an amino acid is added at the A-site. *Peptidyl transferase* activity breaks the link between the first amino acid and its tRNA, and forms a peptide bond with the second amino acid. The catalytic action is due partly to the ribozyme activity of the large subunit rRNA. The ribosome moves along by one triplet so that the second tRNA occupies the P-site. The first tRNA is released from its amino acid, and passes to the *E-site* before being released from the ribosome. A third aminoacyl-tRNA moves into the A-site, corresponding to the next codon on the mRNA. Elongation continues in this way until a STOP codon is encountered (UAG, UAA, UGA). Release factors cleave the polypeptide chain from the final tRNA and the ribosome dissociates into its subunits.

11.4 Regulation of gene expression

The proteins synthesised by microorganisms may be divided broadly into structural proteins required for the fabric of the cell, and enzymes, used to maintain the essential metabolic processes of the cell. It would be terribly wasteful (and possibly harmful) to produce all these proteins incessantly, regardless of whether or not the cell actually required them, so microorganisms, in common with other living things, have mechanisms of *gene*

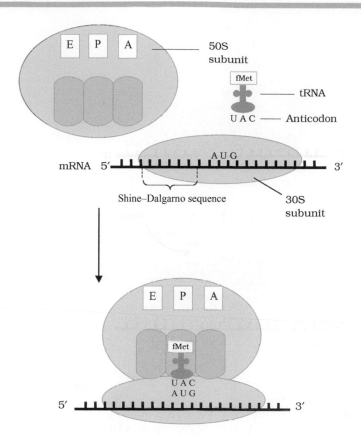

Figure 11.12 Translation: formation of the initiation complex. The small (30S) ribosomal subunit binds to the Shine–Dalgarno sequence, and a formylmethionine initiator tRNA attaches to the AUG codon just downstream. The large ribosomal subunit then attaches in a process requiring energy from GTP, and initiation factors are released.

regulation, whereby genes can be switched on and off according to the cell's requirements.

11.4.1 Induction of gene expression

The synthesis of many enzymes required for the catabolism (breakdown) of a substrate is regulated by enzyme *induction*. If the substrate molecule is not available in the environment, there is little point in synthesising the enzyme needed to break it down. In terms of conservation of cellular energy, it makes more sense for such enzymes to be produced only when they are needed, that is, when the appropriate substrate molecule is present. The substrate

Enzymes whose production can be switched on and off are called *inducible enzymes*. This distinguishes them from *constitutive enzymes*, which are always produced regardless of prevailing conditions.

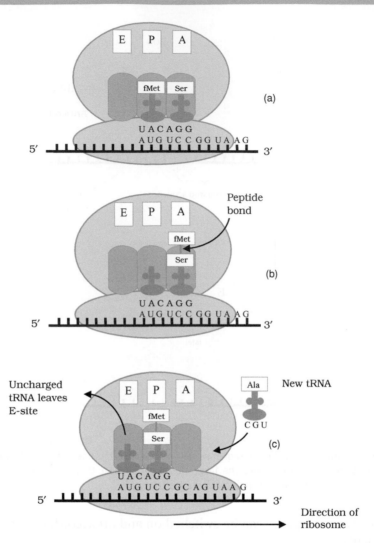

Figure 11.13 Translation: elongation of the peptide chain. (a) Transfer RNA enters the A-site, carrying the amino acid corresponding to the next triplet on the mRNA (serine in the diagram). (b) Peptidyl transferase removes the formylmethionine from its tRNA and joins it to the second amino acid. (c) The ribosome moves along the mRNA by one triplet, pushing the first uncharged tRNA into the E-site, and freeing up the A-site for the next tRNA to enter.

itself therefore acts as the *inducer* of the enzyme's synthesis. An example of this that has been studied in great depth concerns the enzyme β-galactosidase, used by *E. coli* to convert the disaccharide lactose into its constituent sugars:

$$\text{Lactose} \xrightarrow{\quad\quad} \text{Glucose} + \text{Galactose}$$
$$\textbf{β-galactosidase}$$

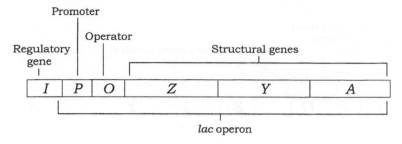

Figure 11.14 The *lac* operon. The *lac* operon comprises three structural genes under the control of a single promoter and operator sequence. The role of the regulatory gene *I* is described in Figure 11.15.

β-Galactosidase has been intensively studied as part of the *lac operon* of *E. coli*. This is made up of three structural genes designated *Z, Y* and *A*, which are clustered together and share a common promoter and terminator (Figure 11.14). The genes code for, respectively, β-*galactosidase*, a *permease* and a *transacetylase*. The permease is necessary for the transport of lactose into the cell, while the role of the transacetylase is not entirely clear, although it is essential for the metabolism of lactose. Grouping the three genes together in this way ensures an 'all-or-nothing' expression of the three proteins. Transcription of these structural genes into their respective mRNAs is initiated by the enzyme RNA polymerase binding to the promoter sequence. However, this is only possible in the presence of lactose; in its absence, a *repressor protein* binds to an operator site, adjacent to the promoter, preventing RNA polymerase binding to the promoter, and therefore preventing mRNA production. Production of the repressor protein is encoded by a *regulator gene* (*I*), situated slightly upstream from the operon (Figure 11.15a).

> A group of functionally related genes involved in the regulation of enzyme synthesis and positioned together at the same locus is called an *operon*. It contains both structural and regulatory genes.

> A *repressor protein* regulates the transcription of a gene by binding to its operator sequence. It is encoded by a regulatory gene.

How then, does the presence of lactose overcome this regulatory mechanism? Allolactose, an isomer of lactose and an intermediate in its breakdown, attaches to a site on the *lac* repressor, thereby reducing the latter's affinity for the operator, and neutralising its blocking effect (Figure 11.15b). The structural genes are then transcribed into mRNA, which is subsequently translated into the three proteins described above, and the lactose is broken down. In the absence of lactose, there are only trace amounts of β-galactosidase present in an *E. coli* cell; this concentration increases some 1000-fold in the presence of lactose. When all the lactose has been consumed, the repressor

(a)

RNA polymerase cannot
bind to its promoter

Repressor protein encoded
by regulatory gene *I*

| *I* | *P* | *O* | *Z* | *Y* | *A* |

(b)

RNA polymerase
binds to promoter

Allolactose binds to repressor,
preventing binding to operator

Transcription

| *I* | *P* | *O* | *Z* | *Y* | *A* |

Figure 11.15   The *lac* operon is inducible. (a) In the absence of the substrate lactose, the *lac* operon is 'switched off', due to a repressor protein encoded by the regulatory gene *I*. The repressor binds to the operator site, preventing the binding of RNA polymerase to the promoter and therefore blocking transcription. (b) Allolactose acts as an inducer by binding the repressor protein and preventing it from blocking the promoter site. Transcription of the three structural genes is able to proceed unhindered.

protein is free to block the operator gene once more, and the needless synthesis of further β-galactosidase ceases.

The *lac* operon can also be induced by isopropyl β-thiogalactoside (IPTG); *E. coli* is not able to break this down, so the genes remain permanently switched on. IPTG is utilised as an inducer in cloning systems involving the expression of the *lac*Z gene on pUC plasmids (see Chapter 12).

The *lac* operon is subject to control by positive as well as negative regulator proteins. Transcription of the operon only occurs if another regulatory protein called *catabolite activator protein* (CAP) is bound to the promoter sequence (see Box 11.5). This is dependent on a relatively high concentration of the nucleotide cAMP, which only occurs when glucose is scarce. The activation of the *lac* operon thus occurs only if lactose is present and glucose is (almost) absent. In the presence of both substrates, the lac operon is only switched on when the glucose has been utilised (diauxic growth; see Box 5.5).

11.4.2   Repression of gene expression

The induction of gene expression, such as we have just described for the *lac* operon, generally relates to catabolic (breakdown) reactions. Anabolic (synthetic) reactions, such as those leading to the production of specific amino acids, by contrast, are often controlled by the *repression* of key genes.

Box 11.5 A choice of substrates

Glucose is central to the reactions of glycolysis (see Chapter 6), and is utilised by *E. coli* with high efficiency, because the enzymes involved are permanently switched on, or *constitutive*. The β-galactosidase required for lactose breakdown, however, must be induced. What happens then, when *E. coli* is presented with a mixture of both glucose and lactose? It would be more efficient to metabolise the glucose, with the ready-to-use enzymes, but from what you've learnt elsewhere in this section (see Figure 11.15b), the presence of lactose would induce formation of β-galactosidase and subsequent lactose breakdown, a less energy-efficient way of going about things. In fact, *E. coli* has a way of making sure that while the readily utilised glucose is present, it takes precedence. It does this by repressing the formation of β-galactosidase, a phenomenon known as *catabolite repression*. Thus the presence of a 'preferred' nutrient prevents the synthesis of enzymes needed to metabolise a less favoured one.

This is because glucose inhibits the formation of cyclic adenosine monophosphate (cAMP), which is required for the binding of the catabolite activator protein (CAP) to its site on the *lac* promoter. When glucose levels drop, more cAMP forms and causes CAP to bind to the CAP binding site. Thus, after a delay, the enzymes needed for lactose catabolism are synthesised, and the lactose is utilised, leading to a diauxic growth curve (see Chapter 5).

Enzyme repression mechanisms operate along similar lines to induction mechanisms, but the determining factor here is not the substrate of the enzymes in question (lactose in our example), but the end-product of their action. The *trp operon* contains a cluster of genes encoding five enzymes involved in the synthesis of the amino acid tryptophan (Figure 11.16). In the presence of tryptophan, the cell has no need to synthesise its own, so the operon is switched off. This is achieved by tryptophan binding to and activating a *repressor protein*, which in turn binds to the operator of the *trp* operon and prevents transcription of the synthetic enzymes. The tryptophan here is said to act as a *corepressor*. As tryptophan is used up and its level in the cell falls, the repressor reverts to its inactive form, allowing transcription of the tryptophan-synthesising enzymes to go ahead unhindered.

11.4.3 Global gene regulation

The systems of gene regulation discussed above apply to individual operons; sometimes, however, a change in environmental conditions necessitates the

(a)   Tryptophan absent

All five enzymes for tryptophan synthesis produced

(b)   Tryptophan absent

No enzymes produced for tryptophan synthesis

Figure 11.16   The *trp* operon. Five structural genes encode enzymes necessary for the synthesis of tryptophan. (a) In the absence of tryptophan, transcription of the operon proceeds unhindered. Although a repressor protein is produced, it is inactive and unable to bind to the operator sequence. (b) Tryptophan activates the repressor by binding to it. This prevents RNA polymerase from binding to the promoter, and transcription is blocked.

regulation of many genes at once. These global regulatory systems respond to stimuli such as oxygen depletion and temperature change, and utilise a number of different mechanisms.

11.5   The molecular basis of mutations

Any alteration made to the DNA sequence of an organism is called a *mutation*. This may or may not have an effect on the phenotype (physically manifested properties) of the organism. It may, for example, enable bacteria to grow without the need for a particular growth supplement, or confer resistance to an antibiotic. Just how this happened was for many years a source of debate. Since the mutant forms only became apparent after the change in conditions (e.g. withdrawal of a nutrient, addition of antibiotic), have some of the bacteria been *induced* to adapt to the new conditions, or are mutant forms arising all the time at a very low frequency, and are merely *selected* by the environmental change? In 1943, Salvador Luria and Max Delbrück devised the *fluctuation test* to settle the matter (see Box 11.6). Results of the fluctuation test together with other evidence led to the understanding that mutations occur *spontaneously* in nature at a very low frequency. As we shall see later on in this section, however, they can also be *induced* by a variety of chemical and physical agents. Any change to the DNA sequence is heritable, thus mutations represent a major source of evolutionary variation. Bacteria make marvellous tools for the study of mutations because of their huge numbers and very short generation times.

Since the DNA sequence of a gene represents highly ordered coded information, most mutations have a detrimental effect on the organism's phenotype, but occasionally a mutation occurs that confers an advantage to an organism, making it better able to survive and reproduce in a particular environment. Mutants that are favoured in this way may eventually become the dominant type in a population, and, by steps like this, evolution gradually takes place.

Mutations occur spontaneously in any part of an organism's genome. Spontaneous mutations causing an inactivation of gene function occur in bacteria at the rate of about one in a million for a given gene at each round of cell division. Most genes within a given organism show similar rates of mutation, relative to their gene size; clearly a larger 'target' will be 'hit' more often than a small one.

11.5.1 How do mutations occur?

Figure 11.17a reminds us how the code in DNA is transcribed into messenger RNA and then translated into a sequence of amino acids. Each time the DNA undergoes replication, this same sequence will be passed on, coding for the same sequence of amino acids. Occasionally, mistakes occur during replication. Cells have repair mechanisms to minimise these errors, but what happens if a mistake still slips through? In Figure 11.17b we can see the effect of one nucleotide being inserted into the strand instead of another. When the next round of replication occurs, the modified DNA will act as a template for a newly synthesised strand, which at this position will be made complementary to the new, 'wrong' base, instead of the original one, and thus the mistake will be perpetuated.

This is an example of the simplest type of mutation, a *point mutation*, where one nucleotide has been substituted by another. The example shown is a *missense mutation*, which has resulted in the affected triplet coding for a different amino acid; this may or may not have an effect on the phenotype of the organism. RNA polymerase, which transcribes the DNA

A *missense mutation* alters the sense of the message encoded in the DNA, and results in an incorrect amino acid being produced at the point where it occurs.

Box 11.6　Settling an argument: the fluctuation test

Luria and Delbrück designed the fluctuation test to show whether resistance in *E. coli* to the bacteriophage T1 was *induced* or occurred *spontaneously*.

Suppose we divide a broth culture of bacteria into two, then inoculate half into a flask of fresh broth, and divide the other half between a large number of smaller cultures in tubes.

E. coli culture

Half of culture divided between 50 small tubes.

Half of culture grown up in a single large flask.

1　2　3　→　50

50 samples plated out onto agar pates

Each sample plated out onto an agar plate.

1　　2　　3　→　50　　　　1　　2　　3　→　50

Colony counts quite similar　　　　Colony counts highly variable

After allowing the bacteria to grow, samples are taken from tubes and flask, and spread onto a selective agar medium (one covered with T1 phage). All the plates deriving from the single bulk culture have roughly the same number of resistant colonies, but the plates resulting from the smaller, tube cultures show very variable colony counts. Although the *average* number of colonies across the 50 tubes is similar to that obtained from the bulk culture, the *individual* plate counts vary greatly, from none on several plates to over 100 on another.

If the phage-resistance was *induced* by the presence of the phage, we would expect all plates in the experiment to give rise to approximately the same number of resistant colonies, as all cultures experienced the same exposure. If, however, the resistant forms were spontaneously arising all the time at a low level in the population, the numbers of resistant colonies would be dependent on when, if at all, a mutation had taken place in a particular tube culture. A mutant arising early in the incubation period would give rise to more resistant offspring and therefore more colonies than one that arose later.

Figure 11.17 Mutations can alter the sense of the DNA message. (a) A short sequence of DNA is transcribed into mRNA and then transcribed into the corresponding amino acids. (b) A single base change from A to T results in a *missense* mutation, as a histidine is substituted for a leucine. (c) A *silent* mutation has altered the DNA (and therefore mRNA) sequence, but has not changed its sense. Both AGG and AGA are mRNA triplets that code for arginine. (d) A *nonsense* mutation has replaced a tryptophan codon with a STOP codon, hence bringing the peptide chain to a premature end.

sequence into mRNA, is unable to tell that an error has occurred, and faithfully transcribes the misinformation. The machinery of translation is similarly 'unaware' of the mistake, and as a consequence, a different amino acid will be inserted into the polypeptide chain. The consequence of expressing a 'wrong' amino acid in the protein product could range from no effect at all to a total loss of its biological properties. This can be understood in terms of protein structure (see Chapter 2), and depends on whether the amino acid affected has a critical role (such as part of the active site of an enzyme), and whether the replacement amino acid has similar or different polar/nonpolar properties. You may recall from the beginning of this chapter that the genetic code is degenerate, and that most amino acids are encoded by more than one triplet; for example, both GAA and GAG code for glutamic acid. This means that some mutations do not affect the amino acid produced; such mutations are said to be *silent*, as in Figure 11.17c. These most commonly occur at the third nucleotide of a triplet.

Another type of point mutation is a *nonsense mutation*. Remember that of the 64 possible triplet permutations of the four DNA bases, three are 'stop' codons, which terminate a polypeptide chain. If a triplet is changed from a coding to a 'stop' codon as shown in Figure 11.17d, then instead of the whole coding sequence being read, transcription to mRNA will end at this point, and a truncated (and probably non-functional) protein will result.

> A *nonsense mutation* results in a 'stop' codon being inserted into the mRNA at the point where it occurs, and the premature termination of translation.

11.5.2   Mutations can add or remove nucleotides

Other mutations involve the *insertion* or *deletion* of nucleotides. This may involve anything from a single nucleotide up to millions. Deletions occur as a result of the replication machinery somehow 'skipping' one or more nucleotides. If the deletion is a single nucleotide, or anything other than a multiple of three, the ribosome will be thrown out of its correct reading frame, and a completely new set of triplet codons will be read (Figure 11.18). This is known as a *frameshift mutation*, and will in most cases result in catastrophic changes to the final protein product. If the deletion is a multiple of three nucleotides, the reading frame will be preserved and the effect on the protein less drastic.

> A *frameshift mutation* results in a change to the reading frame, and an altered sequence of amino acids results downstream of the point where it occurs.

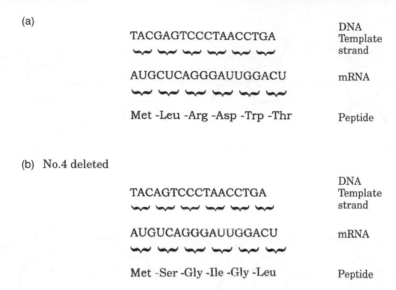

(a)

TACGAGTCCCTAACCTGA   DNA
                                            Template
                                            strand

AUGCUCAGGGAUUGGACU   mRNA

Met -Leu -Arg -Asp -Trp -Thr   Peptide

(b) No.4 deleted

TACAGTCCCTAACCTGA   DNA
                                            Template
                                            strand

AUGUCAGGGAUUGGACU   mRNA

Met -Ser -Gly -Ile -Gly -Leu   Peptide

Figure 11.18   Frameshift mutations. (a) A short sequence of DNA is transcribed into mRNA and then transcribed into the corresponding amino acids. (b) The fourth nucleotide in the sequence is deleted, upsetting the groups of triplets, or reading frame. This alters the sense of the remainder of the message, leading to a completely different sequence of amino acids.

11.5.3   Mutations can be reversed

Just as it is possible for a mutation to occur spontaneously, so it is possible for the nucleotide change causing it to be spontaneously reversed – in other words, a mutant can mutate back to being a *wildtype*. This is known for obvious reasons as a *reverse* or *back mutation*. When this happens, the original genotype and phenotype are restored. Whereas a forward mutation results from *any* change that inactivates a gene, a back mutation is more specific; it must restore function to a protein damaged by a specific mutation. Not surprisingly, given this specificity, the rate of back mutations is much less frequent.

It is possible for the wildtype phenotype to be restored, not through a reversal of the original base change, but due to a second mutation at a different location. The effect of this second mutation is to suppress the effects of the first one. These are called *suppressor* or *second site mutations*. They are double mutants that produce a *pseudowildtype*; the phenotype appears to be wildtype, but the genotype differs.

11.5.4   Mutations have a variety of mechanisms

Why are mistakes made every so often during DNA replication? One source of erroneous base incorporation is a phenomenon called *tautomerism*. Some

(a)

Thymine (keto)   Adenine (amino)   Thymine (enol)   Guanine (keto)

(b)

Cytosine (amino)   Guanine (keto)   Cytosine (imino)   Adenine (amino)

Figure 11.19   Rare forms of nucleotides may cause mispairing. If DNA replication takes place whilst a base is in its rare alternative form, an incorrect base will be incorporated into the newly synthesised second strand. (a) In its normal keto form, thymine pairs with adenine in keeping with the rules of base-pairing. In its rare enol form, however, it preferentially pairs with guanine. (b) Cytosine can likewise assume the rare imino form and mispair with adenine.

of the nucleotide bases occur in rare alternative forms, which have different base-pairing properties. For example, as you will recall from Chapter 2, cytosine normally has an amino group that provides a hydrogen atom for bonding with the complementary keto group of guanine. Occasionally, however (once in about every 10^4–10^5 molecules), the cytosine may undergo a rearrangement called a *tautomeric shift*, which results in the amino group changing to an imino group ($=NH$), and this now behaves in pairing terms as if it were thymine, and therefore pairs with adenine (Figure 11.19a). Similarly, thymine might undergo a tautomeric shift, changing its usual keto form ($C=O$) to the rare enol form (COH). This then takes on the pairing properties of cytosine and pairs with guanine (Figure 11.19b). The result of such mispairing is that at subsequent rounds of replication, one half of the DNA molecules will contain a wrong base pair at that point. The fact that spontaneous mutations occur much less frequently than the rate just quoted is due to the DNA repair mechanisms discussed earlier in this chapter.

11.5.5   Mutations also occur in viruses

As we saw in Chapter 10, the genome of viruses can be composed of either DNA or RNA, and these are subject to mutation just like cellular genomes. The rate of mutation in RNA viruses is much higher than in those containing DNA, about a thousand times higher, in fact. This is significant, because by frequently changing in this way, pathogenic viruses are able to stay one step ahead of host defence systems.

11.5.6   Mutagenic agents increase the rate of mutations

In the early part of the twentieth century, the existence of mutations was appreciated, but attempts to gain a fuller understanding of them were hampered by the fact that they occurred so rarely. In the 1920s, however, it was shown that the mutation rate in both barley plants and the fruit fly *Drosophila* was greatly increased as a result of exposure to *X-rays*. In the following decades, a number of chemical and physical agents were shown also to cause mutations. *Mutagens* such as these raise the general level of mutations in a population, rather than the incidence of mutation in a particular location in the genome. They are extremely useful tools for the microbial geneticist, but need to be treated with great care because they are also mutagenic (and in most cases carcinogenic) towards humans.

> Chemical or physical agents capable of inducing mutations are termed *mutagens*.

Chemical mutagens   Chemical mutagens can be divided into five classes:

1. *Base analogues*. A base analogue is able to 'mimic' one of the four normal DNA bases by having a chemical structure sufficiently similar for it to be incorporated into a DNA molecule instead of the normal base during replication. One such base is 5-bromouracil (5-BU); in its usual keto form, it acts like thymine (which it closely resembles – see Figure 11.20), and therefore pairs with adenine, and is not mutagenic. However 5-BU is capable of tautomerising to the enol form, which, you'll remember, pairs with G, not A. Whereas for thymine the enol form is a rarity, for 5-BU it is more common, hence mispairing with guanine occurs more frequently. So we get the same outcome as with spontaneous mutation – the T-A has been replaced by a C-G – but its frequency is much increased. Several other base analogues, such as 2-aminopurine, which mimics adenine, have similar effects to 5-BU. Mutations brought about by base analogues all result in a purine being replaced by another purine, or a pyrimidine being replaced by another pyrimidine; this type of mutation is called a *transition*.

Figure 11.20 Base analogues mimic DNA base structure. In its keto form, the base analogue 5-bromouracil (5-BU) forms base pairs with adenine, but in the enol form it mispairs with guanine. 5-BU differs from thymine only in the substitution of a bromine atom in place of a methyl group. This makes the occurrence of a tautomeric shift more likely.

2. *Alkylating agents.* As the name suggests, this group of mutagens act by adding an alkyl group (e.g. methyl, ethyl) at various positions on DNA bases. Ethylethanesulphonate (EES) and ethylmethanesulphonate (EMS) are alkylating agents, used as laboratory mutagens. They act by alkylating guanine or thymine at the oxygen atom involved in hydrogen bonding. This leads to an impairment of the normal base-pairing properties, and causes guanine, for example, to mispair with thymine. Substitutions brought about by alkylating agents can be either transitions or *transversions.*

3. *Deaminating agents.* Nitrous acid is a potent mutagen, which alters the base-pairing affinities of cytosine and adenine by replacing an amino group with an oxygen atom. Transitions occur in both directions, A-T → G-C and G-C → A-T. Note that although guanine is deaminated, its base pairing properties are not affected.

4. *Intercalating agents.* These mutagens exert their effect by inserting themselves between adjacent nucleotides in a single strand of DNA. They distort the strand at this site and cause either the addition or, less commonly, deletion of a nucleotide. This leads to frameshift mutations, and often during replication the bound mutagen interferes with new strand synthesis, leading to a gross deletion of nucleotides. *Ethidium bromide*, commonly used for the visualisation of small amounts of DNA in the molecular biology laboratory, is an intercalating agent; it has a planar structure with roughly the same dimensions as a purine-pyrimidine pairing.

5. *Hydroxylating agents.* Hydroxylamine has a specific mutagenic effect, hydroxylating the amino group of cytosine to cause the transition of GC → AT.

Base analogues and certain intercalating agents can only exert their mutagenic effects if they are incorporated into DNA while it is replicating. Others, which depend on altering base pairing by modifying the structure of DNA bases, are effective on both replicating and non-replicating DNA.

Physical mutagens   As mentioned at the start of this section, the first mutagenic agent to be demonstrated was X-radiation; along with ultraviolet light, *X-rays* are the most commonly used physical mutagen. Like other forms of ionising radiation, they cause the formation of highly reactive free radicals. These can bring about changes in base structure as well as gross chromosomal alterations. DNA strands can be broken and reannealed incorrectly, to produce errors in the sequence.

In recent years, we have all become much more aware of the possible perils of excessive exposure to sunlight. *Ultraviolet (UV) light* from the Sun damages DNA in skin cells, which can lead to them becoming cancerous. The specific action is on adjacent pyrimidine bases (usually thymines) on the same strand, which are cross-linked to form a *dimer* (Figure 11.21). This results in a distortion of the helix and interferes with replication. UV light is most effective at wavelengths around 260 nm, as this is the wavelength most strongly absorbed by the DNA bases.

11.5.7   DNA damage can be repaired

All organisms have developed the means to repair damage to their DNA, in addition to the proofreading mechanism described earlier.

Figure 11.21   Thymine dimer formation. Absorption of ultraviolet (UV) light results in the formation of dimers between adjacent thymine residues in the same strand. This can lead to errors being introduced into the DNA sequence. Reproduced from Gardner, EJ, et al. (1991) Principles of Genetics, 8th edn, with permission from John Wiley & Sons.

The most common way of dealing with mutations is by means of a process called *excision repair*, in which enzymes recognise and cut out the altered region of DNA and then fill in the missing bases, using the other strand as a template (Figure 11.22). *Mismatch repair* is used to repair the incorporation of a base that has escaped the proofreading system. In a situation such as this, it is not immediately obvious which is the correct base and which is the mistake, so how does the cell know which strand to replace? In *E. coli*, the old strand is distinguished from the new by the fact that some of its bases are methylated. This only occurs some time after replication, so newly synthesised strands will not have the methyl groups and can thus be recognised.

1.   ─── Mutation

2.   Nicks made in
                                                      phosphate-sugar
                                                      backbone either
                                                      side of mutation

3.   Affected
                                                      section
                                                      removed

4.   New strand
                                                      synthesised
                                                      with correct
                                                      sequence

Figure 11.22   Excision repair. A section of the affected strand either side of the mutation is removed and replaced by the action of DNA polymerase and DNA ligase.

Less frequently, the alteration in DNA structure is simply reversed by *direct repair* mechanisms, the best known of which is *photoreactivation*. This involves an enzyme called *DNA photolyase*, which breaks the bonds formed between adjacent thymine bases by UV light (see above) before replication takes place. Unusually, the photolyase is dependent on visible light (>300 nm) for activation. Another form of direct repair, involving the enzyme methylguanine transferase, allows the reversal of the effects of alkylating agents.

Loss of repair mechanisms can allow mutations to become established that would normally be corrected. A harmful strain of *E. coli* that emerged in the 1980s was shown to have developed its pathogenicity due to a deficiency in its repair enzymes.

11.5.8   Carcinogenicity testing: the Ames test

The great majority of *carcinogenic* substances, that is, substances that cause cancer in humans and animals, are also *mutagenic* in bacteria. This fact has been used to develop an initial screening procedure for carcinogens; instead of the expensive and time-consuming process of exposing laboratory animals (not to mention the moral issues involved), a substance can be tested on bacteria to see if it induces mutations.

The *Ames Test* assesses the ability of a substance to cause reverse mutations in auxotrophic strains of *Salmonella* that have lost the ability to synthesise the amino acid histidine (His−). Rates of back mutation (assessed by the ability to grow in a histidine-free medium) are compared in the presence and absence of the test substance (Figure 11.23). A reversion to His+ at a rate higher than that of the control indicates a mutagen. The strain of *Salmonella* used is also deficient in DNA repair enzymes. This rules out the possibility of a mutation going unnoticed because it was subsequently repaired (false negative). Many substances are procarcinogens, only becoming mutagenic/carcinogenic after metabolic conversion by mammals; in order to test for these, an extract of rat liver is added to the experimental system as a source of the necessary enzymes.

11.6   Genetic transfer in microorganisms

The various mechanisms of mutation described above result in an alteration in the genetic make-up of an organism. This can also occur by *recombination*, in which genetic material from two cells combines to produce a variant different to either parent cell.

In bacteria, this involves the transfer of DNA from one cell (the *donor*) to another (the *recipient*). Because transfer occurs between cells of the same generation (unlike the genetic variation brought about by sexual reproduction in eukaryotes, where it is passed to the next generation), it is sometimes

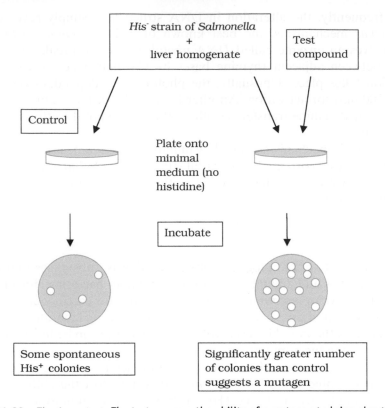

Figure 11.23 The Ames test. The test assesses the ability of a mutagen to bring about reverse mutations in a strain of *Salmonella typhimurium* auxotrophic for histidine. Since it is possible that a small number of revertants may occur due to spontaneous mutation, results are compared with a control plate with no added mutagen.

referred to as *horizontal transfer*. There are three ways in which gene transfer can occur in bacteria, which we shall now explore.

11.6.1   Transformation

Transformation is the simplest of these, and also the first to have been described. We have already referred to the classic experiment of Fred Griffith in 1928, the first demonstration that genetic transfer can occur in bacteria. Griffith had previously demonstrated the existence of two strains of the bacterium *Streptococcus pneumoniae*, which is one of the causative agents of pneumonia in humans, and is also extremely virulent in mice. The *S (smooth)-form* produced a polysaccharide capsule, whilst the *R (rough)-form* did not. These produced recognisably different colonies when grown on a solid medium, but more importantly, differed in their ability to bring about disease in experimental animals. The R-form, lacking the protective capsule,

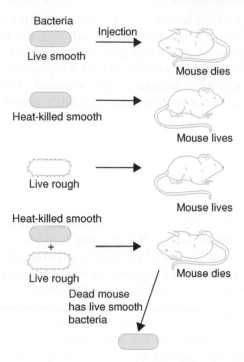

Figure 11.24 Transformation: the Griffith experiment. See text for details. The result of the fourth experiment showed for the first time that genetic material could be passed from one bacterium to another. Reproduced from Reece, RJ (2003) Analysis of Genes and Genomes, with permission from John Wiley & Sons.

was easily destroyed by the defence system of the host. Griffith observed the effects of injecting mice with bacterial cells of both forms; these are outlined here and in Figure 11.24. The results of experiments 1–3 were predictable, that of experiment 4 was startlingly unexpected:

1. Mice injected with live cells of the S-form died, and large numbers of S-form bacteria were recovered from their blood.
2. Mice injected with live cells of the R-form were unaffected.
3. Mice injected with cells of the S-form that had been killed by heating at 60°C were likewise unharmed and no bacteria were recovered from their blood.
4. Mice injected with a mixture of living R-form and heat-killed S-form cells died *and living S-form bacteria were isolated from their blood.*

The S-form bacteria recovered from the mice in the crucial fourth experiment possessed a polysaccharide capsule like other S-forms, and, critically, *were able to pass on this characteristic to subsequent generations.* This finding went against the prevailing view that bacteria simply underwent binary fission, a completely asexual process involving no genetic transfer. Griffith deduced that some as yet unknown substance had passed from the heat-killed S-form cells to some of the living R-forms and conferred on them the ability to make capsules. Not long afterwards it was shown that this process of *transformation* could happen in the test tube, without the involvement of a host animal, and, as we've seen, it was eventually shown that Griffith's 'transforming principle' was DNA.

Box 11.7 Transformation is not due to reverse mutation

At the time of Griffith's experiment, it was already known that the wild-type S-form could mutate to the R-form and vice versa. It might there-fore be argued that the results of his fourth experiment could easily be explained in this way. Griffith's experiment was sufficiently well designed to refute this argument, however, because the bacteria he used were of two different serotypes, meaning that they produced different *types* of capsule (types II and III). Even if the IIR-form cells had mutated back to the wildtype (IIS), they could not have produced the type III capsule Griffith observed in the bacteria he recovered from the mice. The only explanation is that the ability to make type III capsules had been acquired from the heat-killed type III cells.

11.6.2 How does transformation occur?

The uptake of foreign DNA from the environment is known to occur natu-rally in a number of bacterial types, both Gram-positive and Gram-negative, by the acquisition of fragments of *naked DNA* released from dead cells in the vicinity. Being linear and very fragile, the DNA is easily broken into frag-ments, each carrying on average around ten genes. Transformation will only happen at a specific stage in the bacterial life cycle, when cells are in a physio-logical state known as *competence*. This occurs at different times in different bacteria, but is commonly during late log phase. One of the reasons why only a low percentage of recipient cells become transformed is that only a small proportion of them are at any one time in a state of competence. The expres-sion of proteins essential to the transformation process is dependent on the secretion of a *competence factor*.

The exact mechanism of transformation varies somewhat according to species; the process for *Bacillus subtilis* is shown in Figure 11.25. Mere uptake of exogenous DNA is not enough to cause transformation; it must also be integrated into the host genome, displacing a single strand, which is subse-quently degraded. Upon DNA replication and cell division, one daughter cell

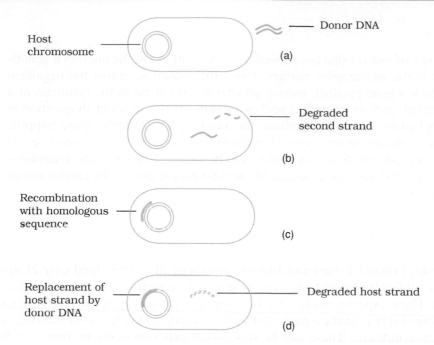

Figure 11.25 Transformation in *Bacillus subtilis.* (a) A fragment of donor DNA becomes bound to the recipient cell surface by means of a DNA-binding protein. (b) After binding, a nuclease contained at the cell surface degrades one strand of the donor DNA, leaving the other strand to be ferried, by other transformation-specific proteins, to the interior of the cell. (c) A fragment of single-stranded DNA aligns with a homologous stretch on the recipient chromosome. (d) The donor fragment becomes integrated by a process of nonreciprocal recombination. At the next cell division, one daughter cell is a transformant, whilst the other retains the parental genotype.

will inherit the parent genotype, and the other the recombinant genotype. In the latter, there is a stable alteration of the cell's genetic composition, and a new phenotype is expressed in subsequent generations.

Uptake of donor DNA from an unrelated species will not result in transformation; this is due to a failure to locate a homologous sequence and integrate into the host's chromosome, rather than an inability to gain entry to the cell.

11.6.3 Induced competence

Transformation is not thought to occur naturally in *E. coli,* but, if subjected to certain treatments in the laboratory, cells of this species can be made to take up DNA, even from a completely unrelated source. This is done by effecting a state of induced or artificial competence, and by introducing the foreign DNA in a self-replicating *vector* molecule, which does not depend on integration into the host chromosome. As we shall see in Chapter 12, this is of enormous significance in the field of genetic engineering.

Box 11.8   *Auxotrophic mutants*

A type of mutant that has proved to be of great use to the microbial geneticist is the *auxotrophic* mutant. Here, the mutation causes the organism to lack a gene product, usually an enzyme, involved in the synthesis of a nutrient such as an amino acid or vitamin. If the nutrient in question is supplied in the culture medium, the auxotroph can survive quite happily, as its deficiency is not exposed. If the nutrient is not provided, however, as in a *minimal medium*, the cells would be unable to grow. Thus microbiologists can detect the existence of an auxotrophic mutant by careful use of selective media.

11.6.4   Conjugation

In 1946, Edward Tatum and Joshua Lederberg (the latter aged only 21 and still a student!) demonstrated a second form of genetic transfer between bacteria. These experiments involved the use of *auxotrophic* mutants, which have lost the ability to make a particular enzyme involved in the biosynthesis of an essential nutrient. These can be recognised experimentally by their inability to grow on a medium lacking the nutrient in question (Figure 11.26). The results of Tatum and Lederberg's experiments suggested that a process was taking place in bacteria akin to sex in eukaryotes, in which there was a recombination of the cells' genetic material (Figure 11.27). Transformation, as demonstrated by Griffith, could not explain their results, since the addition of a DNA-containing extract of one strain to whole cells of the other did not

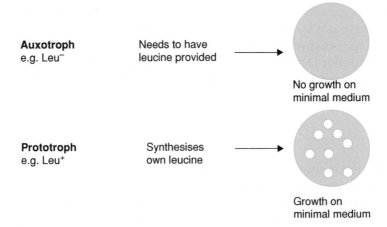

Figure 11.26   Auxotrophs. Auxotrophic mutants provide useful genetic markers. They are unable to synthesise a particular nutrient, and can be detected by their inability to grow on a minimal medium (one containing only inorganic salts and a carbon source such as glucose).

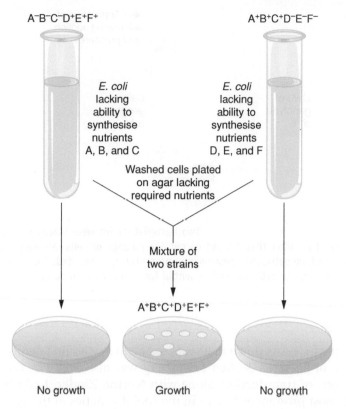

A⁻B⁻C⁻D⁺E⁺F⁺

A⁺B⁺C⁺D⁻E⁻F⁻

E. coli
lacking
ability to
synthesise
nutrients
A, B, and C

E. coli
lacking
ability to
synthesise
nutrients
D, E, and F

Washed cells plated
on agar lacking
required nutrients

Mixture of
two strains

A⁺B⁺C⁺D⁺E⁺F⁺

No growth   Growth   No growth

Figure 11.27   Tatum and Lederberg: sexual recombination in bacteria. Two strains with complementary genotypes (auxotrophic for certain genes, prototrophic for others) are each unable
to grow on a minimal medium. When the two are mixed, however, colonies are formed, indicating that recombinant bacteria prototrophic for all the genes have been formed. The bacteria
are washed before plating out to avoid the carry-over of nutrients from the initial broth to the
minimal medium. Reproduced from Black, JG (1999) Microbiology: Principles and Explorations,
4th edn, with permission from John Wiley & Sons.

result in prototroph formation. This was confirmed in an ingenious experiment in which Bernard Davis showed that Tatum and Lederberg's results
were only obtained if direct cell-to cell contact was allowed (Figure 11.28).

11.6.5   Gene transfer in conjugation is one way only

The process of conjugation was initially envisaged as a fusion of the two partner cells to give a diploid zygote, which subsequently underwent meiosis to
give haploid offspring with modified genotypes. The work of William Hayes,
however, showed that the development of colonies of recombinant cells was
dependent on the survival of only one of the participating strains, the other
strain being required only as a donor of DNA.

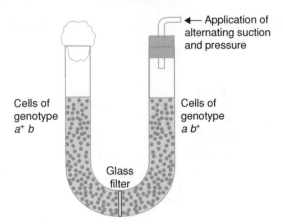

Figure 11.28 Davis's U-tube experiment. Two bacterial strains were placed in two arms of a U-tube, separated by a filter that did not permit the passage of cells. Although suction was used to transfer medium between compartments, no prototrophs resulted. The results provided direct evidence that cell-to-cell contact is essential for conjugation to occur.

Box 11.9   It's that man again!

Not content with going down in history as the man who first demonstrated conjugation in bacteria, Joshua Lederberg was also, in 1952, one of the co-discoverers of transduction, along with Norton Zinder. Lederberg was one of the most prominent figures in microbial genetics in the second half of the twentieth century. He was even responsible for coining the term 'plasmid'!

It became apparent that in *E. coli*, there are two distinct mating types, which became known as F^+ and F^-, depending on whether or not they possessed a plasmid called the *F (fertility) plasmid*. This contains some 30 or 40 genes responsible for its own replication and for the synthesis of a thread-like structure expressed on the cell surface called a *sex pilus*. In a mixture of F^+ and F^- cells, the sex pilus contacts an F^- cell, then contracts, to pull the two cells together. A single strand of plasmid DNA is then passed across a channel made between the two cells, and enters the F^- cell (Figure 11.29). This then serves as a template for the production of the complementary second strand, and similarly the single strand left behind in the donor cell is replicated, leaving us with a double-stranded copy of the F plasmid in both cells. By acquiring the F plasmid, the recipient (F^-) cell has been converted to F^+.

In some cells the F plasmid has become integrated into the main bacterial chromosome and thus loses its ability to replicate independently (Figure 11.30). These are known as *Hfr* (high frequency of recombination) cells. When they are involved in conjugation, genes from the main bacterial

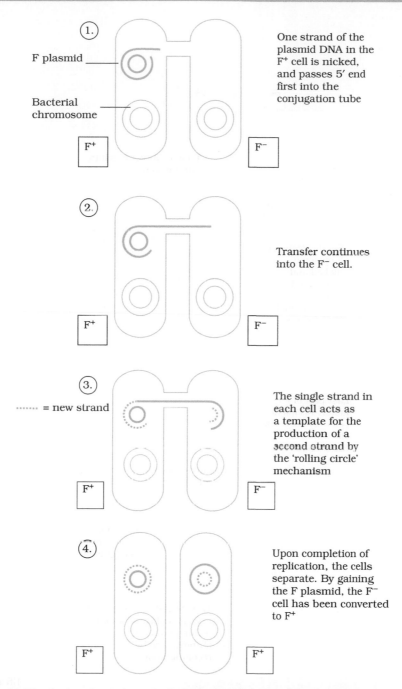

Figure 11.29 Conjugation in bacteria. A single-stranded copy of the F plasmid passes across a conjugation tube from an F⁺ to an F⁻ cell. Both this and the copy left behind act as templates for their own replication, leaving both cells with a complete F plasmid. This means that the recipient cell is converted from F⁻ to F⁺.

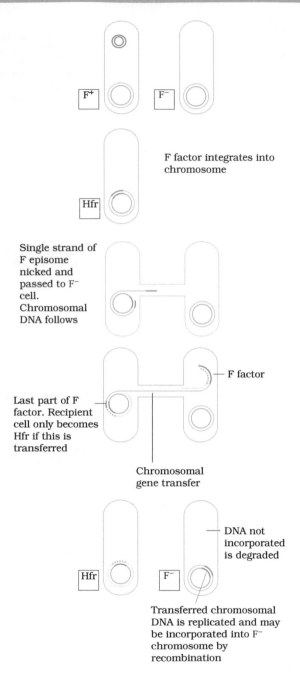

Figure 11.30 Conjugation with an Hfr cell results in transfer of chromosomal genes. **Hfr** cells are formed by the integration of the F factor into the bacterial chromosome. During conjugation, transfer begins part of the way along the F episome, and continues with chromosomal DNA. The amount of chromosomal material transferred depends on how long conjugation is able to proceed. Conjugation may be followed by recombination of transferred chromosomal material with its homologous sequence on the recipient cell's chromosome.

chromosome may be transferred to a recipient cell. The integrated F plasmid behaves just like any other part of the chromosome, although of course it still carries the genes for conjugation and pilus formation. When conjugation occurs, a single strand of Hfr DNA is broken within the F sequence, and is transferred in a linear fashion, carrying behind it chromosomal DNA. Recombination in the recipient cell results in the replacement of a homologous segment of DNA by the transferred material, which is then faithfully replicated in subsequent generations. If transfer is uninterrupted, a process that takes around 100 minutes in *E. coli*, eventually the remaining stretch of F plasmid will enter the F$^-$ cell, bringing up the rear. The fragile nature of the pilus means, however, that transfer is rarely complete, and only a limited portion of the bacterial genome is transferred. This means that those F$^-$ cells receiving DNA from Hfr cells usually remain F$^-$, unlike those in a cross with F$^+$ cells, because the remainder of the F sequence is not transferred.

It soon became clear that this phenomenon afforded a great opportunity to determine the relative positions of genes on the bacterial chromosome. This was done by *interrupted mating* experiments, in which the time allowed for conjugation is deliberately limited by mechanical breakage of the sex pili, and correlated with the phenotypic traits transferred to the recipient cells (Figure 11.31). By these means a *genetic map* of the bacterial chromosome could be developed.

The integration of the F plasmid into the bacterial chromosome is reversible; thus Hfr cells can revert to F$^+$. Excision of the integrated plasmid is not always precise, and sometimes a little chromosomal DNA is removed too. When this happens, the plasmid and the cell containing it are called *F'* (pronounced 'F prime'); transfer of the plasmid to an F$^-$ cell takes with it the extra DNA from the host chromosome. The recipient genome thus becomes partially diploid (*merodiploid*), because for certain genes it has its own copy, plus the 'guest' copy.

11.6.6 Transduction

In the third form of genetic transfer in bacteria, bacteriophages act as carriers of DNA from one cell to another. In order to appreciate the way in which this is done, it is necessary to recall the sequence of events in phage replication cycles discussed in Chapter 10 (see Figure 10.11).

Generalised transduction occurs in virulent phages, that is, those with a lytic life cycle. Sometimes, the enzymes responsible for packaging phage DNA into its protein coat package similarly sized fragments of degraded chromosomal DNA instead (Figure 11.32). Despite containing the wrong DNA, this *transducing phage* particle retains its infectivity, since this is dependent on its protein element. Thus following infection of another bacterial cell, the DNA can be incorporated by recombining with the homologous segment in the recipient cell. Since any chromosomal fragment can be mistakenly

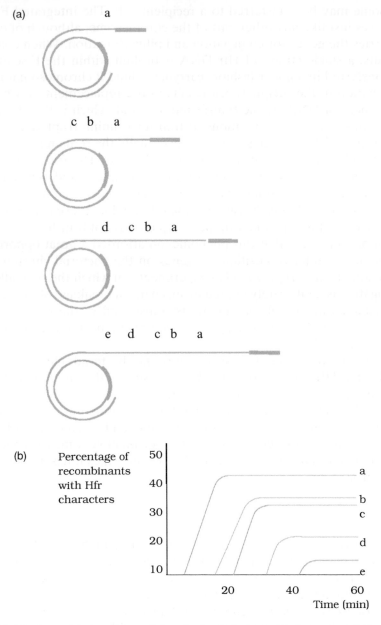

Figure 11.31 Interrupted mating experiments allows the bacterial chromosome to be mapped. (a) The conjugation tube is broken after different time periods, and by plating out onto selective media, the order in which the different genes are transferred can be determined. The time at which different genes are transferred reflects their relative positions on the bacterial chromosome. (b) The graph shows that the first genes to enter the F⁻ cell are those present in the highest proportion of recombinants.

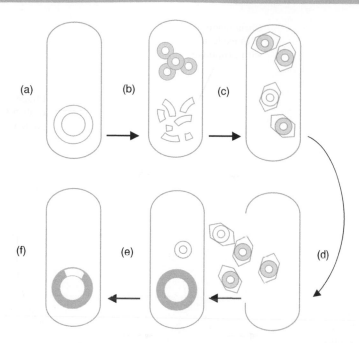

Figure 11.32   Generalised transduction. (a–d) During the lytic cycle of a phage, the host DNA is degraded, and a fragment may be mistakenly packaged into a newly synthesised phage particle. Upon infecting a new host cell (e), the transducing phage releases its DNA; although unable to replicate, this can undergo recombination with a homologous sequence on the host chromosome (f).

packaged in this way (as long as it finds an area of homology), all genes are transferred at a similar (low) frequency.

Specialised transduction results in a much higher efficiency of transfer for specific genes; however, it is limited to genes having a particular chromosomal location. Recall from Chapter 10 that in lysogenic life cycles, the phage DNA is integrated into the host chromosome, and later, perhaps after many rounds of cell division, excised again before re-entering a lytic cycle. If this excision does not happen precisely, some of the adjoining chromosomal DNA, carrying a gene or two, may be incorporated into the phage particle (we saw a similar mechanism in the case of F′ plasmid formation). Upon infecting another cell, the transduced genes would undergo recombination and become incorporated into the recipient's chromosome (Figure 11.33). Although limited to genes in the vicinity of the lysogenic phage's integration, this is a highly efficient form of transfer, since the genes become stably integrated into the host cell.

Transduction experiments, like those involving conjugation, can be used to determine the relative positions of genes on a bacterial chromosome.

Chromosomal
DNA excised
with prophage

Phage
replication and
packaging

Integrated
prophage

Integration of phage
and chromosomal
genes into new host
chromosome

Cell lysis and
release of phages

Figure 11.33   Specialised transduction. During the replication cycle of a lysogenic bacteriophage, phage DNA is incorporated into the host chromosome (see Figure 10.11). When a lytic
cycle resumes and the phage DNA is excised, it may take some surrounding chromosomal DNA
with it. This is packaged into phage particles and infects new host cells, where it is integrated
into the bacterial chromosome. Only genes surrounding the site of phage integration may be
transduced in this way.

11.6.7   Transposable elements

An unusual type of genetic transfer that takes
place at a very low frequency *within* an individual cell involves sequences of DNA called *transposable elements*. These are sections of DNA
that can move from one chromosomal location
to another, and carry the genetic information
necessary to do this. *Insertion sequences* (IS),
are relatively short pieces of chromosomal or
plasmid DNA. The only coding DNA that they

> Transposable   elements
> that also carry genes
> other than those required
> for transposition, such
> as genes for antibiotic
> resistance or toxins, are
> known as *transposons*.

contain is a gene for the enzyme *transposase* (Figure 11.34), which recognises, cuts and re-ligates the insertion sequence anywhere in the bacterial
genome. In so doing, it may interrupt a gene sequence, and thereby inactivate it. Unlike recombination events, no homology is required between the

Figure 11.34 Transposable elements. The insertion sequence IS*1* of *E. coli* is 768 bp in length and is flanked by 23 bp inverted repeat sequences. The IS contains the gene for a transposase enzyme, which catalyses the movement of the insertion sequence from one location to another. By integrating at random points in the genome where there is no sequence homology, IS sequences may disrupt functional genes and give rise to mutations. IR = inverted repeat.

transposable element and the point at which it inserts. This relocation of a transposable element from one place in the genome to another is termed *conservative transposition* ('cut and paste'). In *replicative transposition* ('copy and paste'), the element remains in its original position and a copy is made and inserted elsewhere in the genome. Insertion sequences are flanked by inverted sequences some 9–41 bp in length, which are thought to be essential for the recognition of the sequence by the transposase.

12
Microorganisms in Genetic Engineering

In the last few decades, there has been a revolution in the field of genetics, which has had a profound effect on virtually every other area of biology. This has been due to the development of new techniques that have enabled scientists to analyse and manipulate DNA in a quite unprecedented way. Genetically modified crops, DNA profiling and gene therapy are just three of the many applications made possible by these advances. The subject of 'genetic engineering' is too huge to be discussed here in detail, and indeed it extends into areas far beyond the remit of this book. In this chapter, however, we shall examine some of the ways that microorganisms have contributed to the genetic revolution. As we shall see, their role in the development of new techniques of DNA manipulation since the 1970s has been just as important as their earlier contribution to the elucidation of the structure, role and replication of DNA decades earlier.

The beginnings of genetic engineering can be said to date from the discovery, in the late 1960s, of a class of bacterial enzymes called *restriction endonucleases* (REs). These cut DNA into pieces by making breaks in the sugar-phosphate backbone; in nature, they serve to destroy any foreign DNA that may enter the cell. They do not fragment the DNA in a random fashion, however; their unique usefulness to the molecular biologist lies in the fact that they break the DNA *in a precise and reproducible manner*. They do this by cutting only at specific *recognition sites*, sequences of typically four to six nucleotides (Figure 12.1). Thus, under favourable

> Restriction endonucleases do not destroy the host bacterium's own DNA, because certain nucleotides in the recognition sequence are modified by *methylation*. The REs are unable to cleave the DNA at methylated sites.

Essential Microbiology, Second Edition. Stuart Hogg.
© 2013 John Wiley & Sons, Ltd. Published 2013 by John Wiley & Sons, Ltd.

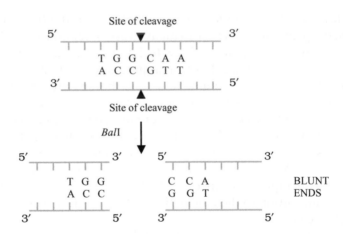

Figure 12.1   Restriction endonucleases cleave DNA at specific nucleotide sequences. They fragment DNA molecules by breaking the sugar-phosphate backbone within a specific sequence of nucleotides. Depending on the site of cleavage, the restriction fragments so produced may be blunt-ended or 'sticky'-ended.

conditions, a particular RE will digest a given piece of DNA in the same places and into an identical collection of fragments, time after time. In the ensuing years, many hundreds of restriction endonucleases have been discovered, many of which recognise different specific sequences, providing biologists with a hugely versatile set of tools for the manipulation of DNA, often likened to pairs of 'molecular scissors'. Not long after REs were first isolated, they were used to create the first synthetic *recombinant DNA* molecule (Figure 12.2). This involved cutting fragments of DNA from different sources, then using DNA ligase to join them together, a process facilitated by using fragments with compatible 'sticky' ends. Remember from

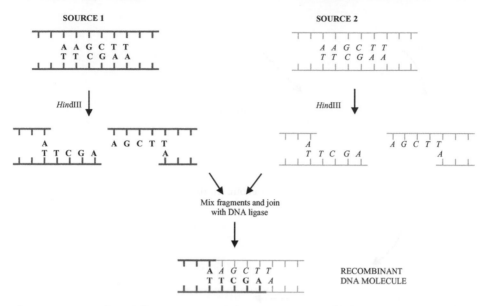

Figure 12.2 DNA from different sources can be joined together. 'Sticky'-ended restriction fragments from one DNA source have single-stranded sequences that are compatible with fragments produced from another source by the same restriction endonuclease (RE). Compatible base pairing attracts the fragments together and the join is made more permanent by the action of DNA ligase.

Chapter 11 that A always pairs with T and C with G; because of this, complementary sequences that come into contact with one another will 'stick' together. DNA, it seems, is DNA, wherever it comes from; consequently DNA from plants, animals, bacteria or viruses can be joined together to create novel sequences undreamed of by Mother Nature.

Of course, a single molecule of our newly recombinant DNA is not much use to us. The important breakthrough came with the development of *cloning* – the ability to produce huge numbers of copies of a given molecule. To do this, two further things are needed: a carrier DNA molecule called a *vector*, and a *host cell* in which it can be replicated.

> *Cloning* is the production of multiple copies of a specific DNA molecule. The term is also used to describe the production of genetically identical cells or even organisms.

Figure 12.3 shows the main steps of a cloning protocol:

- 'Donor' DNA and vector are digested with an RE to provide compatible sticky ends.
- A fragment of donor DNA is spliced into the vector molecule.
- The recombinant vector gains entry to a host cell (usually a bacterium).

Figure 12.3 Gene cloning. DNA fragments obtained by digestion with a restriction endonuclease (RE) can be spliced into a similarly digested vector molecule and transformed into a host bacterial cell. See text for details.

- The vector replicates inside the cell, making further copies of the inserted DNA.
- Host multiplication results in the formation of a clone of cells, all containing the same recombinant plasmid – we now have millions of copies of our donor DNA 'insert'. A collection of such clones is called a *DNA library*. If the original donor DNA was the total DNA of an organism, the library is called a *genomic library* – see later.

Let's look at the role of vectors in a little more detail. The main features required of a cloning vector are:

> A *vector* is a self-replicating DNA molecule used in gene cloning. The sequence to be cloned is inserted into the vector, and replicated along with it.

- *It must be capable of replicating autonomously inside a host cell* – when it does so, any DNA it carries will also be replicated. Vectors make multiple copies of themselves inside the host cell.

- *It must be relatively small* – to facilitate manipulation and entry into a host cell, vectors must not exceed a certain size.

- *It must carry a selectable marker* – since only a proportion of host cells will take up the vector, there must be a means of differentiating them from those that do not. A common way to do this is to use a vector that carries a gene that confers resistance to an antibiotic such as ampicillin. When bacterial cells are

> A *selectable marker* is a gene that allows cells containing it to be identified by the expression of a recognisable characteristic.

plated out on a medium containing the antibiotic, only those that have taken up the vector will be able to form colonies. (The untransformed host strain must, of course, be one that is normally susceptible to the antibiotic.)

- *It must carry a single copy of RE restriction sites* – in order to accommodate a piece of donor DNA, a vector must be cut by a restriction endonuclease in one place only (see Figure 12.3).

12.1 Plasmid cloning vectors

Two main types of vector system are used in cloning, those based on *bacteriophages* and those based on *plasmids* (revisit Chapters 10 and 11, respectively, for a reminder of the main features of these). Naturally occurring examples of these are manipulated so that they possess the above properties. A popular vector in the early days of gene cloning was the plasmid pBR322; Figure 12.4 shows how it satisfies the requirements described above.

Let us consider now what happens, at a molecular level, when donor DNA is ligated into a plasmid vector (Figure 12.5). Sticky ends of the donor fragment form hydrogen bonds with the exposed compatible ends of the opened-up plasmid by complementary base pairing, and DNA ligase consolidates the join. Many plasmids contain engineered sequences called *multiple cloning sites* (MCS) or *polylinkers*; these contain the recognition sequences for several REs, and provide additional flexibility with respect to the restriction fragments that may be accommodated.

The ligation of insert DNA into the cloning site of the plasmid is the desired outcome of the procedure described above; however, it is not the only one. Unless experimental conditions are carefully controlled (there are

Figure 12.4   Plasmid pBR322. One of the earliest plasmid vectors, pBR322 illustrates the major features required for use in gene cloning: an origin of replication (*ori*), selectable markers – genes for resistance to ampicillin (*amp*^{R}) and tetracycline (*tet*^{R}) – and single recognition sites for a number of restriction endonucleases (*Sca*I, *Pvu*I, *Pst*I, *Eco*RI, *Hind*III, *Bam*HI and *Sal*I).

ways of doing this), a more likely outcome is that the two compatible ends of the plasmid will simply 'find' each other again, and rejoin. Since a certain amount of this is inevitable, how are we able to tell the difference between those bacteria that contain a recombinant plasmid (one containing a piece of donor DNA) and those that have taken up a recircularised 'native' plasmid? Since both types will contain the gene for ampicillin resistance, we cannot distinguish them by this means. A strategy commonly used to get around the problem is *insertional inactivation* (Figure 12.6). This clever ploy exploits the fact that we can manipulate DNA, and, for example, insert RE recognition sequences at desired points. If a recognition site occurs in the middle of a gene sequence, and a piece of foreign DNA is inserted at this position, the gene will be interrupted, and unable to produce a functional gene product (i.e. it will be *inactivated*). In the example shown, the gene is *lacZ'*, necessary for the successful expression of the enzyme β-galactosidase. This will only be expressed in those bacteria that contain plasmids in which the gene has remained uninterrupted, that is, those that have *not* taken up an insert. Recombinant plasmids will not be able to express the gene. We can detect whether or not the β-galactosidase has been expressed by growing the bacteria on an artificial substrate, which is converted to a coloured (usually blue) product when acted on by the enzyme. Those bacteria that contain recombinant plasmids are easily identified because disruption of the *lacZ'* gene

The *lacZ'* gene actually only encodes a part of the β-galactosidase enzyme, called the *α-peptide*. The strain of *E. coli* used as host makes an incomplete version of the enzyme, which lacks this portion. Only if the cells contain the plasmid with the *lacZ'* gene can they produce functional β-galactosidase, by *α-complementation*.

Recognition sites
for restriction
endonuclease (RE)

Donor
DNA

Digestion with
RE fragments
donor DNA and
opens up vector

.....G
.....CCTAG

GATCC.....
G.....

Compatible 'sticky ends'

Formation of
recombinant vector
molecule. DNA
ligase consolidates
the join by repairing
the phosphate-sugar
backbone

Figure 12.5 Formation of a recombinant plasmid. **A recombinant plasmid is formed when a
fragment of foreign DNA is taken up and ligated into the plasmid. By cleaving both plasmid and
foreign DNA with the same restriction endonuclease (RE), compatible 'sticky' ends are created,
facilitating the join.**

means that no β-galactosidase is produced, and the colonies formed therefore remain non-pigmented (white).

One problem remains. Remember that our inserted DNA was derived
from the digestion of total (genomic) DNA from the donor organism; this
means that our DNA library will contain recombinant plasmids with a whole

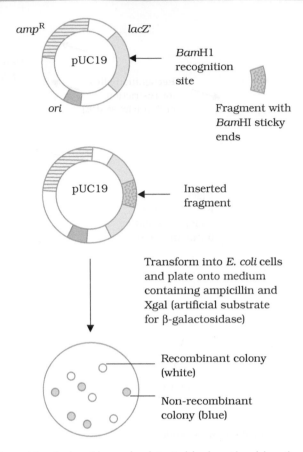

Figure 12.6 Recombinant plasmids can be detected by insertional inactivation. Insertion of foreign DNA is carried out at a site within one of the selectable markers, thus interrupting its gene sequence. Here, a fragment has been inserted into the *Bam*HI site situated within the gene that codes for β-galactosidase. Bacteria transformed with such a plasmid will not produce the functional enzyme, and so can be distinguished from those carrying plasmids with no inserted DNA.

range of fragments from that digestion, and not just the specific one that interests us. How are we able to distinguish this fragment from the others?

A technique called *nucleic acid hybridisation* is used to solve the problem. This once again depends on complementary base pairing, and involves the creation of a *probe*, a short length of single-stranded DNA that is complementary to part of the desired 'target' sequence, and therefore able to seek it out. If searching for the right clone can be likened to looking for a needle in a haystack, then the probe is a powerful 'magnet' that makes it much easier to find the needle. The probe carries a tag or label, so that its location can be determined (Figure 12.7).

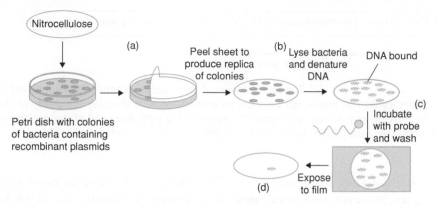

Figure 12.7 Colony probing. (a) A replica of the bacterial colonies is made using a nitrocellulose membrane. (b) Alkali treatment lyses the cells and denatures the DNA, making it single-stranded. (c) Following a period of incubation with a radiolabelled probe to allow hybridisation to take place, the membrane is washed to remove any unbound radioactivity. (d) Position of bound radioactivity is revealed by autoradiography. Comparison with the original master plate reveals the location of colonies carrying specific target sequence. Alternative detection systems that avoid the use of radioactivity such as biotin-streptavidin are now more commonly used. Reproduced from Reece, RJ (2003) Analysis of Genes and Genomes, with permission from John Wiley & Sons.

Once we have identified the clone of bacteria containing plasmids with the insert that interests us, we can grow a pure culture of it and then isolate plasmid DNA. Using the same RE as before, the inserted donor DNA can be removed and purified. We now have enough of this specific DNA sequence (a tiny proportion of the donor organism's total genome) to analyse and manipulate.

Although plasmids are easily isolated and manipulated, their use as cloning vectors is limited by the fact that they tend to become unstable if we attempt to insert much more than about 10 kb of foreign DNA. For inserts larger than this, we must turn to other vector systems.

12.2 Bacteriophages as cloning vectors

The most commonly used bacteriophage vectors are those based on *phage lambda* (λ) (Figure 12.8). The genome of phage λ is 48.5 kb in length; it was the first genome to have its entire sequence determined (in 1982). As drawn conventionally, the genome is linear, and contains 46 genes. At each 5′ end of the double-stranded structure is a 12-base single-stranded sequence known as a *cos* site; these have a complementary base sequence (i.e. they are cohesive, or 'sticky' ends). They can join together to form a double-stranded circular λ molecule, a conformation that is essential for insertion and integration into the host genome (see Chapter 10). The naturally occurring phage is unsuitable as a vector, because being relatively large, it contains multiple copies

Non-essential genes

Figure 12.8 Bacteriophage λ. The central region of the λ genome contains genes involved in its lysogenic cycle. This region can therefore be excised, without affecting the ability of the phage to infect *E. coli* via the lytic cycle. The genes at the right-hand end of the genome as shown have a regulatory function, whilst the structural genes are situated at the left.

of recognition sites for a number of REs. By using genetic manipulation techniques, however, unwanted sites can be removed, allowing the DNA to be 'opened up' at a single location, and fragments of foreign DNA with complementary sticky ends to be ligated.

A further limitation that must be overcome is that in this form, phage λ can only accommodate about another 3 kb of DNA. This is because if the genome exceeds 52 kb, it cannot be packaged properly into the protein head to produce viable phage particles. The arrangement of genes on the phage λ genome offers a solution to this problem. It is known that genes encoding specific functions are clustered together, with genes necessary for the lytic replication cycle being found at the ends of the map as it is conventionally drawn (see Figure 12.8). It is possible, therefore, to remove much of the central part of the λ genome without affecting its ability to infect and lyse its bacterial host.

Insertion vectors have had some of this non-essential material removed, reducing their genome size to around 42 kb, and thus allowing them to carry an insert of up to 10 kb. A single-copy restriction site is opened up, and a fragment with complementary sticky ends inserted with the aid of DNA ligase. Vectors such as λZAPII have a multiple cloning site or polylinker, positioned so that it falls within the *lacZ'* gene. This allows insertional inactivation with blue/white selection to be used for the detection of recombinants as described above for pUC plasmids.

Replacement vectors (Figure 12.9) are able to accommodate larger inserts, because instead of a single copy of a particular RE site, they have two; one situated either side of a disposable central section known as the 'stuffer' fragment. This can be removed by digestion with the appropriate RE and replaced by insert DNA. The vector λEMBL3, which can accommodate inserts of between 9 and 23 kb, has two polylinker sequences containing sites for the restriction endonucleases designated *Sal*I, *Bam*HI and *Eco*RI flanking the stuffer fragment, allowing a variety of restriction fragments to be inserted.

Recombinants in replacement vector systems can be detected by a method that exploits the limitations in the phage's ability to package DNA. Just as too big a genome can't be packaged (see above), neither can one that is too small (<37 kb); consequently, constructs lacking an insert will not result in the formation of plaques.

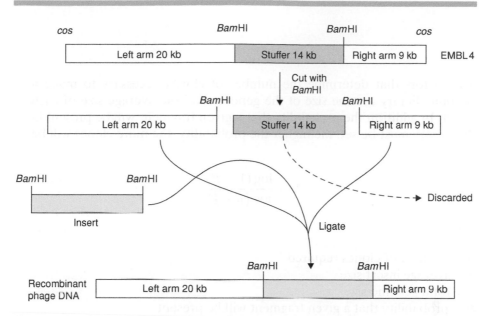

Figure 12.9   Cloning in bacteriophage λ. The removal of non-essential genes in the centre of the genome allows the incorporation of large fragments of foreign DNA. Reproduced from Dale, JW et al. (2002) From Genes to Genomes: Concepts and Applications of DNA Technology, with permission from John Wiley & Sons.

Another naturally occurring bacteriophage of *E. coli* that has been adapted for use as a cloning vector is *phage M13*. This is a single-stranded phage that for part of its replication cycle inside the host cell exists as a double-stranded replicative form, which can be isolated and manipulated like a plasmid. Systems based on M13 have proved popular as a means of obtaining DNA in a single-stranded form for applications such as DNA sequencing using the dideoxy (Sanger) method. M13 vectors contain a multiple cloning site situated within the *lacZ'* gene, allowing blue/white selection of recombinants. The cloning capacity is generally limited to fragments of less than 1.5 kb, but hybrid M13/plasmid vectors (*phagemids*) have been developed that are able to take inserts of up to 10 kb.

As we have seen, a large collection of cloned DNA fragments is called a DNA library. If the original donor DNA comprised the entire genome of an organism, then the collection of clones, which between them contain an entire donor genome, is known as a *genomic library*. We could create a library of the entire *E. coli* genome (total size 4.8×10^6 bp), for example, by having just over 700 clones with an average insert size of 20 kb (see Box 12.1 for calculation). For a more complex genome, however, we would either need to have very many more clones, or to increase the average fragment size. For this reason, further types of cloning vectors have been developed, to allow the cloning of larger fragments. *Cosmids* are entirely artificial creations, and have no naturally occurring counterpart. They incorporate beneficial features

Box 12.1   A large number of clones are required to cover an entire genome

The factors that determine the number of clones necessary to make a genomic library are the size of the genome and the average size of each insert. In addition the probability that a given fragment will be present in the library must be set: the higher the probability, the more clones will be necessary.

$$N = \frac{\log (1 - P)}{\log \left(1 - \frac{a}{b}\right)}$$

where:

N = number of clones required
a = average insert size
b = total genome size
P = probability that a given fragment will be present

Example

Calculate the number of clones required for a 95% probability of a given fragment to be present in a genomic library of *E. coli* (genome size 4.8×10^6) with an average insert size of 20 kb:

$$\begin{aligned} N &= \log (1 - 0.95) / \log (1 - 4.167 \times 10^{-3}) \\ &= \log 0.05 / \log (0.9958) \\ &= 712 \text{ clones} \end{aligned}$$

This may seem a manageable number, but remember that the genome of *E. coli* is very small compared to that of higher organisms (e.g. *Homo sapiens*: 3×10^9 bp).

of both plasmids and phage vectors, and may be capable of accommodating insert fragments of more than 45 kb. Cosmids are essentially plasmids that contain the *cos* sites from λ phage, with the plasmid supplying the necessary origin of replication, restriction sites and selectable marker. As we've seen, λ DNA will be packaged into phage heads if the *cos* sites are 37–52 kb apart; however, the only parts of the DNA recognised by the packaging enzymes are the *cos* sites, so any DNA can be used to fill the intervening sequence. The recombinant DNA is packaged into phage particles by a process called *in vitro packaging*, and introduced into the host *E. coli*. Lacking phage genes, cosmids do not lead to lysis of the host cells and plaque production, but are instead replicated as if they were plasmids. Growth of host cells on a selective medium allows the detection of transformants, namely those that have taken up the cosmid. There is no need to select for recombinants, because non-recombinants are too small to be packaged into the phage heads.

12.3 YACs, BACs and PACs

We saw in the preceding section how the use of cosmid vectors extended the size of clonable inserts to around 45 kb. Many eukaryotic genes, however, are bigger than this, and so cannot be cloned in a single sequence using such a system. In addition, the ability to clone large DNA fragments is essential for the physical mapping of complex eukaryotic genomes such as that of humans (3×10^9 bp). *Yeast artificial chromosomes* (YACs) have been developed to accommodate insert fragments of up to 1 Mb (1000 kb). They contain key sequences from a yeast chromosome that ensure its stable replication; these include an origin of replication (the autonomous replication sequence, *ARS*) the *CEN* sequence from around the centromere, and the telomeres at the end of the chromosome. When placed in *Saccharomyces cerevisiae* cells, the presence of these sequences allows the YAC to replicate along with the natural chromosomes. The fragment to be cloned is inserted between the two arms of the YAC (Figure 12.10). Each arm carries a selectable marker; it is important to have two, to ensure that each construct comprises both a right and a left arm, and not two of the same. Insertional inactivation of a third selectable marker (situated around the point of insertion) allows the detection of recombinants.

> The *centromere* is the central region of the chromosome that ensures correct distribution of chromosomes between daughter cells during cell division, whilst *telomeres* are important in preserving the stability of the tips of each chromosome arm.

Bacterial artificial chromosomes (BACs) allow the cloning of fragments as large as 300 kb in length into *E. coli* host cells, although 100–150 kb is more routinely used. They are based on the naturally occurring F plasmid of *E. coli*; recall from our description of bacterial conjugation in Chapter 11 that the F plasmid can pick up considerable lengths of chromosomal DNA, when it becomes known as F''. BACs have the advantage over YACs that they are easier to manipulate, and inherently more stable. *Phage P1-derived artificial chromosomes* (PACs) are another class of vector developed for use in *E. coli*, with a comparable capacity to that of BACs.

12.4 Expression vectors

Sometimes, the aim of a cloning experiment is not just to obtain large amounts of a specific gene, but for the gene to be expressed. This involves using the host cell as a sort of 'factory', to manufacture the specific protein

Figure 12.10   Yeast artificial chromosomes (YACs) are used to clone large fragments of DNA. Cleavage with the restriction endonucleases (REs) *Bam*HI and *Sna*BI leads to the formation of two arms, between which fragments of foreign DNA of several hundred kilobase may be ligated. The final construct contains a centromere, telomere sequences and an origin of replication, and is able to act like a natural chromosome. Reproduced from Brown, TA (1995) Gene Cloning, 3rd edn, with permission from Taylor & Francis Group.

encoded by the cloned gene. One of the earliest applications of genetic engineering was the production of human insulin in *E. coli* (Figure 12.11). Insulin is needed in considerable quantities for the treatment of diabetics; for years it was obtained from the pancreas of pigs and cattle, but this had several disadvantages including immunological complications and the risk of viral contamination. Insulin generated by recombinant means is free from these problems. Many proteins can now be produced in this way by microorganisms at a rate several times that of the normal host cell.

In order for a gene to be expressed, it must have specific nucleotide sequences around it that act as signals for the host cell's transcription/

The artificial genes

Synthesis of insulin protein

Figure 12.11 Production of recombinant human insulin. Human insulin comprises two short peptide chains, 21 and 30 amino acid residues in length. Because they are so short, the nucleotide sequence of their genes can be predicted from their amino acid sequence, and synthetic genes produced. Each is cloned under the influence of the *lac* promoter and downstream of part of the *lacZ'* gene. By being produced as fusion proteins with β-galactosidase, the insulin chains are protected against being degraded by the *E. coli* cell.

translation machinery (promoter, ribosomal binding site and terminator – see Chapter 11). Since these sequences differ between, say, humans and *E. coli*, the bacterial RNA polymerase will not recognise the human sequences, and therefore although a human gene may be cloned in *E. coli* using a simple

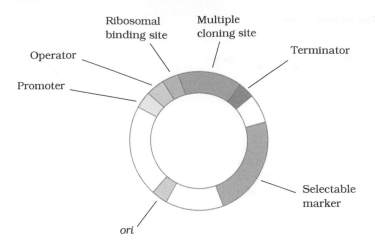

Figure 12.12 A cassette vector. In order for a foreign gene to be transcribed and translated successfully, it must be surrounded by the appropriate expression sequences. In a cassette vector, these sequences flank the restriction endonuclease (RE) recognition sites at which the foreign DNA is inserted. Reproduced from Reece, RJ (2003) Analysis of Genes and Genomes, with permission from John Wiley & Sons.

vector, it will not be expressed. If, however, the human gene could be inserted so that it was under the control of the *E. coli* expression signals, then transcription should take place. Specially designed vectors that provide these signals are called *expression vectors*. The choice of promoter sequence is particularly important; often, a strong (i.e. very efficient) promoter is selected, so as to maximise the amount of protein product obtained. Genes whose protein products are naturally synthesised in abundance are likely to have such promoters. It is often helpful to be able to regulate gene expression; *inducible* promoters can be switched on and off by the presence of certain substances. The *lac* promoter (which controls the *lacZ* gene) is an example of this. *Cassette vectors* have promoter, ribosomal binding site and terminator sequences clustered together as a discrete unit, with a single recognition site for one or more REs being situated just downstream of the promoter (Figure 12.12).

The small size of the insulin molecule (and gene) and the size of the potential market made it a prime early target for production by recombinant DNA technology. Most insulin used in the treatment of diabetes nowadays is produced in this way. Systems based on *E. coli* have also been used to synthesise other small human proteins with therapeutic potential such as human growth hormone, γ-interferon and tumour necrosis factor (TNF).

> The polypeptide products of translation may require *post-translational modification* before functional protein is produced. Examples include phosphorylation, acetylation and glycosylation.

Bacteria, however, are not suitable host cells for the production of many other human proteins such tissue plasminogen activator (TPA) or blood clotting factor VIII, due to the size and complexity of their genes. This is because many proteins of complex eukaryotes are subject to *post-translational modifications;* this does not occur in prokaryotes, so bacteria such as *E. coli* are not equipped with the cellular machinery to make the necessary modifications to human proteins.

Another obstacle to the cloning of such proteins concerns a fundamental difference in the way that prokaryotic and eukaryotic systems convert the message encoded in DNA to messenger RNA (see Chapter 11). Prokaryotes lack the means to remove *introns*, so if a human gene, for example, is expressed, the whole of the primary transcript will be translated, instead of just the coding sequences, leading to a non-

> *Complementary DNA (cDNA)* is DNA produced using mRNA as a template, using the enzyme reverse transcriptase. It represents only the coding sequences of the parent DNA molecule.

functional protein. This problem can be circumvented by cloning not the entire gene, but its *cDNA*, that is, just those DNA sequences that are transcribed into mRNA and subsequently translated into amino acid sequences. This can be done by isolating mRNA, then using reverse transcriptase to make a DNA copy. In the case of proteins such as insulin, the very small size enabled artificial genes to be synthesised, based on their known amino acid sequences.

12.5 Eukaryotic cloning vectors

A number of problems are associated with the expression of complex eukaryotic genes in bacterial cloning systems, some of which have just been outlined. In order to avoid these, such genes may be expressed in eukaryotic cloning systems, the model system for which is the yeast *S. cerevisiae*. This holds many attractions for the experimenter:

- it is safe and easy to handle in the laboratory;
- it grows at higher density than *E. coli*;
- it grows much faster than other eukaryotic cells in culture (e.g. cultured mammalian cells).

Several types of vector have been developed for use in yeasts. These are summarised in Table 12.1. The choice of which type to use will depend on how important are factors such as yield and long-term stability to a particular application. Most yeast vectors are *shuttle vectors*, that is, vectors that can be replicated in more than one type of host cell (in this case, yeast and *E. coli*). Initial cloning is more easily done in *E. coli*, then the recombinant vector is transferred to yeast cells, which can carry out functions such as protein

Table 12.1 Characteristics of yeast cloning vectors

Plasmid type	Abbreviation	Features
Yeast episomal plasmid	YEp	Contains 2 μm circle (naturally occurring yeast plasmid) origin of replication and whole of pBR322 sequence. Can integrate into yeast chromosomal DNA or replicate independently. Multiple copies per cell
Yeast integrative plasmid	YIp	Bacterial plasmid containing yeast selectable marker. Stably integrates into yeast chromosome. Unable to exist independently in yeast. Low transformation rate and single copy number
Yeast replicative plasmid	YRp	Carries yeast chromosomal origin of replication (*ARS*). High copy number. Unstable
Yeast centromere plasmid	YCp	Contains centromere sequence – ensures stability, but low (single) copy number

folding and glycosylation that are not possible in bacteria. Shuttle vectors must contain selectable markers and an origin of replication (or means of integration) for both cell types.

12.6   Viruses as vectors in eukaryotic systems

Several proteins of clinical or commercial interest are too complex to be able to express using microbial host cells, even eukaryotic ones, and are only properly produced in mammalian systems. Vectors based on animal viruses such as SV40, adenoviruses and vaccinia virus have been successfully developed for use in these. Vaccinia has been particularly valuable in the development of recombinant vaccines.

Viruses of humans such as adenoviruses and retroviruses have also been tested as vectors in the exciting technique of *gene therapy*, which attempts to ameliorate the effects of genetic disorders by introducing the 'correct' form of the defective gene into the patient's cells. Here it is important to ensure stable integration of the inserted DNA into the host chromosome.

A large virus that infects insects, the *baculovirus*, has been found to be a highly efficient vector for the large-scale expression of eukaryotic proteins in cultured insect cells. The rate of expression is much higher than in cultured mammalian cells, and the necessary protein folding and post-translational modifications are correctly executed.

12.7 Cloning vectors for higher plants

The most important single tool for the genetic engineering of plants is the *Ti plasmid*. This is found naturally in the soil bacterium *Agrobacterium tumefaciens,* which infects plants at wound sites, and leads to a condition called *crown gall disease*. The important feature of this plasmid is that part of it, called the *T-DNA*, can integrate into the host plant's chromosomes, and be expressed along with host genes (Figure 12.13). Geneticists were quick to spot the potential of the Ti plasmid, replacing tumour-forming genes with foreign genes, and having them expressed in plant tissues. The recombinant *A. tumefaciens* is used to infect protoplasts, which can be

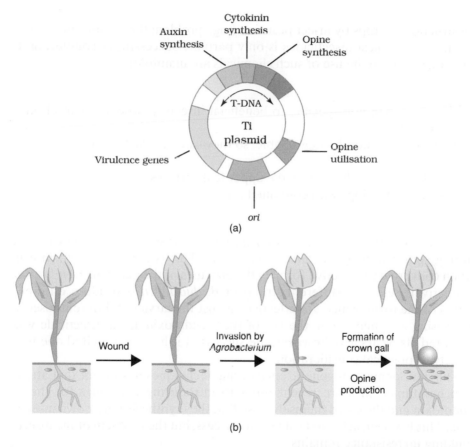

(a)

(b)

Figure 12.13   *Agrobacterium tumefaciens* and the Ti plasmid. (a) The T-DNA contains genes for tumour production and the synthesis of opines, unusual amino acid derivatives that serve as nutrients for *A. tumefaciens*. (b) Crown gall formation by *A. tumefaciens*. The T-DNA integrates into the DNA of the host, where its genes are expressed. This ability has been exploited in order to transfer foreign genes into plant cells. Reproduced from Reece, RJ (2003) Analysis of Genes and Genomes, with permission from John Wiley & Sons.

regenerated into a whole plant, every cell of which will contain the integrated foreign gene. The Ti plasmid system has been used in the successful transfer of genes for resistance to insects and herbicides into economically significant crop plants.

Plant viruses have very limited usefulness as vectors. Only the *caulimoviruses* and the *geminiviruses* have DNA rather than RNA as their genomic material, and a variety of problems, including instability of inserts and narrow host range, have been encountered with the use of these.

12.8   Applications of gene cloning in the microbial world

Destruction of crops by insect pests is a huge problem throughout the world, the use of chemical insecticides is only partially successful in countering it. The drawbacks to the use of such chemicals are manifold:

- they are often non-specific, so beneficial insects as well as harmful ones are killed;
- they are often non-biodegradable, so they can have lasting environmental effects;
- aerial spraying only reaches the upper leaf surfaces;
- resistance develops with continued use.

A form of natural insecticide does exist; it is a crystalline protein produced during sporulation by *Bacillus thuringiensis*. This is highly toxic to insects when converted to its active form by the enzymes of the gut. This δ-*endotoxin* is relatively selective; different strains of *B. thuringiensis* produce different forms of the toxin, which are effective against the larvae of different insects. Surprising as it may seem, the use of the δ-endotoxin as an insecticide was patented 100 years ago; however, the success of it has been limited due to a variety of practical considerations.

The development of genetic engineering techniques has meant that attention has turned in more recent times to introducing the genes for the δ-endotoxin into the crops themselves, so that they synthesise their own insecticide. This has been achieved with some success, but the problem of the insects building up resistance remains.

As a spin-off from the Human Genome Project, the genetic make-up of many microorganisms has been elucidated. One of these is *Photorhabdus luminescens*, which encodes a toxin lethal to the two species of mosquito responsible for the spread of malaria, and it is hoped that determining the sequence of the gene will lead to applications in insect control.

12.9   DNA microarrays

DNA microarrays utilise an extension of hybridisation technology that permits the simultaneous screening of tens of thousands of DNA sequences. The technique has a number of applications, such as comparing how a gene is expressed in different tissues, individuals or species, or detecting subtle differences in the same sequence among a large number of individuals or species. A specific application in the microbial world enables investigators to determine the degree of microbial diversity in an environmental sample by using a commercially manufactured 'chip' only a few square centimetres in size, which carries immobilised segments of 16S RNA genes from over a million bacterial taxa. The DNA from the sample under investigation is applied, and a fluorescence detection system identifies those sequences to which it has hybridised. An advantage of the system over other methods of assessing microbial diversity is that it identifies all species that give a positive hybridisation, rather than just the dominant members of a mixed population. In an early study using this method, samples of air taken over two US cities were shown to have a level of microbial diversity almost as great as that found in soil.

12.10   Polymerase chain reaction (PCR)

First described in the mid-1980s by Kary Mullis, PCR is probably the most significant development in molecular biology since the advent of gene cloning. PCR allows us to amplify a specific section of a genome (e.g. a particular gene) millions of times from a tiny amount of starting material (theoretically a single molecule!). The impact of this powerful and highly specific process has been felt in all areas of biology and beyond. Medicine, forensic science and evolutionary studies are but three areas where PCR has opened up new possibilities over the last 20 years or so.

> The *polymerase chain reaction (PCR)* selectively replicates a specific DNA sequence by means of in vitro enzymatic amplification.

To appreciate how PCR works, you will need to understand the role of the enzyme DNA polymerase in DNA replication, as outlined in Chapter 11. This is the enzyme, you'll recall, that when provided with single-stranded DNA and a short primer, can direct the synthesis of a complementary second strand. Figure 12.14 illustrates the three steps in the PCR process:

> A **primer** is a short nucleotide sequence that can be extended by DNA polymerase. In PCR, synthetic primers are designed to flank either side of the target sequence.

* *Denaturation:* by heating to 95°C, the DNA is separated into single strands, providing a template for the DNA polymerase.

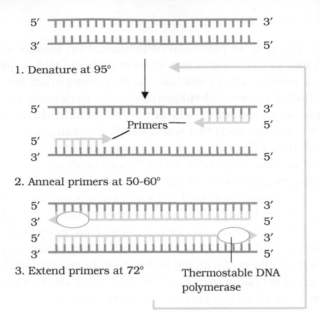

5'   ┬┬┬┬┬┬┬┬┬┬┬┬┬┬┬┬┬┬┬┬┬┬┬┬   3'
3'   ┴┴┴┴┴┴┴┴┴┴┴┴┴┴┴┴┴┴┴┴┴┴┴┴   5'

1. Denature at 95°

5'   ┬┬┬┬┬┬┬┬┬┬┬┬┬┬┬┬┬┬┬┬┬┬┬┬   3'
                  Primers—   5'

5'   ┴┴┴┴┴
3'   ┴┴┴┴┴┴┴┴┴┴┴┴┴┴┴┴┴┴┴┴┴┴┴┴   5'

2. Anneal primers at 50-60°

5'   ┬┬┬┬┬┬┬┬┬┬┬┬┬┬┬┬┬┬┬┬┬┬┬┬   3'
3'   ⬭┴┴┴┴┴┴┴┴┴┴┴┴┴┴┴┴┴┴┴┴┴┴   5'
5'   ┴┴┴┴┴┴┴┴┴┴┴┴┴┴┴┴┴┴┴┴┴⬭   3'
3'   ┴┴┴┴┴┴┴┴┴┴┴┴┴┴┴┴┴┴┴┴┴┴┴┴   5'

3. Extend primers at 72°   Thermostable DNA
                                polymerase

Figure 12.14   The polymerase chain reaction (PCR). Millions of copies of a specific sequence of DNA can be made by the polymerase chain reaction. Samples are subjected to repeated cycles of denaturation, annealing of primers and extension by a themostable DNA polymerase (for details see text). At the end of the first cycle (shown in the diagram), each of the four strands can serve as a template for replication in cycle two. After 30 cycles, over a billion copies of the target sequence will have been made.

- *Primer annealing:* at a lower temperature (typically 50–60°C, the exact value depends on the primer sequence), short, single-stranded primers are allowed to attach. Primers are synthetic oligonucleotides with a sequence complementary to the regions flanking the region we wish to amplify. One primer anneals to each strand, at either side of the target sequence.
- *Polymerase extension:* at around 72°C, the DNA polymerase extends the primer by adding complementary nucleotides and so forms a second strand.

The net result of this process is that instead of one double-stranded molecule, we now have two. We now come to the key concept of PCR; if we raise the temperature to 95°C again to start another cycle, we will have not two, but four single-stranded templates to work on, each of which can be converted to the double-stranded form as before. After 20 such cycles we should have, in theory, over *one million* copies of our original molecule! Typical PCR protocols run for 30–35 cycles. All this can be achieved in just a couple of hours; the temperature cycling is carried out by a programmable microprocessor-controlled machine called a *thermal cycler*.

PCR has widespread applications in microbiology, as in other fields of biology, but in addition, microorganisms play a crucial role in the process itself. If you've read the above description of PCR carefully, something should have struck you as being not quite right: how can the DNA polymerase (a protein) tolerate being repeatedly heated to over 90°C, and what sort of an enzyme can work effectively at 72°C? The answer is that the DNA polymerase used in PCR comes from thermophilic bacteria such as *Thermus aquaticus*, a species found naturally in hot springs. The optimum temperature for *Taq polymerase* is 72°C, and it can tolerate being raised to values considerably higher than this for short periods. The availability of this enzyme meant that it was not necessary to add fresh enzyme after each cycle, and the whole process could be automated, a key factor in its subsequent phenomenal success.

V
Microorganisms in the Environment

13
Microbial Associations

During the course of this book we have stressed on more than one occasion that microorganisms exist in nature not on their own as pure cultures but in, on or alongside numerous other organisms, microbial or otherwise, with which they may have to compete in the never-ending struggle for survival. In a number of cases, this coexistence may extend beyond merely sharing the same environmental niche; some microorganisms form a close physical association with another type of organism, from which special benefits may accrue for one or both parties. Such an association is termed *symbiosis* ('living together': Table 13.1). Three general forms of symbiotic relationship may be defined:

> *Symbiosis* is sometimes taken to mean a relationship between different organisms from which both participants derive benefit. We use the term in its broader sense, as described in the text.

- *Parasitism*: an association from which one partner derives some or all of its nutritional requirements by living either in or on the other partner (the *host*), which usually suffers some harm as a result.
- *Mutualism*: an association from which both participants derive benefit. The relationship is frequently obligatory, that is, both are dependent upon the other for survival. Mutualism that is not obligatory is sometimes called *protocooperation*.
- *Commensalism*: an association from which one participant (the commensal) derives benefit, and the other is neither benefited nor harmed. The relationship is not usually obligatory.

Microorganisms may be associated with plants, animals or other types of microorganism in any of these types of symbiosis (Tables 13.2, 13.3 and 13.4).

Essential Microbiology, Second Edition. Stuart Hogg.
© 2013 John Wiley & Sons, Ltd. Published 2013 by John Wiley & Sons, Ltd.

Table 13.1 Types of symbiotic association. Participants in symbiosis may derive benefit, harm or neither from the association. + denotes benefit, x denotes harm and – denotes neither

Association	Species A	Species B
Mutualism	+	+
Protocooperation	+	+
Commensalism	–	+
Parasitism	x	+

Table 13.2 Microorganism–animal associations

Microorganism	Animal	Type of relationship
Anaerobic bacteria	Ruminants	Mutualism
Flagellated protozoans	Termites	Mutualism
Sulphur-oxidising bacteria	*Riftia* (marine tube worm)	Mutualism
Luminescent bacteria	Fish, molluscs	Mutualism
Bacteria, yeasts	Honey guide (bird)	Mutualism
Fungus	Leaf-cutter ants	Protocooperation
Resident bacteria of skin, large intestine, etc.	Humans	Commensalism

Table 13.3 Microorganism–plant associations

Microorganism	Plant	Type of relationship
N_2-fixing bacteria	Legumes	Mutualism
Mycorrhizal fungi	Various	Mutualism
Agrobacterium tumefaciens	Various	Parasitism (crown gall disease)
Acremonium (fungus)	Grass	Mutualism

Table 13.4 Microorganism-microorganism associations

Microorganism	Microorganism	Type of relationship
Fungi	Alga/blue-green	Mutualism (lichen)
Amoebas, flagellates	Methanogenic archaea	Mutualism

13.1   Microbial associations with animals

Termites are insects belonging to the order Isoptera that are found particularly in tropical regions. Their well-known ability to destroy trees and wooden structures such as buildings and furniture is due to a resident population of flagellated protozoans in their hindgut, which are able to break down cellulose. Termites lack the enzymes necessary to do this, and would thus starve to death if the protozoans were not present. In return, however, they are able to provide the protozoans with the anaerobic conditions they require to ferment the cellulose to acetate, carbon dioxide and hydrogen. The acetate is then utilised as a carbon source by the termites themselves.

In addition to the protozoans, anaerobic bacteria resident in the hindgut also play an important role in the metabolism of the termites. Acetogenic and methanogenic species compete for the carbon dioxide and hydrogen produced by the protozoans. The acetogenic bacteria contribute more acetate for the termite

> The total global amount of methane production by termites is comparable to that generated by ruminants.

to use, whilst methanogens produce significant amounts of methane. Some methanogens exist as endosymbionts within the protozoans. In other types of termite, no resident population of cellulose digesters is present. Instead, the termite ingests a fungus, which provides the necessary cellulolytic enzymes.

Another example of a host's staple diet being indigestible without the assistance of resident microorganisms is provided by the brightly coloured African bird known as the honey guide. The honey guide eats beeswax, and relies on a two-stage digestion process by bacteria (*Micrococcus cerolyticus*) and yeast (*Candida albicans*) to render it in a usable form.

At the bottom of the deepest oceans, around geothermal vents, live enormous (2 m or more) tube worms belonging to the genus *Riftia*. These lack any sort of digestive system, but instead contain in their body cavity a tissue known as the *trophosome*. This comprises vascular tissue plus cells packed with endosymbiotic bacteria. These are able to generate ATP and NADPH by the oxidation of hydrogen sulphide generated by volcanic activity, and fix carbon dioxide via the Calvin cycle, providing the worm with a supply of organic nutrients. Hydrogen sulphide is transported to the trophosome from the worm's gill plume by a form of haemoglobin present in its blood (Figure 13.1).

Warm-blooded animals including humans play host in their lower intestinal tract to vast populations of bacteria. Although some of these are capable of producing useful metabolites such as vitamin K, most of them live as *commensals*, neither benefiting nor harming their host. It could be argued, however, that the very presence of the resident intestinal microflora acts as an important defence against colonisation by other, pathogenic, microorganisms, thus making the association more one of mutualism.

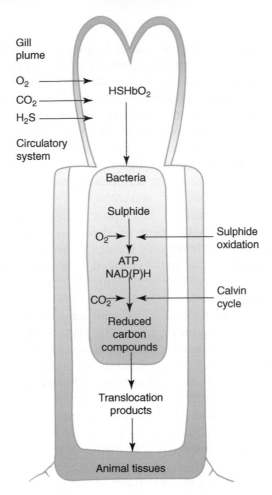

Figure 13.1 Symbiosis in *Riftia*, the giant tube worm. **Found in deep-sea hydrothermal vents,** *Riftia* acts as host to sulphur-oxidising bacteria. Energy and reducing power derived from sulphide oxidation are used to fix CO_2 via the Calvin cycle and provide the worm with organic carbon. Reproduced from Prescott, LM, et al. (2002) Microbiology, 5th edn, with permission from McGraw-Hill.

A number of bacteria, viruses, fungi, protozoans and even algae act as pathogens in animals, and cause millions of human deaths every year. Examples of diseases caused by each group are discussed in Chapter 15.

13.2 Microbial associations with plants

The roots of almost all plants form mutualistic associations with fungi, known as *mycorrhizae*, which serve to enhance the uptake of water and mineral

Figure 13.2 Endomycorrhizae. Section through a plant root colonised by an endomycorrhizal fungus. Note the spreading 'treelike' arbuscles.

nutrients, especially phosphate, by the plants. The beneficial effect of a mycorrhizal association is particularly noticeable in soils with a low phosphorus content. In return, the plant supplies reduced carbon in the form of carbohydrates to the fungi. Unlike other plant–microorganism interactions that occur in the rhizosphere, mycorrhizal associations involve the formation of a distinct, integrated structure comprising root cells and fungal hyphae. In *ectomycorrhizae* the plant partner is always a tree; the fungus surrounds the root tip, and hyphae spread between (but do not enter) root cells. In the case of the more common *endomycorrizae*, the fungal hyphae actually penetrate the cells by releasing cellulolytic enzymes.

> A *mycorrhiza* is a mutually beneficial association between plant roots and a species of fungus.

Arbuscular mycorrhizal fungi (AMF) belong to the phylum Glomeromycota, and are found in most plant types, including 'lower' plants (mosses, ferns). They form highly branched *arbuscules* within the root cells that gradually

> The **rhizosphere** is the region around the surface of a plant's root system.

lyse, releasing nutrients into the plant cells (Figure 13.2). The plants benefit particularly from the increased uptake of phosphorus that results from the association, whereas the AMF receives a supply of hexose sugars. In contrast to pathogenic fungi, mycorrhizal fungi are often rather non-specific in their choice of 'partner' plant.

The ability of crop plants to thrive is frequently limited by the supply of available nitrogen; although there is a lot of it in the atmosphere, plants are unable to utilise it, and instead must rely on an inorganic supply (both naturally occurring and in the form of fertilisers). As we saw in Chapter 7,

Figure 13.3   Nitrogen-fixing bacteria form root nodules in legumes. (a) Nitrogen-fixing *Rhizobium* proliferate inside root cells of leguminous plants as bacteroids. (b) Root nodule formation.

however, certain bacterial species are able to 'fix' atmospheric nitrogen into a usable form. Some of these, notably *Rhizobium* spp., form a mutualistic relationship with leguminous plants such as peas, beans and clover, converting nitrogen to ammonia, which the legume can incorporate into amino acids. In return, the bacteria receive a supply of organic carbon, which they can use as an energy source for the fixation of nitrogen.

The free-living *Rhizobium* enters the plant via its root hairs, forming an infection thread and infecting more and more cells (Figure 13.3). Normally rod-shaped, they proliferate as irregularly shaped *bacteroids*, densely packing the

> *Root nodules* are tumour-like growths on the roots of legumes, where nitrogen fixation takes place.

cells and causing them to swell, forming *root nodules*. *Rhizobium* requires oxygen as a terminal electron acceptor in oxidative phosphorylation, but as you may recall from Chapter 7, the nitrogenase enzyme that fixes the nitrogen is sensitive to oxygen. The correct microaerophilic conditions are maintained by means of a unique oxygen-binding pigment, *leghaemoglobin*. This is only synthesised by means of a collaboration between both partners. Nitrogen fixation requires a considerable input of energy in the form of ATP (16 molecules for every molecule of nitrogen), so when ammonia is in plentiful supply the synthesis of the nitrogenase enzyme is repressed.

Farmers have long recognised the value of incorporating a legume into a crop rotation system; the nodules left behind in the soil after harvesting the crop appreciably enhance the nitrogen content of the soil.

Legumes are not the only plants able to benefit from the nitrogen-fixing capabilities of bacteria. The water fern *Azolla*, which grows prolifically in the paddy fields of southeast Asia, has its nitrogen supplied by the blue-green bacterium *Anabaena*. When the fern dies, it acts as a natural fertiliser for the rice crop. *Anabaena*

> Unlike higher plants, ferns do not possess true roots, stems and leaves. The structure equivalent to a leaf is called a *frond*.

does not form root nodules, but takes up residence in small pores in the *Azolla* fronds. Nitrogen fixation takes place in heterocysts, specialised cells whose thick walls slow down the rate at which oxygen can diffuse into the cell, providing appropriate conditions for the oxygen-sensitive nitrogenase.

The alder tree (*Alnus* spp.) is able to grow in soils with a poor nitrogen content due to its association in root nodules with the nitrogen-fixing acti-nomycete *Frankia*. The filamentous *Frankia* solves the problem of nitroge-nase's sensitivity to oxygen by compartmentalising it in thick-walled vesicles at the tips of its hyphae, which serve the same function as the heterocysts of *Anabaena*.

Several genera of bacteria live inside the cells of insects and other arthro-pods. Perhaps the best known of these are members of the genus *Wolbachia*, whose relationship with their host ranges from parasitism to obligate mutual-ism. *Wolbachia* bacteria have the ability to alter the reproductive capabilities of some hosts by specifically infecting reproductive cells. The result can be males being killed off, males turned into females, or parthenogenesis (asex-ual reproduction of female with no male involvement). *Wolbachia* also infect species of nematode worms, including some serious parasites of humans. Since in some cases the nematode is completely dependent on *Wolbachia* for survival, some disease-control strategies target the bacterium with antibi-otics, rather than trying to kill off the worms directly. *Carsonella ruddii* is an endosymbiont of psyllids, which are sap-sucking insects. It provides its host with amino acids that are missing from the psyllid's sugar-rich but protein-poor diet. *C. ruddii* has the smallest genome known so far, just 159 662 base pairs, encoding a mere 182 proteins. It lacks many essential genes such as those responsible for DNA replication and key metabolic pathways, and is thus entirely dependent on its host for survival. A significant proportion of its small genome is devoted to amino acid synthesis. Since it lacks so many of the genes necessary for independent existence, some argue that *C. ruddi* has a status closer to that of an organelle rather than that of an organism in its own right.

Many microorganisms, particularly bacteria and yeasts, are to be found living as harmless commensals on the surface structures of plants such as leaves, stems and fruits.

> Organisms that grow on the surface of a plant are called *epiphytes*. They fre-quently live as commen-sals.

13.2.1 Plant diseases

Plant disease may be caused by viruses, bacteria, fungi or protozoans. These frequently have an impact on humans, especially if the plant affected is a commercially important crop. Occasionally the effect on a human popula-tion can be catastrophic, as with the Irish famine of the 1840s brought about by potato blight. A number of microbial diseases of plants are listed in Table 13.5.

Table 13.5 Some microbial diseases of plants

Causative agent	Type of microorganism	Host	Disease
Heterobasidion	Fungus	Pine trees	Heart rot
Ceratocystis	Fungus	Elm trees	Dutch elm disease
Puccinia graminis	Fungus	Wheat	Wheat rust
Phytophthora infestans	Water mould	Potato	Potato blight
Erwinia amylovora	Bacterium	Apple, pear tree	Fire blight
Pseudomonas syringae	Bacterium	Various	Chlorosis
Agrobacterium	Bacterium	Various	Crown gall disease
Tobacco mosaic virus	Virus	Tobacco	Tobacco mosaic disease

We have already encountered the soil bacterium *Agrobacterium tumefaciens* in Chapter 12, where we saw how it has been exploited as a means of genetically modifying plants. *A. tumefaciens* is useful for introducing foreign DNA because it is a natural pathogen of plants, entering wounds and causing *crown gall disease*, a condition characterised by areas of uncontrolled growth, analogous to tumour formation in animals. This proliferation is caused by the expression within the plant cell of genes that encode the sequence for enzymes involved in the synthesis of certain plant hormones. The genes are carried on the T-DNA, part of an *A. tumefaciens* plasmid, which integrates into a host chromosome. Also on the T-DNA are genes that code for amino acids called opines. These are of no value to the plant, but are utilised by the *A. tumefaciens* as a food source.

13.3   Microbial associations with other microorganisms

The most familiar example of mutualism between microorganisms is that of *lichens*, which comprise a close association between the cells of a fungus (usually belonging to the Ascomycota) and a photosynthetic alga or cyanobacterium. Although many different fungal species may take part in lichens, only a limited number of algae or cyanobacteria do so. Lichens are typically found on exposed hard surfaces such as rocks, tree bark and the roofs of buildings, and grow very slowly at a rate of a millimetre or two per year. They often occupy particularly harsh environments, from the polar regions to the hottest deserts. The photosynthetic partner usually exists as a layer of cells scattered among fungal hyphae (Figure 13.4). Often unicellular, it fixes carbon dioxide as organic matter, which the heterotrophic fungus absorbs and utilises.

> A *lichen* is a mutually beneficial association between a fungus and an alga or cyanobacterium (blue green).

Algal cell

Fungal hyphae

Figure 13.4 Algae and fungi combine to form lichens. Algal cells are embedded among the fungal hyphae just below the surface, where light is able to penetrate. Organic carbon and oxygen produced by photosynthesis are used by the fungus, while it provides water, minerals and shelter for its algal partner.

The fungal member provides anchorage and supplies inorganic nutrients and water, as well as protecting the alga from excessive exposure to sunlight.

Although lichens are tolerant of extremes of temperature and water loss, they have a well-known sensitivity to atmospheric pollutants such as the oxides of nitrogen and sulphur. Their presence in an urban setting is therefore a useful indicator of air quality. Lichens were used for many years as a source of brightly coloured dyes for the textile industry and also used in the perfume industry. The dye used in litmus paper is derived from a lichen belonging to the genus *Roccella*.

It should be stressed that lichens are not just a mixture of fungal and algal cells. They are distinctive structures with properties not possessed by either of their component species. Indeed, the relationship between the two partners of a lichen is so intimate that the composite organisms are given taxonomic status, and many thousands of species of lichen have been identified. Their classification is based on their fungal component.

As noted in the definitions at the beginning of this chapter, a mutualistic relationship need not always be as intimate and essential to both partners as it is in a lichen. The sulphate-reducing bacterium *Desulfovibrio*, for example, can obtain the sulphate and organic substrates it needs from the photosynthetic purple sulphur bacterium *Chromatium*, which in turn receives the carbon dioxide and hydrogen sulphide it requires (Figure 13.5). This protocooperation is not essential to either bacterium, however, as each is able to satisfy its requirements by alternative means. In another bacterial relationship, the gut bacteria *E. coli* and *Enterococcus faecalis* join forces to utilise arginine. As shown in Figure 13.6, neither can usefully metabolise this amino acid on its own, but neither is dependent on the cooperation.

Commensal relationships, in which one partner benefits and neither suffer, are common among microorganisms; such associations are rarely obligatory. A common basis for microbial commensalism is for one partner to benefit as a coincidental consequence of the normal metabolic activities of the other.

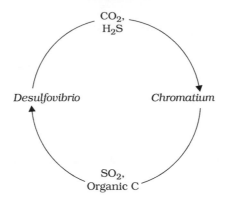

Figure 13.5 Protocooperation in bacteria. The sulphate-reducing *Desulfovibrio* and the sulphur-oxidising *Chromatium* can supply each other with the raw materials required for energy production. Neither is completely dependent on the association.

Thus one may excrete vitamins or amino acids that can be utilised by the other, or a facultative anaerobe may assist its obligate anaerobe neighbour by removing oxygen from the atmosphere, thus providing the conditions for the latter to grow.

The nature of a role played by an organism in a symbiotic relationship may alter according to the prevailing conditions. In a soil already rich in nitrogen, for example, the legume derives no benefit from the *Rhizobium*, which is then more accurately classed as a parasite, as it continues to utilise organic carbon produced by the plant. In humans, harmless gut symbionts such as *E. coli* can become *opportunistic pathogens,* and cause infections if introduced to an inappropriate site such as a wound or the urinary tract.

Box 13.1   *Once bitten, twice shy*

An interesting example of mutualistic association concerns the endophytic (= 'inside plant') fungus *Acremonium*. It derives reduced carbon compounds and shelter by living within the tissues of the grass, *Stipa robusta*, and in return deters animals from grazing on it. It does this by producing various alkaloids that, if ingested in sufficient amounts, are powerful enough to send a horse to sleep for several days! The horse clearly does not relish the experience, as it avoids the grass thereafter. The *Acremonium* passes to future generations through the seeds, so the relationship between plant and fungus is perpetuated. The nickname of 'sleepy grass' is self-explanatory!

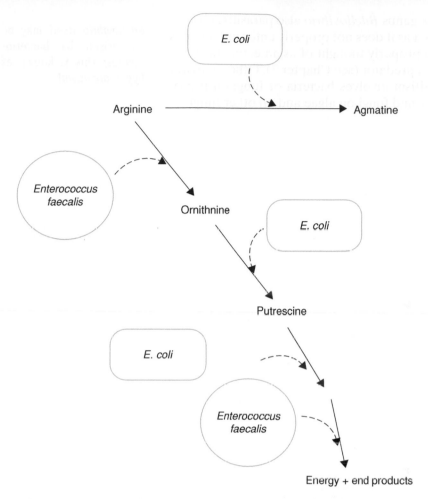

Figure 13.6 Protocooperation can make available an otherwise unutilisable substrate. Individually, neither *Enterococcus faecalis* nor *E. coli* is able to utilise arginine; however, working together they can convert it into putrescine, which can then be metabolised further by either organism to produce energy.

A limited number of microorganisms exist by living *parasitically* inside another. Viruses (see Chapter 10) are all obligate endoparasites that form an association with a specific host. This host may be microbial: bacteria, fungi, protozoans and algae all act as hosts to their own viruses. Viruses of bacteria are termed *bacteriophages* and are only able to replicate themselves inside an actively metabolising bacterial cell (see Chapter 10 for bacteriophage replication cycles). An unusual bacterium belonging

> An *endoparasite* fully enters its host and lives inside it. An *ectoparasite* attaches to the outside.

to the genus *Bdellovibrio* also parasitises bacteria, but as it does not properly enter the cell, it is more properly thought of as an ectoparasite or even a predator (see Chapter 7). Other nonviral parasitism involves bacteria or fungi on protozoans, and fungi on algae and on other fungi.

> *Bdellovibrio* itself may be parasitised by bacteriophages; this is known as *hyperparasitism*!

14
Microorganisms in the Environment

At various points in this book we have referred to the different environments in which particular microorganisms are to be found. Like other living organisms, they live as components of *ecosystems*, and we shall consider the three main types of ecosystem – terrestrial, freshwater and marine – later in this chapter.

> Living organisms, together with their physical surroundings, make up an *ecosystem*.

Firstly, however, we must turn our attention again to the subject of energy relations in living things. In Chapter 4, we looked at the different ways in which microorganisms can derive and utilise energy from various sources. We now need to put these processes into a global perspective. All organisms may be placed into one of three categories with respect to their part in the global flow of energy:

- *(Primary) Producers*: autotrophs that obtain energy from the Sun or chemical sources (e.g. green plants, photosynthetic bacteria, chemolithotrophic bacteria). They use the energy to synthesise organic material from carbon dioxide and water.
- *Consumers*: heterotrophs that derive energy through the consumption of other organisms (producers or other consumers). They may serve as a link between the primary producers and the decomposers.
- *Decomposers*: organisms that break down the remains and waste products of producers and consumers, obtaining energy and releasing nutrients, including CO_2, that can be reused by the producers.

Essential Microbiology, Second Edition. Stuart Hogg.
© 2013 John Wiley & Sons, Ltd. Published 2013 by John Wiley & Sons, Ltd.

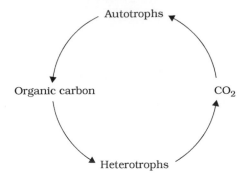

Figure 14.1   The carbon cycle. Autotrophs fix CO_2 as an organic compound, which heterotrophs convert back to CO_2. The recycling of carbon satisfies the requirements of both nutritional types.

Natural systems exist in a balance. Carbon and all the other elements that make up living things are subject to repeated *recycling*, so that they are available to different organisms in different forms. Think back to Chapter 6, in which we discussed how algae, green plants and certain bacteria capture light energy, then use it to synthesise organic carbon compounds from carbon dioxide and water. What happens to all this organic carbon? It doesn't just accumulate, but is recycled by other living things, which convert it back to carbon dioxide by respiration. This can be seen in its simplest form in Figure 14.1. Many other elements such as sulphur, nitrogen and iron are similarly transformed from one form to another in this way, by a cyclic series of reactions. Microorganisms are responsible for most of these reactions, oxidising and reducing the elements according to their metabolic needs. The continuation of life on Earth is dependent on the cycling of finite resources in this way.

14.1   The carbon cycle

A more detailed scheme of the carbon cycle is shown in Figure 14.2. Both aerobic and anaerobic reactions contribute to the cycle. The numbers in parentheses in the following description refer to the pathways in Figure 14.2.

Atmospheric CO_2 is fixed into organic compounds by plants, together with phototrophic and chemoautotrophic microorganisms (1). These compounds undergo cellular respiration, and CO_2 is returned to the atmosphere (2). The carbon may have been passed along a food chain to consumers before this

> The *carbon cycle* is the series of processes by which carbon from the environment is incorporated into living organisms and returned to the atmosphere as carbon dioxide.

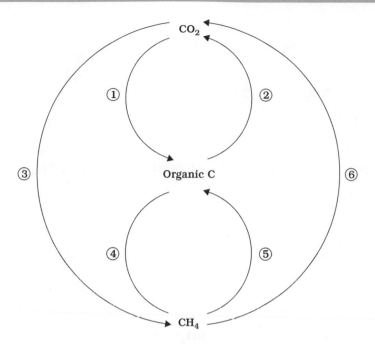

Figure 14.2 The carbon cycle: a closer look. Carbon is circulated as one of three forms, carbon dioxide, methane and organic compounds. Different organisms are able to utilise each form for their own metabolic requirements, converting it in the process to one of the others. Numbered arrows refer to reactions described in the text.

occurs. Carbon dioxide is also produced by the decomposition of dead plant, animal and microbial material by heterotrophic bacteria and fungi.

Methanogenic bacteria produce methane from organic carbon or CO_2 (3, 4). This in turn is oxidised by methanotrophic bacteria; carbon may be incorporated into organic material or lost as CO_2 (5, 6).

14.2 The nitrogen cycle

Nitrogen is essential to all living things as a component of proteins and nucleic acids. Although elemental nitrogen makes up three-quarters of the Earth's atmosphere, only a handful of life forms are able to utilise it for metabolic purposes. These are termed nitrogen-fixing bacteria, and incorporate the nitrogen into ammonia (Figure 14.3, reaction 1):

$$N_2 + 8e^- + 8H^+ + 16ATP \rightarrow 2NH_4^+ + 16ADP + 16Pi$$

The nitrogenase enzyme complex responsible for the reaction is very sensitive to oxygen, and is thought to have evolved early in the Earth's history,

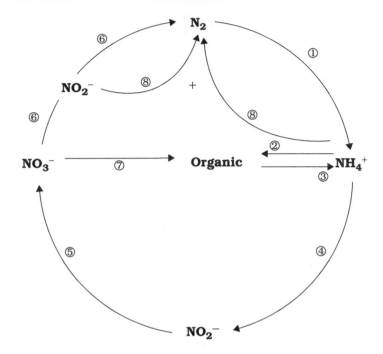

Figure 14.3 The nitrogen cycle. See the text for further details of reactions. Numbered arrows refer to reactions described in the text.

when the atmosphere was still largely oxygen-free. Many nitrogen-fixing bacteria are anaerobes; those that are not have devised ways of keeping the cell interior anoxic. *Azotobacter* spp., for example, utilise oxygen at a high rate, so that it never accumulates in the cell, and so does not inactivate the nitrogenase. Many cyanophytes (blue-greens) carry out nitrogen fixation in thick-walled heterocysts, which help maintain anoxic conditions.

Some nitrogen-fixing bacteria such as *Rhizobium* infect the roots of leguminous plants such as peas, beans and clover, where they produce nodules and form a mutually beneficial association.

Ammonia produced by nitrogen fixation is assimilated as amino acids, which can then form proteins and feed into pathways of nucleotide synthesis (2). Organic nitrogen in the form of dead plant and animal material plus excrement re-enters the environment, where it undergoes *mineralisation* (3) to an inorganic form at the hands of a range of microorganisms. This involves the deamination of amino acids to their corresponding organic acid. This process of mineralisation may occur aerobically or anaerobically, in a wide range of microorganisms:

> The process by which microorganisms convert organic matter to an inorganic form such as CO_2, CH_4, NH_4^+ is termed *mineralisation*.

$$2 \text{ HC-NH}_3^+ + O_2 \quad \xrightarrow{\textbf{Deamination}} \quad 2 \text{ C=O} + 2 \text{ NH}_4^+$$

Alanine   Pyruvate

The process of *nitrification*, by which ammonia is oxidised stepwise firstly to nitrite and then to nitrate, involves two different groups of bacteria (4, 5:

$$NH_4^+ \rightarrow NO_2^-$$
$$NO_2^- \rightarrow NO_3^-$$

The nitrate thus formed may suffer a number of fates. It may act as an electron acceptor in anaerobic respiration, becoming reduced to nitrogen via a series of intermediates including nitrite (6). This process of *denitrification* occurs in anaerobic conditions such as waterlogged soils. Alternatively, it can be reduced once again to ammonia and be converted to organic nitrogen (7).

Denitrification is the reduction, under anaerobic conditions, of nitrite and nitrate to nitrogen gas.

A final pathway of nitrogen cycling has only been discovered relatively recently. It is known as *anammox* (anaerobic ammonia oxidation), and is carried out by members of a group of Gram-negative bacteria called the Planctomycetes (see Chapter 7). The reaction, which can be represented thus:

Anammox is the formation of nitrogen gas by the anaerobic oxidation of ammonia and nitrite.

$$H_4^+ + NO_2^- \rightarrow N_2 + 2H_2O (8)$$

has considerable potential in the removal of nitrogen from wastewater.

14.3 The sulphur cycle

Sulphur is found in living organisms in the form of compounds such as amino acids, coenzymes and vitamins. It can be utilised by different types of organisms in several forms; Figure 14.4 shows the principal components of the sulphur cycle.

In its elemental form, sulphur is unavailable to most organisms; however, certain bacteria such as *Acidithiobacillus* are able to oxidise it to sulphate (1), a form that can be utilised by a much broader range of organisms (see Chapter 7):

$$2S + 3O_2 + 2H_2O \rightarrow 2H_2SO_4$$

Figure 14.4   The sulphur cycle. See the text for further details of reactions. Numbered arrows refer to reactions described in the text.

Powdered sulphur is often added to alkaline soils in order to encourage this reaction and thereby reduce the pH.

Sulphate-reducing bacteria convert the sulphate to hydrogen sulphide gas (2) using either an organic compound or hydrogen gas as electron donor:

$$8H^+ + SO_4^{2-} \rightarrow H_2S + 2H_2O + 2OH^-$$

These bacteria are obligate anaerobes, and the process is termed *dissimilatory* sulphate reduction.

Plants are also able to utilise sulphate, incorporating it into cellular constituents such as the amino acids methionine and cysteine (3) (*assimilatory* sulphate reduction).

When the plants die, these are broken down, again with the release of hydrogen sulphide (4) (see mineralisation, earlier).

$$
\begin{array}{ccc}
\text{CH}_2\text{HS} & & \text{CH}_3 \\
| & & | \\
\text{HC-NH}_3^+ + \text{H}_2\text{O} & \longrightarrow & \text{C=O} + \text{NH}_4^+ + \text{H}_2\text{S} \\
| & & | \\
\text{COO}^- & & \text{COO}^- \\
& & \\
\text{Cysteine} & & \text{Pyruvate}
\end{array}
$$

Green and purple photosynthetic bacteria and some chemoautotrophs use hydrogen sulphide as an electron donor in the reduction of carbon dioxide, producing elemental sulphur and thus completing the cycle (5):

$$H_2S + CO_2 \rightarrow (CH_2O)_n + S^0$$

Phosphorus exists almost exclusively as phosphate; however, this is cycled between soluble and insoluble forms. This conversion is pH-dependent, and if phosphate is only present in an insoluble form, it will act as a limiting nutrient. This explains the sudden surge in the growth of plants, algae and cyanobacteria in water bodies when a source of soluble phosphate (typically fertiliser or detergent) enters a watercourse.

14.4 The microbiology of soil

In the following section it will be necessary to generalise, and treat soil as a homogeneous medium. In fact, it is no such thing; its precise make-up is dependent upon the underlying geology, and the climatic conditions both past and present. In addition, the microbial population of a soil will vary according to the amount of available water and organic matter, and different organisms colonise different strata in the soil.

> The *topsoil* is the top few centimetres of a soil, characterised by its high content of organic material in various stages of decomposition. It is distinguished from the succeeding layers underneath it, termed the subsoil, parent layer and bedrock.

The organic content of a soil derives from the remains of dead plants and animals. These are broken down in the soil by a combination of invertebrates and microorganisms (mainly bacteria and fungi) known as the *decomposers*. Their action results in the release of substances that can be used by plants and by other microorganisms. Much organic material is easily degraded, while the more resistant fraction is referred to as *humus*, and comprises lignin together with various other macromolecules. The humus content of a soil, then, is a reflection of how favourable (or otherwise) conditions are for its decomposition; the value usually lies between 2 and 10%

> *Humus* is the complex organic content of a soil, comprising complex materials that remain after microbial degradation.

by weight. The inorganic fraction of a soil derives from the weathering of minerals. Microorganisms may be present in soils in huge numbers, mostly attached to soil particles. Their numbers vary according to the availability of suitable nutrients. Bacteria (notably actinomycetes) form the largest fraction of the microbial population, together with much smaller numbers of fungi, algae and protozoans. Published values of bacterial numbers range from overestimates (those that do not distinguish between living and dead cells) and underestimates (those that depend on colony counts and therefore exclude those organisms we are not yet able to grow in the laboratory – 99% of species according to some experts!). Suffice to say that many millions (possibly billions) of bacteria may be present in a single gram of topsoil. In spite of being present in such enormous numbers, bacteria only represent a minute percentage of the volume of most soils. Fungi, although present in much smaller numbers than bacteria, form a higher proportion of the soil biomass, due to their greater size. The majority of soil microorganisms are aerobic heterotrophs, involved in the decomposition of organic substrates, thus microbial numbers diminish greatly the further down into the soil we go, away from organic matter and oxygen. The proportion of anaerobes increases with depth, but unless the soil is waterlogged, they are unlikely to predominate.

Other factors affecting microbial distribution include pH, temperature and moisture. Broadly speaking, neutral conditions favour bacteria, while fungi flourish in mildly acidic conditions (down to about pH 4), although extremophiles survive well outside these limits. Actinomycetes favour slightly alkaline conditions. Bacterial forms occurring commonly in soils include *Pseudomonas*, *Bacillus*, *Clostridium*, *Nitrobacter* and the nitrogen-fixing *Rhizobium* and *Azotobacter*, as well as cyanobacteria such as *Nostoc* and *Anabaena*. Commonly found actinomycetes include *Streptomyces* and *Nocardia*. As we shall see in Chapter 17, actinomycetes are notable for their secretion of antimicrobial compounds into their surroundings. This provides an example of how the presence of one type of microorganism in a soil population can influence the growth of others, forming a dynamic, interactive ecosystem. In addition, bacteria may serve as prey for predatory protozoans, and secondary colonisers may depend on a supply of nutrients from, for example, cellulose degraders. Important fungal genera common in soil include the familiar *Penicillium* and *Aspergillus*; these not only recycle nutrients by breaking down organic material, but also contribute to the fabric of the soil, by binding together microscopic soil particles. Soil protozoans are mostly predators, which ingest bacteria or protists such as yeasts or unicellular algae. All the major forms of protozoans may be present (flagellates, ciliates and amoebas), moving around the water-lined spaces between soil particles. Algae are of course phototrophic, and are therefore to be found mostly near the soil surface, although it will be recalled from Chapter 9 that some forms are capable of heterotrophic growth, and may thus survive further down.

The surface of soil particles is a good natural habitat for the development of *biofilms*, complex microenvironments comprising microbial cells held together in a polysaccharide matrix. The microorganisms themselves produce the polysaccharide, which also allows the passage of nutrients from the environment. Biofilms can form on almost any surface, and are often to be found in rapidly flowing waters. Microorganisms with different metabolic properties may occupy different positions in a biofilm, with, for example, phototrophic species near to the surface.

Although we have emphasised the importance of organic matter in soil ecosystems, microorganisms may also be found growing on or even within rocks. The growth of such organisms, together with the action of wind and rainfall, contribute to the weathering of rocks.

14.5 The microbiology of freshwater

The microbial population of freshwater is strongly influenced by the presence or absence of oxygen and light. A body of water such as a pond or lake is stratified into zones (Figure 14.5), each having its own characteristic microflora, determined by the availability of these factors. The *littoral* zone is the region situated close to land where the water is sufficiently shallow for sunlight to penetrate to the bottom. The *limnetic* zone occupies the same depth, but is in

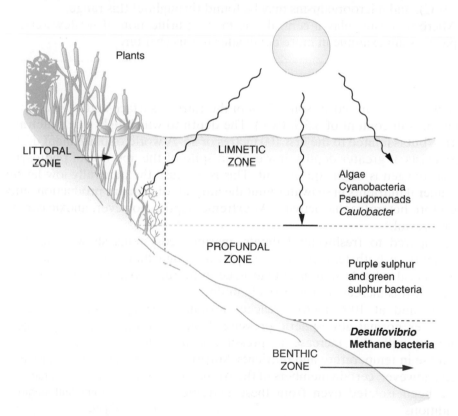

Figure 14.5 Vertical zonation in a lake or pond. Representative organisms are indicated for each zone. Reproduced from Black, JG (1999) Microbiology: Principles and Explorations, 4th edn, with permission from John Wiley & Sons.

open water, away from the shore. The *profundal* zone occupies deeper water, where sunlight is unable to penetrate, and finally the *benthic* zone comprises the sediment of mud and organic matter at the bottom of the pond or lake.

Oxygen is poorly soluble in water (9 mg/litre at 20°C), so its availability is often a limiting factor in determining the microbial population of a body of water. Oxygen availability in lakes and ponds is closely linked to oxygenic photosynthesis and therefore, indirectly, to the penetrability of light. Phototrophs such as algae and blue-greens are limited to those regions where light is able to penetrate. Oxygen is absent or very limited in the benthic zone, where anaerobic forms such as the methanogenic bacteria are to be found. Another factor influencing microbial populations is the organic content of the water; if this is high, the growth of decomposers will be encouraged, which will in turn deplete the oxygen. This is much less of an issue in rivers and streams, where physical agitation of the water generally ensures its continued oxygenation.

The temperature of freshwater ecosystems ranges between extremes (<0 to 90°C), and microorganisms may be found throughout this range.

Microorganisms play a central role in the purification of wastewaters, a topic we shall examine in more detail later in this chapter.

14.6   The microbiology of seawater

The world's oceans cover some 70% of the Earth's surface and have a fairly constant salt content of 3.5% (w/v). The depth to which light can penetrate varies, but is limited to the first 100 metres or so. A world of permanent darkness exists at greater depths; however, in spite of the absence of photosynthesis, oxygen is often still present. This is because the generally low levels of mineral nutrients in seawater limit the amount of primary production, and therefore heterotrophic activity. At extreme depths, however, anoxic conditions prevail.

Compared to freshwater habitats, marine ecosystems show much less variability in both temperature and pH, although there are exceptions to this general rule. A more pertinent issue in marine environments is that of *pressure*; this increases progressively in deeper waters, and at 1000 metres reaches around 100 times normal atmospheric pressure. Concomitant with this increase in pressure is a decrease in temperature and nutrients. Surprisingly, however, certain members of the Archaea have been isolated even from these extreme conditions.

> *Phytoplankton* is a collective term used to describe the unicellular photosynthesisers, which include cyanobacteria, dinoflagellates, diatoms and single-celled algae.

In contrast to terrestrial ecosystems, where plants are responsible for most of the energy fixation via photosynthesis, marine primary production is largely microbial, in the shape of members of the *phytoplankton*. As we have seen, such forms are restricted to those zones where light is able to penetrate. Also found here may be protozoans and fungi that feed on the phytoplankton. Because of the high salt concentration of sea-water, the bacteria that are typically found in such environments differ from those in freshwater. In the last decade or so, the presence of *ultramicrobacteria* has been detected in marine ecosystems at relatively high densities; these are around one-tenth of the size of 'normal' bacteria. Marine bacteria are of necessity halophilic. Anaerobic decomposing bacteria inhabit the benthic zone, carrying out reactions similar to those that occur in fresh-water sediments, whilst the profundal zone is largely free of microbial life.

> *Ultramicrobacteria* are bacteria that are much smaller than normal forms, and some are able to pass through a 0.22 μm filter. They may represent a response to reduced nutrient conditions.

14.7 Detection and isolation of microorganisms in the environment

As we emphasised in the last chapter, microorganisms rarely, if ever, exist in nature as pure cultures but rather as members of mixed populations. Methods are required, therefore, for the detection and isolation of specific microbial types from such mixtures. The traditional method of isolation is the use of an enrichment culture, as described in Chapter 4. As examples, aerobic incubation with a supply of nitrite would assist in the isolation of nitrifying bacteria such as *Nitrobacter* from mud or sewage, whilst a minimal medium containing $FeSO_4$ at pH 2 would encourage the isolation of *Acidithiobacillus ferrooxidans* from a water sample.

We now know, however, that there are many types of microorganism in the environment that have so far resisted all attempts to culture them in the laboratory (often referred to as *viable but non-culturable*). Modern molecular techniques have helped us to identify the existence of a much broader range of bacteria and archaea than had previously been thought to exist. The extreme sensitivity of such methods means that we are able to demonstrate the presence of even a single individual of a particular bacterium in a mixed population. One such technique is called *fluorescence in situ hybridisation* or *FISH*. This uses an oligonucleotide probe comprising a short sequence of single-stranded DNA or RNA that is unique to a particular microorganism, attached to a fluorescent dye. The microorganisms are fixed to a glass slide and incubated with the probe. The rules of base pairing in nucleic acids mean that the probe will seek out its complementary sequence, and cells carrying this sequence can be visualised under a fluorescence microscope. The most commonly used 'target' is ribosomal RNA, since this shows sequence variation from one microbial type to another, and because there are multiple copies within each cell, providing a stronger response. The polymerase chain reaction (PCR, see Chapter 12) is another valuable tool in the identification of specific nucleic acid sequences. Other methods, not dependent on DNA, include the use of fluorescence-labelled antibodies raised against specific microorganisms.

> *Antibodies* are proteins produced by the immune systems of higher animals in response to infection by a foreign organism; their main characteristic is their extreme specificity, thus they can be used to locate a specific protein.

14.8   Beneficial effects of microorganisms in the environment

The central role played by microorganisms in the recycling of essential elements on a global scale has already been stressed in this chapter. Many of their natural activities are exploited by humans for their own benefit. Some form the basis of industrial processes such as those used in the food and drink industries and are considered in Chapter 18, while the application of others is essentially environmental. Notable among these is the harnessing of natural processes of *biodegradation* to treat the colossal volumes of liquid and solid wastes generated by our society. These are reviewed briefly in the following section.

> *Biodegradation* is the term used to describe the natural processes of breakdown of matter by microorganisms.

14.8.1   Solid waste treatment: composting and landfill

Those of us who live in the industrialised nations are often said to belong to a 'throwaway society'. On average, each of us generates around 2 tonnes of solid waste material per year, and all of this must be disposed of in some way! Most of it ends up in landfill sites – huge holes in the ground where refuse is deposited to prevent it being a hazard. The non-biodegradable components (metals, plastics, rubble, etc.) remain there more or less indefinitely; however, over a period of time biodegradable material (food waste, textiles, paper, etc.) undergoes a decomposition process. The rate at which this happens is dependent on the nature of the waste and the conditions of the landfill, but can take several decades. Aerobic processes give way to anaerobic ones and a significant result of the latter is the generation of methane. Modern landfill sites incorporate systems that remove this to prevent it being a fire/explosion hazard, and may put it to good use as a fuel source.

Many householders separate organic waste items such as vegetable peelings and grass cuttings and use them to make *compost*. This practice, apart from providing a useful gardening supplement, also substantially reduces the volume of material that has to be disposed of by other means (see earlier). We have already mentioned the role of microorganisms in the recycling of carbon in the biosphere; these same processes serve to degrade the organic waste, especially the cellulose, resulting in a considerable reduction of the bulk. Fungi and bacteria, particularly actinomycetes, break down the organic matter to produce CO_2, water and humus, a relatively stable organic end product. Compost is not really a fertiliser, since its nitrogen content is not high, but it nevertheless provides nutrients to a soil and generally helps to improve its condition. Composting is carried out on a large scale by local authorities using the waste generated in municipal parks and gardens.

14.8.2 Wastewater treatment

The aim of wastewater treatment is the removal of undesirable substances and hazardous microorganisms in order that the water may safely enter a watercourse such as a river or stream. Further purification procedures are required before it can be used as drinking water. Wastewater treatment is fundamental to any developed society, and greatly reduces the incidence of waterborne diseases such as cholera. Wastewater may come from domestic or commercial sources; highly toxic industrial effluents may require pre-treatment before entering a water treatment system. *Sewage* is the term used to describe liquid wastes that contain faecal matter (human or animal).

The effectiveness of the treatment process is judged chiefly by the reduction of the wastewater's *biochemical oxygen demand* (BOD). This is a measure of the amount of oxygen needed by microorganisms to oxidise its organic content. A high BOD leads to the removal of oxygen from water, a certain indicator of pollution.

Wastewater treatment usually occurs in stages, the first of which (primary treatment) is purely physical, and involves the removal of floating objects followed by sedimentation, a process that removes up to one-third of the BOD value. Secondary treatment involves microbial oxidation, leading to a substantial further reduction in BOD. This may take one of two forms, both of which are aerobic, the traditional *trickling filter* and the more recent *activated sludge* process (Figure 14.6). In the former, the wastewater is passed slowly over beds of stones or pieces of moulded plastic. These develop a biofilm comprising bacteria, protozoans, fungi and algae, and the resulting treated water has its BOD reduced by some 80–85%. Activated sludge facilities achieve an even higher degree of BOD reduction. Here the wastewater is aerated in tanks that have been seeded with a mixed microbial sludge. The main component of this is the bacterium *Zoogloea,* which secretes slime, forming aggregates called *flocs*, around which other microorganisms such as protozoans attach. Some of the water's organic content is not immediately oxidised, but becomes incorporated into the flocs. After a few hours' residence

Figure 14.6   Wastewater treatment achieves a reduction of the biochemical oxygen demand of the water by primary (physical) and secondary (biological) treatment.

in the tank, the sludge is allowed to settle out, and the treated water passes out of the system. Before being discharged to a watercourse, it is treated with chlorine to remove any pathogenic microorganisms that may remain.

The principal operating problem encountered with activated sludge is that of *bulking*. This is caused by filamentous bacteria such as *Sphaerotilus natans*; it results in the sludge not settling properly and consequently passing out with the treated water.

Both secondary treatment processes result in some surplus sludge, which undergoes *anaerobic digestion*, resulting in the production of methane and CO_2. The methane can be used as a fuel to power the plant, and any remaining sludge is dewatered and used as a soil conditioner. Care must be taken in this context, however, that the sludge does not contain toxic heavy metals.

14.8.3   Bioremediation

One of the biggest problems facing the developed world in the early twenty-first century is that of pollution of the environment. Our dependence on the products of the chemical industries has resulted in the production of vast amounts of toxic waste material. One way of dealing with such (mostly organic) waste is to encourage the growth of bacteria and fungi that are able to oxidise the pollutants, a process known as *bioremediation*. Elsewhere in this book we have seen how microorganisms are able to utilise an enormous range of organic compounds as carbon sources. Many organisms can metabolise not only naturally occurring substances, but also synthetic ones, making them valuable allies in the process of bioremediation. Often the most effective microorganisms to use are those found living naturally at the contaminated site, since they have demonstrated a naturally developed ability to survive the toxic effects of the pollutant, although in other cases specially adapted or genetically modified bacteria may be introduced (*bioaugmentation*). Examples of the use of microorganisms include the treatment of toxic waste sites, chemical spills, pesticides in groundwater and oil spills. One of the first large-scale attempts at bioremediation came in the aftermath of the *Exxon Valdez* disaster in 1989, when thousands of tons of crude oil were released off the coast of Alaska. Depending on the circumstances, bioremediation procedures may occur *in situ*, or the contaminated soil or water may be removed to a specialist facility for treatment.

> *Bioremediation* is the use of biological processes to improve a specific environment, such as by the removal of a pollutant.

> *Bioaugmentation* is the deliberate introduction of specific microorganisms to an environment in order to assist in bioremediation.

14.9 Harmful effects of microorganisms in the environment

The natural processes of bioconversion that are so important in the global recycling of elements may have unwanted consequences for humans. One of these is *acid mine drainage*, a frequently encountered problem in mining regions. Bacterial oxidation of mineral sulphides, particularly the ubiquitous iron pyrite, leads to the release of a highly acidic leachate into streams and rivers. This also contains toxic dissolved metals such as copper and also ferric iron. When it mixes with stream water, the pH is raised sufficiently for the iron to precipitate as unsightly orange ferric hydroxides, blanketing the stream bed and wiping out plant and animal life. The main culprits in the formation of acid mine drainage are sulphur-oxidising bacteria, notably *Acidithiobacillus ferrooxidans;* as we shall see in Chapter 18, under controlled conditions this same organism can also provide economic benefits to the mining industry by extracting valuable metals from low-grade ores.

Another area in which environmental microorganisms can have detrimental effects is that of *biodeterioration*, whereby economically important materials such as wood, paper, textiles, petroleum and even metals and concrete may be subject to damage by a range of microorganisms, mainly fungi and bacteria.

> *Biodeterioration* is the damage caused to materials of economic importance due to biological (mainly microbial) processes.

The most important microorganisms in the biodeterioration of wood are members of the Basidiomycota. Wood is only susceptible to fungal attack when its moisture level reaches around 30%. The major component of wood that is subject to microbial attack is *cellulose*, although some microorganisms can also degrade *lignin*. There are two main forms of rot: *white rot*, which involves the degradation of lignin as well as cellulose, and *brown rot*, in which the lignin is unaltered.

The dry rot fungus *Serpula lacrymans* produces thick strands of hyphae called *rhizomorphs*, which it uses to conduct water and nutrients from damper areas. These are very strong, and able to travel over brickwork and masonry barriers. *S. lacrymans* can generate water as a metabolic end-product and thus, once established, is able to grow even on dry wood. Dry rot flourishes in areas of static dampness such as badly ventilated, uninhabited properties.

Since cellulose is also an important component of paper and textiles, its breakdown is clearly of great economic importance. Degradation by fungi and, to a lesser extent, bacteria, results in a loss of strength of the material in question. The paper-making process provides warm, wet conditions rich in nutrients, ideal for microbial growth, which can clog up machinery and discolour the finished product. A variety of biocides are used in an effort to minimise microbial contamination.

The discoloration referred to above raises the point that biodeterioration of a material need not necessarily affect its physical or chemical make-up; aesthetic damage can lessen the economic value of a material by altering its appearance. The blackening of shower curtains by moulds growing on surface detritus, familiar to generations of students, is another example of this!

VI
Medical Microbiology

15
Human Microbial Diseases

In Chapter 1 we saw how, in the decades either side of 1900, advances in microbiology meant that a causative organism could be assigned to most of the major infectious diseases (see Table 1.1). The present chapter is devoted to human microbial diseases, although it must be appreciated that only a relatively superficial coverage is possible in an introductory text such as this. We shall consider some of the mechanisms used by microorganisms to cause disease, together with modes of disease transmission, before describing some representative examples of diseases caused by different types of microorganisms.

As we have seen, only a small proportion of the vast numbers of microbial species give rise to infections; these are termed *pathogens*. An *infectious disease* can be defined as a change in the state of an individual's health brought about by the invasion and colonisation of the body by pathogens, which may be bacteria, viruses, fungi or protists. It is worth making clear the distinction between infection and disease in this context. An *infection* is the colonisation, multiplication, invasion or persistence of a pathogen in or on a host, whereas the term *infectious disease* is only used when this results in significant harm.

A *primary infection* is one where there is a clear invasion and multiplication of the microorganisms leading to local tissue damage. A *secondary infection* occurs when there is an invasion of microorganisms subsequent to a primary infection, such as bacterial pneumonia following a viral lung infection. An *opportunistic infection* is one that occurs only when the host's defence systems have been compromised in some way; the infection of a tissue injury by normal host microflora would be an example of this, or the infection of an AIDS patient by normally harmless microorganisms due to the person's suppressed immune system. *Accidental infections* are caused by organisms that are not normally associated with humans but that in unusual circumstances can cause disease. For example, the anaerobe *Clostridium tetani* is widespread

Essential Microbiology, Second Edition. Stuart Hogg.
© 2013 John Wiley & Sons, Ltd. Published 2013 by John Wiley & Sons, Ltd.

in soils and only gains access to the body as a result of being introduced via a deep puncture or scratch, where it can cause the potentially fatal condition of tetanus.

The ability of an organism to cause disease is called its *pathogenicity*, and the degree of its pathogenicity is its *virulence*. Many factors affect the virulence of an organism, including the numbers of organisms involved, the route of entry, the state of competence of the host's defences and *virulence factors* of the pathogen. Virulence can be quantified by determining the number of bacteria required to kill 50% of a group of experimental animals within a defined time period, the *lethal dose 50* (LD_{50}). The virulence of a pathogen may be closely related to factors such as its mode of transmission and ability to live outside the host. If a pathogen depends on direct contact for transmission, it would not be to its benefit to damage its host so extensively that it was unable to contact others and continue transmission of the disease. The common cold, for example, has a low virulence; sufferers are able to carry on with their lives, and in doing so continue to come into contact with other people and pass the virus on to them. When a pathogen is not dependent on host mobility, the effect on the host can be more drastic; despite the death of the host, the pathogen may still flourish if it can be transmitted effectively by other means, such as by water or an insect vector (see examples later). Similarly, those pathogens that are unable to survive outside their hosts and therefore depend on direct contact transmission will again be dependent on the survival of their host and tend to be less virulent. Those that can readily survive in the environment can afford to cause serious damage or death to their host because they are not dependent on it for their survival. The causative agents of diseases such as tuberculosis and diphtheria can survive for months without entering a human host. It is also important to realise that the virulence of a given organism is not a constant, but may change under the influence of a range of environmental factors.

To be able to induce an infectious disease, a pathogen must be able to:

- travel/be carried to the host;
- attach to, colonise or invade the host;
- resist host defences and multiply or complete its life cycle within the host;
- cause damage to the host by chemical or mechanical means.

We now considering each of these in turn.

15.1   Transmission

Transmission of a pathogen may be direct (involving person-to-person contact) or indirect.

Direct transmission involves person-to-person transmission without the involvement of any intermediate, such as in the case of the common cold, influenza and hepatitis A, as well as sexually transmitted diseases. Pathogens transmitted in this way are often very sensitive to desiccation and other environmental factors and cannot survive away from the body for even a short time, hence the need for direct transmission. Also classed as directly transmitted are skin pathogens such as staphylococci that cause boils and pimples, and fungi that give rise to ringworm, for example. These, however, are more resistant to environmental factors, hence contact literally from person to person is not always necessary. Organisms that are spread by droplets of saliva or mucus as a result of coughing, sneezing or even laughing are also regarded as being directly transmitted. Droplets generated in this way travel less than a metre through the air and the organisms are not regarded as being truly airborne. A special form of direct transmission is the passing of a pathogen from mother to unborn or newborn child (vertical transmission). AIDS, toxoplasmosis and congenital syphilis can all be transmitted in this way.

Indirect transmission involves an intermediate of some sort. *Fomites* are inanimate objects on which pathogens can remain viable until contacted by a second individual. Common examples include towels, razors, syringe needles and children's toys. Athlete's foot and tetanus are both transmitted by fomites. Microbial diseases can also be transmitted by living intermediates called *vectors*, such as insects or cats and dogs. Sometimes the pathogen simply hitches a ride on the surface of the vector, which thus plays an entirely passive role; for example, flies can carry gastrointestinal pathogens of humans such as *Shigella* on their feet and bodies after visiting a source of faeces. The rat flea, on the other hand, which transmits *Yersinia pestis*, the causative agent of plague, does so by actually ingesting the bacteria with a blood meal and then passing them on when feeding on another host. Both of these are examples of *mechanical transmission*. In *biological transmission,* the pathogen undergoes some part of its life cycle inside the vector; an example of this is the malaria protozoan *Plasmodium*, which multiplies and accumulates inside the mosquito that transmits it.

Common vehicle transmission refers to those pathogens that are transmitted by means of an inanimate *reservoir of disease* used by both the pathogen and its prospective host. Contaminated water (e.g. with cholera) is the most obvious example, but air, blood transfusions, food and medications are others. To be regarded as truly *airborne transmission*, the droplet or dust particle carrying the pathogen needs to travel over a metre from the disease source to the host. Some microorganisms can be carried in small droplets of mucus called droplet nuclei. These result from the evaporation of droplets from sneezes and the like and are so light that they can stay airborne over long distances. Tuberculosis and measles are largely airborne transmitted diseases. Finally, microbial infections may be transmitted through food that is contaminated with a pathogen such as *Salmonella* or Norovirus, which causes diarrhoea and food poisoning.

15.2 Attachment and colonisation

Having reached the host, the first thing a potential invader of the body will come across is an epithelial surface of some description – the skin, or the lining of the various entrances to the body, that is, the alimentary, respiratory and genitourinary tracts. These represent a formidable series of physical and chemical barriers that must be overcome before the invader can gain entry to the body.

15.2.1 Skin

Skin forms an effective barrier to invading organisms for a number of reasons:

- Few organisms can penetrate the intact skin because the outer layer is made up of thick, closely packed keratinised cells, resistant to enzymatic attack.
- Any organisms that do manage to adhere will soon be removed as the process of desquamation continually removes the outer cells.
- The skin is mildly acidic (pH 5–6) due to the production of fatty acids such as oleic acid from the breakdown of lipids; this inhibits the growth of many microorganisms. The low pH of sweat has a similar effect.
- Lysozyme, which breaks the NAG–NAM bond in peptidoglycan, is to be found in secretions such as tears and saliva.
- The relative dryness of the skin slows down microbial growth but it flourishes in damper areas such as the groin, under arms and between the toes.
- Normal, harmless skin microflora compete for nutrients and attachment sites and will in many cases act antagonistically against invading pathogens.

15.2.2 Mucous membranes

The alimentary, respiratory and genitourinary tracts have mucous membranes whose stratified epithelium and secretions of mucus provide a protective covering that resists penetration. In the respiratory tract, particles greater than 10 μm in size are trapped by the hairs and cilia of the nasal lining, whereas smaller particles are deposited on the mucous surface; particles such as bacteria that are trapped in this way are then removed by mechanisms such as coughing and ciliary movement. Mucus is made up of long, branched acidic carbohydrates called *mucins,* which are good at trapping and sticking to bacteria and preventing them from attaching to receptors on the epithelial cells. In the alimentary tract, saliva washes bacteria down to the acidic stomach, which destroys most bacteria and their toxins. Toxins of *Staphylococcus* and *Clostridium* are exceptions to this, and cysts of protozoans such as *Cryptosporidium* can also survive. Microorganisms that have survived the acid environment of the stomach, for example by being protected by a layer of food, pass to the small intestine, where bile, pancreatic enzymes and enzymes from intestinal secretions may destroy them. The process of peristalsis and the normal loss of columnar epithelium lining the small intestine help to remove invading organisms, and in addition, specific immune proteins in the alimentary tract prevent bacterial attachment. In extreme cases of microbial ingestion and toxin production, a vomiting reflex may be triggered. In the large intestine there is a large resident microflora, which is important in preventing invaders from getting established. It does this by competing for attachment sites and nutrients and by producing inhibitory metabolic products. A layer of mucus here also performs a protective function.

> The bacterial count of a healthy stomach is as few as 10 cells per ml!

The kidneys, ureters and bladders of mammals are normally sterile, as is the urine itself. Some bacteria are killed because of the relatively acid pH of the urine, and also because of the presence of urea and other metabolic products such as uric and hippuric acids. In the kidney itself, conditions are so hypertonic that few microorganisms are able to survive. Potential pathogens are flushed from the urinary tract some 4–10 times daily by urination. In the female genitourinary tract, acid-tolerant lactobacilli degrade glycogen from the vaginal epithelium to form lactic acid, producing a pH of 3–5, which is unfavourable to most organisms.

A final, less obvious opening to the body is represented by the eye. The *conjunctiva* is a specialised mucus-secreting epithelial membrane that lines the inner surface of each eyelid and the exposed surface of the eye itself. This is protected by the secretions in tears, which, as well as keeping the conjunctiva moist, are rich in lysozyme and other antimicrobial substances.

In order to give rise to an infection, pathogens need either to colonise the cell surface or to penetrate within it. To do this, they need to overcome the host's defence mechanisms described above. The routes by which they achieve this are called *portals of entry*. Many pathogens have a preferred portal of entry, and are only able to cause an infection if that route is used (Table 15.1). Streptococci, for example, can cause pneumonia if they invade

Table 15.1   Portals of entry for human pathogens

Mucous membranes:
   respiratory tract
   gastrointestinal tract
   genitourinary tract
   eyes
Breaks in host barriers, mainly skin (parenteral route)
Placenta

the respiratory tract, but if they are swallowed there are generally no harmful effects. Other pathogens, such as *Mycobacterium tuberculosis* can gain entry by a variety of routes.

The virulence of an organism depends on a combination of its own properties and the ability of the host to resist it. Pathogenic bacteria often possess *virulence factors* – structural or physiological features that enhance their disease-causing capabilities by helping with attachment to the host or evasion of its defences. Virulence factors may also take the form of toxins that actually cause the adverse effects.

Many bacteria and viruses gain access to the body by penetrating mucous membranes lining the respiratory, gastrointestinal and genitourinary tracts. The most commonly invaded of these is the respiratory mucosa by, for example, the causative agents of the common cold, influenza, pneumonia, or whooping cough. Microorganisms ingested with food or drink or from dirty fingers enter the gastrointestinal tract, but most are destroyed by enzymatic secretions and the low pH of the stomach. Those that survive, either by being protected inside a food bolus or through sheer weight of numbers, can cause gastroenteritis, cholera, typhoid and dysentery. They are eliminated in the faeces and passed to the next host via contaminated water, poor food hygiene, etc. The route of entry for most urinary infections and sexually transmitted diseases is via the genitourinary tract. Microorganisms may also be introduced through a breakage of the skin or mucous membranes, from cuts, bites or burns.

15.2.3   How do pathogens penetrate the mucosa?

Some bacteria release extracellular enzymes, sometimes known as 'spreading factors', that facilitate entry to host cells. *Hyaluronidase* breaks down hyaluronic acid, a key component of the matrix that holds cells together, by digesting it; bacteria such as *Streptococcus pyogenes* and *Staphylococcus aureus* are able to penetrate more deeply into a tissue by passing through the intercellular spaces. *Collagenase* is another enzyme that breaks down connective tissue, in this case collagen.

Coagulase, produced by *Staphylococcus aureus,* works by accelerating the clotting of blood. Plasma leaking out of vessels into tissues will be clotted; this can have the effect of coating the pathogens with fibrin and protecting them from the host's immune cells. This explains the isolated nature of *S. aureus* infections in boils and pimples. Another bacterial enzyme, streptokinase (produced by *Streptococcus pyogenes*), has the opposite effect; it dissolves blood clots, releasing trapped bacteria and allowing them to spread to other tissues. It does this by activating the production of plasmin, which digests fibrin.

Once inside a host, most bacteria are able to attach themselves to the host's epithelia, frequently with a high degree of specificity, both for host and type of tissue. This attachment takes place by means of cell surface molecules called *adhesins*. These are proteins or glycoproteins found on surface structures such as fimbriae, pili or capsules. The term is sometimes extended to describe the structures themselves.

Most adhesins only allow the pathogen to attach to certain cells or tissues, because these carry specific receptors for them. *Campylobacter* spp. and *Vibrio cholerae*, for example, both have adhesins on their flagella that bind to specific receptors on the intestinal lining, whereas *Streptococcus mutans*, which causes tooth decay, attaches via its capsule to tooth enamel. The possession of fimbriae and capsules also makes the bacterium more difficult for the host's phagocytic cells to deal with. As described in Chapter 10, viruses attach by binding to specific receptors on the surface of their host cells.

Host macrophages form part of the front line of defence against pathogens. They function by internalising the invaders by phagocytosis, and then causing their destruction by fusing with lysosomes, membrane-bound packets of potent destructive enzymes. Some pathogens, however, have developed ways of avoiding lysis, so the process of internalisation actually aids their invasion of the host.

> Certain pathogenic strains of *E. coli* such as O157:H7 are able to synthesise a protein that acts as a receptor for their own adhesin. The protein, known as Tir (translocated intimin receptor), is secreted into the membrane of the cells lining the small intestine, where it enables the adhesin intimin, which is expressed on the bacterial surface, to bind.

15.3   Bacterial toxins

Disease is rarely caused simply by the presence of microorganisms, even in large numbers, although there are exceptions to this, such as when a crucial vessel or heart valve becomes blocked due to sheer weight of bacterial numbers.

In most cases, the pathogens exert their effect by producing *toxins*. A toxin is a substance, generally a metabolic product of the pathogen, which causes damage to the host. Bacterial toxins can be conveniently divided into two types, *exotoxins* and *endotoxins* (Table 15.2).

- *Exotoxins* are soluble proteins released from living bacteria into host tissues; they may be transported away from the site of infection to produce damage at distant sites.
- *Endotoxins* form part of the bacterial cell surface and are mainly released when the cell is lysed. They are lipopolysaccharides and are limited to Gram-negative bacteria.

Table 15.2 Features of endotoxins and exotoxins

Exotoxins	Endotoxins
Proteins	Lipopolysaccharides
Secreted from bacterial cell	Integral part of bacterial outer membrane
Denatured by temperatures >60°C	Stable to autoclaving
Usually site-specific	Systemic
Toxic at very low concentrations	Only toxic at much higher concentrations
Not fever-inducing	Fever-inducing
Strongly immunogenic	Weakly immunogenic

15.3.1 Exotoxins

Exotoxins are among the most potent toxic substances known, per unit weight. They are mostly produced by Gram-positive bacteria, together with a few Gram-negative. They have no known function for the bacteria, and have been shown to be non-essential for survival by deleting the genes coding for them without ill effects. The site of action of most exotoxins is restricted to certain cells or cell receptors; we can categorise exotoxins on the basis of this site specificity, and will now consider some examples in more detail.

Tetanus and *botulinum* toxins are both produced by members of the genus *Clostridium (C. tetani and C. botulinum)*, and both are examples of *neurotoxins*. They are also both *A-B toxins*, that is, they comprise two subunits joined by covalent bonding, the B component binding to a cell surface receptor, allowing the toxic A subunit to be transferred into the host cell where it exerts its effect. The A subunit acts as a proteolytic enzyme, preventing neurotransmitter release.

The botulinum toxin is one of the most toxic substances known; it has been calculated that a milligram of pure botulinum toxin would be enough to kill over 10 000 people. It acts by binding to presynaptic membranes of stimulatory motor neurons, preventing the release of the neurotransmitter acetylcholine. This means that muscles do not receive an excitatory signal and therefore cannot contract, leading to flaccid paralysis. Without adequate treatment around 30% of patients can die within 2–3 days from respiratory failure or cardiac arrest. Botulinum toxin is used clinically to relieve the effects of conditions in which there is uncontrolled muscular contraction. One such condition is blepharospasm, in which uncontrolled contractions of the muscles around the eyes may render the patient functionally blind. Injection of minute amounts of

'Botox' injections, much in vogue in certain circles as a cosmetic treatment, involve low doses of *C. botulinum* exotoxin. By acting as a muscle relaxant, they are intended to reduce the facial wrinkles that develop with the passing of time.

botulinum toxin every few months results in the blocking of nerve impulses to the muscles, allowing the eyes to be opened.

Tetanus toxin is another A-B toxin that affects the nervous system. It is transported back through the motor neurons to the spinal cord, where it binds specifically to the surface of *inhibitory* neurons. These normally work by releasing glycine, which acts as an inhibitory neurotransmitter, preventing the release of acetylcholine and therefore inhibiting muscle contraction. If the glycine release is inhibited, there will be a continual release of acetylcholine and uncontrolled muscle contraction. This is what the tetanus toxin *tetanospasmin* does. One manifestation of this is lockjaw, a prolonged spasm restricting movement of the mouth, and if respiratory muscles are involved, death can result from asphyxiation.

Thus although the two toxins both have serious and even fatal outcomes, the mechanisms by which they work are in a sense the exact opposite of each other. An important difference between them at another level is that while *C. tetani* grows within its host and secretes its toxin, botulism is caused by the ingestion of food contaminated with toxin, rather than of the organism itself. For this reason, antibiotic therapy for foodborne botulism would be ineffective and treatment focuses instead on neutralising the exotoxin. Botulism is therefore not strictly speaking an infectious disease, but an *intoxication*.

Enterotoxins cause a huge secretion of water and electrolytes from the intestinal epithelium into the lumen of the intestine, resulting in the symptoms of diarrhoea. They are produced by organisms such as *Salmonella enteroviridis, Clostridium perfringens* and *Vibrio cholerae*. Cholera is a disease we'll be looking at in more detail later in this chapter.

A third main group of exotoxins are the *cytotoxins*. They are a heterogeneous group of soluble proteins that are secreted by a range of bacteria, and share in common the feature of either killing host cells or affecting their functions. An example is the *diphtheria* toxin, which has the distinction of being the first exotoxin to be discovered. It is yet another A-B toxin, with the B subunit binding to the host cell surface, and the A subunit entering to have its toxic effect as before. The mode of action of the A subunit is different again, however. It acts by blocking transfer of amino acids from tRNA to the growing polypeptide chain, and so disrupting protein synthesis (see Chapter 11). The actual effect is on a protein called elongation factor 2. The bacterium in question, *Corynebacterium diphtheriae*, only produces the diphtheria toxin if it has been infected by a bacteriophage known as phage β. This carries the *tox* gene responsible for encoding the diphtheria toxin. Some cytotoxins work by disrupting host membranes, causing cell lysis. They do this enzymatically, or by the insertion of protein channels into the host cytoplasmic membrane. *Leukocidins* are cytotoxins produced by certain streptococci and staphylococci that are capable of killing host white blood cells, specifically macrophages and neutrophils. Since these are phagocytic cells and form an important part of the host's immune system, this affects the host's ability to defend itself against microbial attack. Leukocidins act by inserting into the host cell membrane to form a pore. *Haemolysins* lyse red blood cells by a similar mechanism. Their presence can be demonstrated in the laboratory by areas of clearing around the colonies of growth when the bacteria are cultured on blood agar. α-Haemolysis results in a greenish halo around the bacteria due to lysis of the red blood cells and partial breakdown of haemoglobin, whereas β-haemolysis involves complete breakdown of haemoglobin, giving a clear area. Destruction of red cells is not on a scale to have any harmful effects for the host, but the breakdown of haemoglobin provides the bacteria with a source of iron, essential for bacterial growth. Red cells are not the only cells attacked by haemolysins, but provide an easy method of detection. Streptococci and staphylococci both produce haemolysins.

Unlike endotoxins, exotoxins are strongly *immunogenic*, that is, they stimulate the immune system to produce antibodies (*antitoxins*) against them, providing the host with immunity against future infections by the same organism. Most individuals who suffer with recurrent *Clostridium difficile*-associated diarrhoea are only able to produce low levels of antibodies against the *C. difficile* exotoxins TcdA (an enterotoxin) and TcdB (a cytotoxin).

Toxoids are toxins that have been altered so that they have lost their toxicity but retained their immunogenicity, and are administered in order to develop immunity against the toxin. Tetanus injections contain the tetanus toxoid.

15.3.2   Endotoxins

You will recall from Chapter 3 that the cell wall of Gram-negative bacteria has a so-called outer membrane, comprising a bilayer of phospholipid and lipopolysaccharide, and that the lipid A component of the latter can act as an endotoxin. Many Gram-negative bacteria including *Salmonella* and *Yersinia pestis* owe their toxic properties to their endotoxin. It is the lipid A component that contributes nearly all the toxic properties of the lipopolysaccharide. Other Gram-negative bacteria may contain a version of lipopolysaccharide lacking specific parts of the lipid A and as a result are not toxic. Endotoxins are produced when Gram-negative bacteria such as *E. coli*, *Shigella* and *Salmonella* die and their cell walls are fragmented.

Endotoxins produce the same symptoms when released into the bloodstream, regardless of the specificity of the O-polysaccharide component, although they may differ in their degree. Macrophages are stimulated to produce inflammatory *cytokines* known collectively as endogenous pyrogens, and the net effect is to induce *inflammation, intravascular coagulation, haemorrhage* and *shock* (Gram-negative septicaemia or septic shock). In serious cases this can lead to

> *Cytokines* are small proteins released from cells of the immune system that are involved in cellular communication and regulation.

multiple organ failure and death. As mentioned above, endotoxins are poorly immunogenic compared with exotoxins; their immunogenicity is due to the O-polysaccharide fraction. Antibiotics taken to combat Gram-negative infections work by lysing the bacterial cells; in so doing, they may release endotoxins and thereby cause a temporary worsening of the condition.

It is important that when we receive medical treatment in the form of antibiotics and other drugs, they are free of endotoxins, otherwise some of the adverse effects just described could result. To this end rigorous standards are observed and sensitive tests have been developed for the detection of small amounts of endotoxin contamination. One of these is based on the fact that trace amounts of endotoxin will react with the clotting protein from the amoebocytes of *Limulus*, the horseshoe crab. In the commercial assay, the presence of endotoxin causes the clotting protein to precipitate, resulting in changes in turbidity, which can be measured spectrophotometrically. The *Limulus* amoebocyte lysate (LAL) assay is sensitive down to picogram quantities of lipolysaccharide.

15.3.3 Superantigens

Superantigens are bacterial proteins that produce a set of symptoms very similar to those of septic shock, but by a different mechanism. They activate a large proportion of the immune system's T-lymphocytes in a non-specific way and elicit a massive inflammatory reaction due to an overproduction of cytokines. This may result in the same symptoms of diarrhoea and vomiting, fever or potentially fatal systemic shock. Toxic shock syndrome (*Staphylococcus aureus*) and scarlet fever (*Streptococcus pyogenes*) are both caused by bacterial superantigens.

15.3.4 Siderophores

The final class of virulence factors we're going to consider are *siderophores*. Both bacteria and their hosts require iron for growth and metabolism, and animals have developed mechanisms of withholding it from their tissue fluids, making conditions unfavourable for invading bacteria. Siderophores are structurally diverse molecules that share the common feature of high affinity for ferric ions. An environment deficient in iron stimulates the transcription of bacterial genes that encode enzymes involved in the synthesis of siderophores. In some cases the affinity of siderophores may be so high as to solubilise iron already tied up as lactoferrin or transferrin and transfer it to the bacteria. Once the iron-siderophore

> Much of the iron in the blood is not available to bacteria because it is bound up as either haemoglobin (in red cells) or transferrin (in plasma). Similarly iron in other body fluids is tied up as lactoferrin.

Table 15.3 Some bacterial diseases of humans. The table shows the genera responsible for some important bacterial diseases of humans. Note that many of the diseases listed may be caused only by a particular species within the genus

	Genus	Disease
Gram-positive	Staphylococcus	Impetigo, food poisoning, endocarditis, bronchitis, toxic shock syndrome
	Streptococcus	Pneumonia, pharyngitis, meningitis, scarlet fever, dental caries
	Enterococcus	Enteritis
	Bacillus	Anthrax
	Clostridium	Tetanus, botulism, gangrene
	Corynebacterium	Diphtheria
	Listeria	Listeriosis
	Mycobacterium	Leprosy, tuberculosis
	Propionibacterium	Acne
	Mycoplasma	Pneumonia, vaginosis
Gram-negative	Salmonella	Salmonellosis
	Escherichia	Gastroenteritis
	Shigella	Dysentery
	Neisseria	Gonorrhoea, meningitis
	Bordetella	Whooping cough
	Legionella	Legionellosis
	Pseudomonas	Infections of burns
	Vibrio	Cholera
	Campylobacter	Gastroenteritis
	Helicobacter	Peptic ulcers
	Haemophilus	Bronchitis, pneumonia
	Treponema	Syphilis
	Chlamydia	Pneumonia, urethritis, trachoma

complex reaches the bacterial surface it binds to a specific receptor protein, whose synthesis is also stimulated by the lack of environmental iron. The iron is then released into the bacterial cell and reduced to the Fe^{2+} form.

15.4 Bacterial diseases in humans

The remainder of this chapter is devoted to a discussion of representative examples of diseases caused by different classes of microorganisms. We start with bacterial diseases, with an emphasis on the different ways they can be transmitted. Table 15.3 summarises the principal bacterial diseases of humans.

15.4.1   Waterborne transmission: cholera

Cholera is most commonly contracted as a result of ingesting contaminated water containing faecal material. It is a form of gastroenteritis caused by a toxin released by *Vibrio cholerae*. The bacteria attach by means of adhesins to the intestinal mucosa, where, without actually penetrating the cells, they release the cholera exotoxin. This comprises an 'A' and several 'B' subunits; the former is the active component, while the latter attach to epithelial cells by binding to a specific glycolipid in the membrane. This allows the passage of the 'A' subunit into the cell, where it causes the activation of an enzyme called adenylate cyclase (Figure 15.1). This results in uncontrolled production of cyclic AMP, causing active secretion of Cl^- and HCO_3^- ions into the intestinal lumen, followed by large volumes of water in an effort to maintain ionic balance.

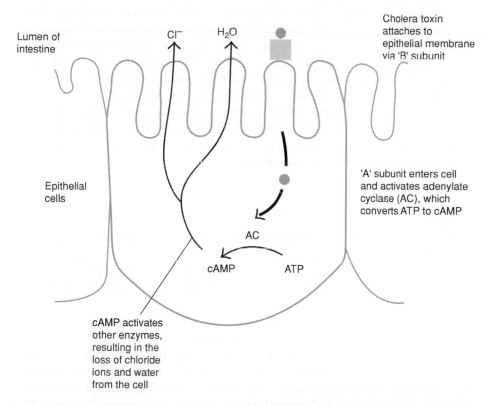

Figure 15.1   The action of the cholera exotoxin. Activation of the enzyme adenylate cyclase results in elevated levels of cyclic AMP and the secretion of electrolytes out of the epithelial cells lining the intestine and into the lumen. This is followed by water loss, resulting in debilitating dehydration.

The outcome of this is huge fluid loss (10 litres or more per day) through profuse and debilitating diarrhoea. In the young, old and sick, death through dehydration and salt depletion can follow within a few hours. If proper liquid and electrolyte replacement therapy is available, recovery rates are very high, but left untreated, mortality is 50% or more. Although it is now very rare in the developed world, cholera remains a major killer in less developed countries, claiming over 100 000 lives each year. It is easily preventable by means of clean water supplies and improved sanitation; however, when these services break down, for example in times of war or after natural disasters, cholera outbreaks quickly follow; indeed, the most severe outbreak of recent times occurred in the wake of the Haiti earthquake of 2010.

15.4.2   Airborne transmission: 'strep' throat

Streptococcal pharyngitis, commonly known as strep throat, is one of the commonest bacterial diseases of humans, being particularly common in children of school age. The primary means of transmission is by the inhalation from coughs and sneezes of respiratory droplets containing *Streptococcus pyogenes* (β-haemolytic type A streptococci), although other routes (infected handkerchiefs, kissing) are possible. The primary symptoms are a red and raw throat (and/or tonsils), accompanied by headaches and fever. *S. pyogenes* attaches to the throat mucosa, stimulating an inflammatory response and secreting virulence factors that destroy host blood cells. Although self-limiting within a week or so, strep throat should be treated with penicillin or erythromycin as more serious streptococcal diseases such as *scarlet fever* and *rheumatic fever* may follow if it is left untreated.

15.4.3   Contact transmission: syphilis

Causative organisms of sexually transmitted diseases such as *Neisseria gonorrhoeae* (gonorrhoea) and *Treponema pallidum* (syphilis) are extremely sensitive to the effects of environmental factors such as UV light and desiccation. They are therefore unable to live outside their human host, and rely for transmission on intimate human contact.

The spirochaete *T. pallidum* enters the body through minor abrasions, generally on the genitalia or mouth, where a characteristic lesion called a *chancre* develops. The disease may proceed no further than this, but if *T. pallidum* enters the bloodstream and passes around the body, the more serious secondary stage develops, lasting some weeks. Following a latent period of several years, around half of secondary syphilis cases go on to develop into the tertiary stage of the disease, whose symptoms may include mental retardation, paralysis and blindness. Congenital syphilis is caused by *T. pallidum* being passed from a mother to her unborn child.

The primary and secondary stages of syphilis are readily treated by penicillin; however, the tertiary stage is much less responsive to such therapy.

15.4.4 Vector-borne transmission: plague

A limited number of bacterial diseases reach their human hosts from their main host, usually another species of mammal, via an insect intermediary.

Plague (bubonic plague, the Black Death) has been responsible for the deaths of untold millions of people in terrible epidemics such as the ones that wiped out as much as one-third of the population of Europe in the Middle Ages. It is caused by the Gram-negative bacterium *Yersinia pestis*, whose normal host is a rat, but which can be spread to humans by fleas. The bacteria pass to the lymph nodes, where they multiply, causing the swellings known as buboes. *Y. pestis* produces an exotoxin, which prevents it from being destroyed by the host's macrophages; instead, it is able to multiply inside them. From the lymph nodes, the bacteria spread via the bloodstream to other tissues such as the liver and lungs.

Once established in the lungs (pneumonic plague), plague can spread from human to human by airborne transmission in respiratory droplets. Plague is extremely infectious, and left untreated has a high rate of fatality, particularly for the pneumonic form of the disease. Early treatment with streptomycin or tetracycline, however, is largely successful. Improved public health measures and the awareness of the dangers of rats and other rodents have meant that confirmed cases of plague are now relatively few.

15.5 Viral diseases in humans

Viruses are responsible for some of the most serious infectious diseases to affect humans. Some important examples are listed in Table 15.4, and two are discussed in a little more detail below.

15.5.1 Airborne transmission: influenza

Influenza is a disease of the respiratory tract caused by members of the *Orthomyxoviridae*. Transmission occurs as a result of inhaling airborne respiratory droplets from an infected individual. Infection by the influenza virus results in the destruction of epithelial cells of the respiratory tract, leaving the host open to secondary infections from bacteria such as *Haemophilus influenzae* and *Staphylococcus aureus*. It is these secondary infections that are responsible for the great majority of fatalities due to influenza. Generally, sufferers from influenza recover completely within 10–14 days, but some people, notably the elderly and those with chronic health problems, may develop complications such as pneumonia.

Table 15.4 Some important viruses of humans

Virus	Family	Disease	Genome type
Adenovirus	*Adenoviridae*	Respiratory infections	dsDNA
Ebola virus	*Filoviridae*	Haemorrhagic fever	(−)ssRNA
Epstein–Barr	*Herpesviridae*	Infectious mononucleosis	dsDNA
Hepatovirus A	*Picornaviridae*	Hepatitis A	(+)ssRNA
Herpes simplex Type I	*Herpesviridae*	Cold sores	dsDNA
Herpes simplex Type II	*Herpesviridae*	Genital warts	dsDNA
Human immunodeficiency virus (HIV)	*Retroviridae*	Acquired immunodeficiency syndrome (AIDS)	(+)ssRNA*
Human papillomavirus	*Papovaviridae*	Warts	dsDNA
Influenza virus	*Orthomyxoviridae*	Influenza	(−)ssRNA
Lassa virus	*Arenaviridae*	Lassa fever	(−)ssRNA
Morbillivirus	*Paramyxoviridae*	Measles	(−)ssRNA
Norovirus	*Calciviridae*	Enteritis	(+)ssRNA
Paramyxovirus	*Paramyxoviridae*	Mumps	(−)ssRNA
Polio virus	*Picornaviridae*	Poliomyelitis	(+)ssRNA
Rabies virus	*Rhabdoviridae*	Rabies	(−)ssRNA
Rhinovirus	*Picornaviridae*	Common cold	(+)ssRNA
Rotavirus	*Reoviridae*	Enteritis	dsRNA
Rubella virus	*Togaviridae*	German measles	(+)ssRNA
Smallpox virus	*Poxviridae*	Smallpox	dsDNA
Varicella-zoster	*Herpesviridae*	Chicken pox, shingles	dsDNA
Yellow fever virus	*Flaviviridae*	Yellow fever	(+)ssRNA

*The genome of HIV, like that of other retroviruses, also has a DNA phase; see Chapter 10.

The influenza virus has an envelope, and a segmented (−)sense ssRNA genome (Figure 15.2). The envelope contains two types of protein spike, each of which plays a crucial role in the virus's infectivity:

- *Neuraminidase* is an enzyme that hydrolyses sialic acid, thereby assisting in the release of viral particles.
- *Haemagglutinin* enables the virus to attach to host cells by binding to epithelial sialic acid residues. It also helps in the fusion of the viral envelope with the cell membrane.

Both types of spike act as antigens, proteins that stimulate the production of antibodies in a host. One of the reasons that influenza is such a successful virus is that the 'N' and 'H' antigens are prone to undergoing changes (*antigenic shift*) so that the antigenic 'signature' of the virus becomes altered, and host immunity is evaded. Different strains of the influenza virus are given a code denoting which variants of the antigens they carry; the subtype N_1H_1, for example, caused both the 1918 'Spanish flu' and the 2009 'swine flu' pandemics, while H_5N_1 was responsible for the 'bird flu' outbreak in SE Asia in 2003–4.

Figure 15.2 The influenza virus. The RNA segments are bound to protein, forming a nucleocapsid, and are surrounded by further protein. The two types of spike assist in the attachment and penetration of the virus into its host.

15.5.2 Transmission by water or food: viral gastroenteritis

Most people will be only too familiar with the symptoms of gastroenteritis – sickness, diarrhoea, headaches and fever. The cause of this gastroenteritis may be bacterial (e.g. *Salmonella*) or viral. The major cause of the viral form is the human rotavirus, which, together with the norovirus, is responsible for the majority of reported cases.

rota = 'a wheel', describing the distinctive appearance of this type of virus.

The viruses damage the villi in the upper part of the intestinal tract, affecting normal ion transport, and resulting in the characteristic water loss. Transmission of gastroenteritis is via the faecal-oral route, that is, by the ingestion of faecally contaminated food or water. Poor hygiene practice or contaminated water supplies are usually to blame for the perpetuation of the cycle. Normally, the condition is self-limiting, lasting only a couple of days; the normal treatment is fluid replacement therapy. In areas where clean water supplies are not available, however, the outcome can be much more serious. In less developed countries, the condition is a major killer; it is the principal cause of infant mortality, and the cause of some five to ten million deaths per year.

15.5.3 Vector-borne transmission

Some human viral diseases are spread by means of arthropods such as ticks and mosquitoes that feed on human blood. The viruses are known as *arboviruses* (*ar*thropod-*bo*rne viruses), and are mostly enveloped RNA viruses. Note, however, that the arboviruses do not form a discrete taxonomic grouping, but simply share a means of transmission. Diseases caused by arboviruses include dengue fever, yellow fever, and forms of viral encephalitis such as West Nile encephalitis and St Louis encephalitis.

15.5.4 Latent and slow (persistent) viral infections

After an infection has passed, a virus may sometimes remain in the body for long periods, causing no harm. It may be reactivated, however, by stress or some change in the individual's health, and initiate a disease state. Well-known examples of *latent viral infections* are cold sores and shingles, both caused by members of the herpesvirus family. A virus of this sort will remain with an individual throughout their lifetime.

Whereas latent virus infections are characterised by a sudden increase in virus production, in *persistent (slow) infections* the increase is more gradual, building up over several years. Such infections have a serious effect on the target cells, and are generally fatal. An example is the measles virus, which can re-manifest itself after many years in a rare condition called subacute sclerosing panencephalitis (SSPE).

15.5.5 Viruses and cancer

A number of chemical and physical agents are known to trigger the uncontrolled proliferation of cells that characterise cancers, but in the last two decades it has become clear that at least six types of human cancer can be virally induced. How do cells lose control of their division, and how are viruses able to bring this about? It is now known that cells contain genes called *proto-oncogenes*, involved in normal cell replication. They are normally under the control of other, *tumour-suppressor genes*,

> An *oncogene* is a gene associated with the conversion of a cell to a cancerous form.

but these can be blocked by proteins encoded by certain DNA viruses. When this happens, the proto-oncogene functions as an *oncogene*, and cell division is allowed to proceed uncontrolled. Retroviruses have a different mechanism; they carry their own, altered, version of the cellular oncogene, which becomes integrated into the host's genome and leads to uncontrolled cell growth. Retrovirus oncogenes are thought to have been acquired originally from human (or animal) genomes, with the RNA transcript becoming incorporated into the retrovirus particle.

15.5.6 Emerging and re-emerging viral diseases

As a result of changes in the pathogen or in the host population, completely new infectious diseases may arise, or we may experience the reappearance of diseases previously considered to be under control. These are known as, respectively, *emerging* and *re-emerging* infections. Social, economic and climatic changes and changed patterns of human population movement may all create the conditions that are often respon-

> A *zoonosis* is a disease normally found in animals, but transmissible to humans under certain circumstances.

sible for the development of such infections. Frequently, emerging virus infections are *zoonotic* in origin, that is, they are transferred to humans from animal reservoirs. The human immunodeficiency virus, for example, is thought to have developed from a similar virus found in certain types of monkey.

While the first edition of this book was in preparation, there was a sharp reminder of the ever-present threat of emerging viral diseases, in the form of a new viral disease called *severe acute respiratory syndrome* (SARS). The outbreak of this disease began in Guangdong province in southern China in November 2002. The Chinese authorities were heavily criticised for not reporting the extent of the outbreak until some three months later, by which time cases were appearing in many parts of the world, illustrating the role of increased intercontinental travel in the spread of disease in the modern era. At its peak in April 2003, over 1000 new cases of SARS were being reported per week. The cause of SARS was quickly identified as a member of the *Coronaviridae* (single-stranded RNA viruses). Transmitted by droplets from coughs and sneezes, it produces flulike symptoms, but has a mortality rate of around 4%. Strict public health measures were brought into force, including restrictions in flights to and from affected areas, and the number of reported cases began to subside. By the time the outbreak was contained, SARS had claimed over 900 lives, mostly in China and Hong Kong, but with a number of deaths occurring as far afield as Canada and South Africa.

15.5.7 Virus vaccines

Smallpox, once the scourge of millions, was in 1979 the first infectious disease to be declared successfully eradicated. This followed a worldwide campaign of vaccination by the World Health Organization over the previous decade, and was made feasible by the fact that humans are the only reservoir for the virus. *Vaccination* is a preventive strategy that aims to stimulate the host immune system, by exposing it to the infectious agent in question in an inactivated or incomplete form. There are four main classes of virus vaccines.

Attenuated (= 'weakened') vaccines contain 'live' viruses, but ones whose pathogenicity has been greatly reduced. The aim is to mimic an infection in order to stimulate an immune response, but without bringing about the disease itself. A famous example of this type of vaccine is the polio vaccine developed by Albert Sabin in the 1960s. The cowpox virus used by Edward

Jenner in his pioneering vaccination work in the late eighteenth century was a naturally occurring attenuated version of the smallpox virus.

Inactivated vaccines contain viruses that have been exposed to a denaturing agent such as formalin. This has the effect of rendering them non-infectious, while at the same time retaining their ability to stimulate an immune response. Vaccines directed against influenza are of this type.

Subunit vaccines depend on the stimulation of an immune response by just a part of the virus. Since the complete virus is not introduced, there is no chance of infection, so vaccines of this type have the attraction of being very safe. Subunit vaccines are often made using recombinant DNA technology (see Chapter 12); the first example to be approved for human use was the hepatitis B vaccine, which consists of part of the protein coat of the virus produced in specially engineered yeast cells.

DNA vaccines are also the product of modern molecular biology techniques. DNA encoding virus antigens is directly injected into the host, where it is expressed and triggers a response by the immune system. Whilst showing great potential, vaccines of this type have not so far been approved for use in humans.

15.6   Protists and disease

When microbial diseases are discussed, it is generally bacteria and viruses that come to mind; however, a significant number are caused by eukaryotic pathogens, chiefly protozoans. Some important protozoan diseases such as cryptosporidiosis and toxoplasmosis are particularly prevalent in immunocompromised individuals such as those with AIDS. In some parts of the world, notably the tropics and subtropics, infections by protozoans are endemic; it is estimated that over two million people die each year from malaria alone. With international travel now commonplace, however, cases of travellers from Western countries returning home with these diseases are on the increase; around 2000 cases of malaria are now reported annually in the United Kingdom.

> A number of cases of 'airport malaria' have been documented, in which the unfortunate victim has been bitten by a 'stowaway' mosquito at an airport in Europe or the USA.

Each of the four categories of protozoans mentioned in Chapter 9 contains representatives capable of causing disease in humans (Table 15.5). Some of the better known examples are described in the following pages, illustrating the range of infectious cycles that exist among protozoan parasites. Some parasitic protozoans complete their life cycle in a single host, whilst others require more than one host to complete the different stages of their life cycle. The host in which the sexual stage of the life cycle is completed is called the *definitive host*, and the one with the asexual stage the *intermediate host*.

Table 15.5 Some major protozoal diseases of humans

Protozoan	Disease	Mode of infection
Mastigophora		
Trypanosoma brucei	African trypanosomiasis	Insect vector (tsetse fly)
Trypanosoma cruzi	Chagas' disease	Insect vector (kissing bug)
Giardia lamblia	Giardiasis	Faecally contaminated water
Leishmania	Leishmaniasis	Insect vector (sandfly)
Trichomonas vaginalis	Vaginitis	Sexual contact
Ciliophora		
Balantidium coli	Balantidial dysentery	Faecally contaminated water
Apicomplexa		
Cryptosporidium	Cryptosporidiosis	Contaminated water
Cyclospora cayetanensis	Cyclosporosis	Contaminated water
Plasmodium spp.	Malaria	Insect vector (mosquito)
Toxoplasma gondii	Toxoplasmosis	Ingestion of oocysts
Sarcodina		
Entamoeba histolytica	Amoebic dysentery	Faecally contaminated water

15.6.1 Malaria

Malaria, toxoplasmosis and cryptosporidiosis are all caused by members of the Apicomplexa group of protozoans (Chapter 9). Their life cycles vary considerably and are described below; however, all depend on being able to penetrate host cells by means of their apical complex of specialised organelles. Four species of *Plasmodium* cause malaria in humans – *P. falciparum*, *P. vivax*, *P. malariae* and *P. ovale* – with the first two of these being responsible for around 80% of cases. In addition, a fifth species, *P. knowlesi*, which infects macaque monkeys, has also been shown to infect humans. The life cycle of *Plasmodium* species involves both asexual and sexual reproduction, which take place respectively in the mammalian host and mosquitoes of the genus *Anopheles* (Figure 15.3). Within an hour of being introduced into the human bloodstream by a mosquito taking a blood meal, sporozoites invade liver cells, and initiate the infection. The subsequent invasion and bursting of red blood cells every two or three days causes the classic malaria symptoms of periodic chills and fever. Some of the released merozoites colonise other red blood cells, perpetuating the so-called erythrocytic cycle. Some develop into gametocytes, which are ingested by another mosquito when it bites the infected person. From here the sexual stage of the life cycle proceeds, resulting in a zygote that develops into an oocyst, the only diploid stages of *Plasmodium*. Meiosis produces large numbers of sporozoites, which travel to the mosquito's salivary gland, ready to recommence the cycle.

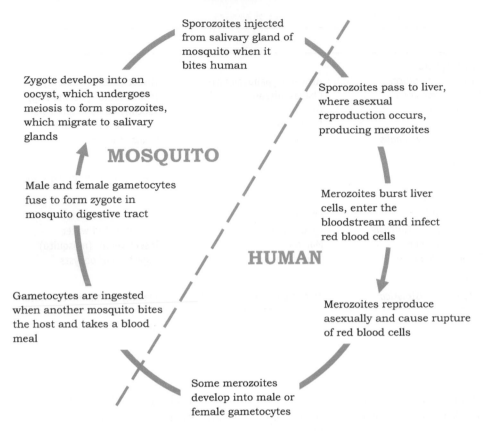

Figure 15.3 The *Plasmodium* parasite that causes malaria has two hosts and is transmitted by the *Anopheles* mosquito. Asexual reproduction (schizogeny) takes place in the human host; the sexual cycle occurs in the mosquito, following ingestion of blood containing gametocytes.

15.6.2 Toxoplasmosis

Another apicomplexan, *Toxoplasma gondii*, the causative agent of toxoplasmosis, has a life cycle that involves at least two mammalian hosts. The primary host is always a member of the cat family, whilst birds and mammals such as mice, farming livestock or humans act as intermediate hosts. Huge numbers of oocysts resulting from the fusion of gametocytes in the cat's digestive system are released in the animal's faeces, from where they are ingested by the intermediate host. Once inside the intermediate host, motile crescent-shaped tachyzoites are released from the oocysts. These are able to penetrate host cells, reproduce asexually and kill the cells by bursting them. Some tachyzoites give rise to tissue cysts or pseudocysts containing thousands of bradyzoites, usually in the heart, brain or skeletal muscle. Being intracellular parasites, they are protected from the effects of circulating antibodies. The life cycle is completed when a cat eats an infected bird or mouse containing bradyzoites, or ingests oocytes from faeces. Humans become infected either directly as a result of contact with pets (or the soil they use as a toilet), or indirectly by the ingestion of poorly cooked meat (usually pork or lamb) from an infected farm animal. Most infected people remain largely asymptomatic, but some may develop serious conditions including encephalitis and hepatitis, with immunocompromised individuals being particularly at risk. Congenital toxoplasmosis arises when *T. gondii* crosses the placenta to infect the foetus; it can result in spontaneous abortion or a range of severe birth defects.

15.6.3 Cryptosporidiosis

Cryptosporidiosis is a disease of the gastrointestinal tract caused by *Cryptosporidium* species, mostly *C. parvum* and *C. hominis*. The latter is only found in humans, whereas *C. parvum* also infects other species, primarily cattle. Thus for a single disease we can have transmission that is either *anthroponotic* (person to person) or *zoonotic* (animal to person). In each case, transmission is via faecal contamination of a drinking water supply. Infection of humans starts with the ingestion of oocysts; these can remain viable for long periods in the environment, and it is thought that as few as ten may be enough to initiate an infection. Once ingested, they make their way to the small intestine, where they penetrate into epithelial cells. Reproduction of *Cryptosporidium* may be either sexual or asexual, and results in the release of more oocysts in the faeces. In healthy individuals cryptosporidiosis causes diarrhoea that tends to be self-limiting over a period of weeks, but in AIDS patients it can be irreversible and often fatal.

Oocysts of *Cryptosporidium* are resistant to chlorination, and filtration or ultraviolet light treatment are necessary to ensure their removal from water supplies. In 1993, a huge outbreak of cryptosporidiosis occurred in Milwaukee, United States, due to a failure in the filtration system of a local water treatment plant. Over 400 000 people were affected, around half of the total population served by the plant. Fifty-four deaths were attributed to the outbreak, most of whom were AIDS patients.

15.6.4 Leishmaniasis

Leishmaniasis is a disease of the tropics and subtropics and is caused by species of the kinetoplasmid protozoan *Leishmania*. Dogs and rodents act as the reservoir for the parasite. It is spread from animal to human and also from human to human by female sandflies. When a sandfly takes a blood meal from the infected host, ingested amastigotes develop in the fly's gut into promastigotes, which divide and pass to the proboscis, whence they are passed with saliva to the next host. Host macrophages

> The *reservoir* of an infection is the site or organism where the pathogen is normally found.

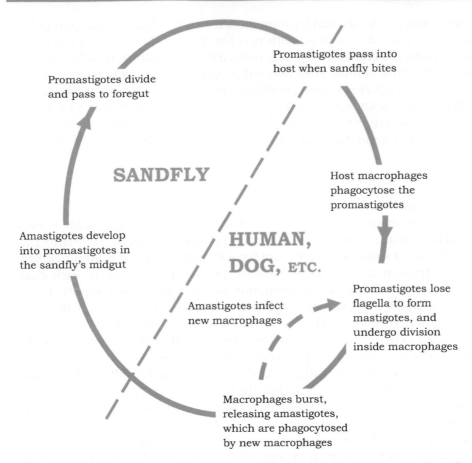

Promastigotes pass into
host when sandfly bites

Promastigotes divide
and pass to foregut

SANDFLY

Host macrophages
phagocytose the
promastigotes

Amastigotes develop
into promastigotes in
the sandfly's midgut

HUMAN,

DOG, ETC.

Promastigotes lose
flagella to form
mastigotes, and
undergo division
inside macrophages

Amastigotes infect
new macrophages

Macrophages burst,
releasing amastigotes,
which are phagocytosed
by new macrophages

Figure 15.4   Leishmaniasis is transmitted by sandflies. Females of several genera of sandflies act as vectors for the *Leishmania* parasite between mammalian hosts.

phagocytose the *promastigotes*, which lose their flagella to become non-motile *amastigotes;* these undergo rapid division by binary fission until the macrophage bursts (Figure 15.4). The amastigotes so released are then engulfed by more macrophages, and the process continues until the reduction in macrophage numbers affects the efficiency of the host's immune response. Leishmaniasis is usually cutaneous, involving ulceration of the skin, and may range from minor and self-limiting to severe and permanently disfiguring. In some cases the amastigotes may spread to vital organs, especially the liver and spleen (visceral leishmaniasis). This condition has a very high mortality rate if not treated. The nature of the disease depends on the species of *Leishmania*, and the degree to which the immune system is compromised.

Trypanosoma brucei, the causative agent of African sleeping sickness, is also transmitted by a type of blood-sucking insect, in this case the tsetse fly. Although details of the life cycle and the tissues affected differ from leishmaniasis, the general cycle of infection is similar, with the insect vector transmitting the parasite from one human to another through its bite. The infectious form of *T. brucei* develops in the salivary glands of the intermediate host, the tsetse fly, and is passed to the human host when a bite punctures the skin. Here, it eventually reaches the central nervous system by way of the blood or lymphatic systems. Inflammation of the brain and spinal cord results in the characteristic lethargy, coma and eventual death of the patient. Another species of *Trypanosoma, T. cruzi*, is the cause of Chagas' disease, sometimes known as American trypanosomiasis.

15.6.5 Amoebic dysentery

Another common cause of gastrointestinal disease in humans is the amoeba *Entamoeba histolytica*. Ingestion of cysts is usually via faecal contamination of water or food. Unlike its free-living relatives, *E. histolytica* is unable to reproduce outside its host. Excystation occurs in the small intestine, with trophozoites (free-living amoebas) passing to the large intestine. Here they may stay, causing no harm to the host, who remains asymptomatic. Although unaffected, such individuals can still pass cysts to another person in their faeces, either through person-to-person contact (poor personal hygiene) or via the environment, where cysts may survive for many weeks. Infection with *E. histolytica* may develop into amoebic dysentery, when trophozoites invade the intestinal mucosa, resulting in diarrhoea with bloody, mucous stools, and sometimes colitis or appendicitis. In extreme cases, trophozoites can spread throughout the body, leading to abscesses in the liver and other major organs. Amoebic dysentery is second only to malaria as the leading cause of death from protozoan infections; as many as 100 000 fatalities occur each year, mostly in areas where water sanitation is poor or human excrement is used as a fertiliser.

Another amoeba to cause disease in humans is *Acanthamoeba*, a common inhabitant of soil and fresh water that can cause keratitis of the eye as well as amoebic cephalitis.

> Amoebic keratitis used to be rare but its incidence has increased greatly since the 1980s, and is nearly always associated with the use of contact lenses. Failure to comply with strict cleaning and disinfecting protocols is the most common cause.

15.7 Fungal diseases in humans

A limited number of fungi are pathogenic to humans (Table 15.6). Mycoses (sing. *mycosis*) in humans may be cutaneous or systemic; in the latter, spores generally enter the body by inhalation, but subsequently spread to other organ systems via the blood, causing serious, even fatal disease.

Cutaneous mycoses are the most common fungal infections found in humans, and are caused by fungi known as *dermatophytes*, which are able

Table 15.6 Some fungal diseases of humans

Disease	Fungus
Histoplasmosis	*Histoplasma capsulatum*
Blastoplasmosis	*Blastomyces dermatitidis*
Cryptococcosis	*Cryptococcus neoformans*
Cutaneous mycoses	*Trichophyton* spp.
Pneumocystis pneumonia	*Pneumocystis jirovecii*
Candidiasis ('thrush')	*Candida albicans*
Aspergillosis	*Aspergillus fumigatus*

to utilise the keratin of skin, hair or nails by secreting the enzyme keratinase. Popular names for such infections include ringworm and athletes' foot. They are highly contagious, but not usually serious conditions.

Systemic mycoses can be much more serious, and include conditions such as histoplasmosis and blastomycosis. The former is caused by *Histoplasma capsulatum*, and is associated with areas where there is contamination by bat or bird excrement. It is thought that the number of people displaying clinical symptoms of histoplasmosis represents only a small proportion of the total number infected. If confined to the lungs, the condition is generally self-limiting, but if disseminated to other parts of the body such as the heart or central nervous system, it can be fatal. The causative agents of both diseases exhibit dimorphism; they exist in the environment as mycelia but convert to yeasts at the higher temperature of their human host.

A number of fungi act as opportunistic pathogens, including *Aspergillus fumigatus* (aspergillosis), *Candida albicans* (candidiasis or 'thrush') and *Pneumocystis jirovecii* (previously named *P carinii*) (pneumonia). The latter is found in a high percentage of AIDS patients, whose immune defences have been compromised. The causative organism was previously considered to be a protozoan, and has only been classed as a fungus relatively recently, as a result of DNA/RNA sequence evidence. It lives as a commensal in a variety of mammals, and is probably transmitted to humans through contact with dogs.

The incidence of opportunistic mycoses has increased greatly since the introduction of antibiotics, immunosuppressants and cytotoxic drugs. Each of these either suppresses the individual's natural defences, or eliminates harmless microbial competitors, allowing the fungal species to flourish. *Microsporidiosis* is an opportunistic infection, mainly of the intestine, caused by members of

> A recent study isolated species of the potentially pathogenic *Exophiala* fungus from the rubber seal of over half of domestic dishwashers tested; they were able to tolerate the harsh conditions of temperature and salt concentrations that exist when the appliance is in operation.

the Microsporidia (see Chapter 8). It is mainly found in immunocompromised individuals, especially those with AIDS but also recipients of organ transplants. Many fungi produce natural *mycotoxins*; these are secondary metabolites, which, if consumed by humans, can cause food poisoning, which can sometimes be fatal. Certain species of mushrooms ('toadstools'), including the genus *Amanita*, contain substances that are highly poisonous to humans. Other examples of mycotoxin illnesses include ergotism (see Box 8.2) and

> *Secondary metabolites* are produced by a microorganism after the phase of active growth has ceased. Such substances are usually not required for essential metabolic or cell maintenance purposes. Examples include toxins, pigments and most antibiotics.

aflatoxin poisoning. Aflatoxins are carcinogenic toxins produced by *Aspergillus flavus,* which grows on stored peanuts. In the early 1960s, the turkey industry in the UK was almost crippled by 'turkey X disease', caused by the consumption of feed contaminated by *A. flavus.*

> *Carcinogenic* = cancer-causing.

It is thought likely that all animals are parasitised by one fungus or another. Extraordinary though it may seem, there are even fungi that act as predators on small soil animals such as nematode worms, producing constrictive hyphal loops that tighten, immobilising the prey.

15.8 Algal diseases of humans

In Chapter 9 we mentioned that shellfish can accumulate toxins produced by dinoflagellates, causing disease when consumed by humans. A range of such toxins is produced by different dinoflagellates, which act on specific cellular targets and lead to different symptoms, including gastrointestinal and neurological symptoms. *Paralytic shellfish poisoning* (PSP) is caused by toxins called saxitoxins, produced principally by dinoflagellates of the genus *Alexandrium.* They act by blocking sodium channels in nerve cells, leading to tingling, numbness and paralysis. If symptoms extend to respiratory paralysis, the condition can be fatal if not treated in time.

Ciguatera is a form of food poisoning caused by consuming fish from tropical regions that have accumulated potent neurotoxins called ciguatoxins from dinoflagellates, notably *Gambierdiscus toxicus.* The toxin accumulates as it passes up the food chain until it reaches fish such as snapper and barracuda. Ciguatoxins also affect the opening of sodium channels, but in a different way, thus leading to different clinical symptoms. Like saxitoxins they are heat-stable, and so are not destroyed by cooking.

16

The Control of Microorganisms

Prior to Lister's pioneering work with antiseptics, around four out of every ten patients undergoing an amputation failed to survive the experience, such was the prevalence of infections associated with operative procedures, and yet, only 40 years later, this figure had fallen to

> An *antiseptic* is a chemical agent of disinfection that is mild enough to be used on human skin or tissues.

just 3 in 100. This is a dramatic demonstration of the fact that in some situations it is necessary for us to destroy, or at least limit, microbial growth, because of the undesirable consequences of the presence of microorganisms, or their products.

Control of microorganisms can be achieved by a variety of chemical and physical methods. *Sterilisation* is generally achieved by using physical means such as heat, radiation and filtration. Agents that destroy bacteria are said to be *bactericidal*. Chemical methods, whilst effective at *disinfection*, are generally not reliable for achieving total sterility. Agents that inhibit the growth and reproduction of bacteria without bringing about their total destruction are described as *bacteriostatic*.

16.1   Sterilisation

One of the oldest forms of antimicrobial treatment is heating, and in most cases this remains the preferred means of sterilisation, provided that it does not cause damage to the material in question. The benefits of boiling drinking water have been known at least since the fourth

> *Sterilisation* is the process by which all microorganisms present on or in an object are destroyed or removed.

Essential Microbiology, Second Edition. Stuart Hogg.
© 2013 John Wiley & Sons, Ltd. Published 2013 by John Wiley & Sons, Ltd.

century BC, when Aristotle is said to have advised Alexander the Great to order his troops to take this precaution. This of course was many centuries before the existence of microorganisms had been demonstrated or perhaps even suspected.

Boiling at 100°C for 10 minutes is usually enough to achieve sterility, provided that organisms are not present in high concentrations; in fact most bacteria are killed at about 70°C. If, however, endospores of certain bacteria (notably *Bacillus* and *Clostridium*) are present, they can resist boiling, sometimes for several hours. As we saw in Chapter 7, the causative agents of some particularly nasty conditions, such as botulism and tetanus, are members of this group. In order to destroy the heat-resistant endospores, heating beyond 100°C is required, and this can be achieved by heating under pressure in a closed vessel (Table 16.1). A typical laboratory treatment is 15 minutes at a pressure of 103 kPa (15 lb/in^2 (psi)), raising the temperature of steam to 121°C. This is carried out in an *autoclave*, which is, to all intents and purposes, a large-scale pressure cooker (Figure 16.1). Air is driven out of the system so that the atmosphere is made up entirely of steam; the desired temperature will not be reached if this is not achieved (Figure 16.2). Large loads, large volumes of liquids, and loads that contain trapped air, such as blankets or bedding, may need a longer treatment time in order for the heat to penetrate throughout. The problem of trapped air can be overcome with a *prevacuum autoclave*, which removes air from the chamber prior to sterilisation, thereby reducing the time needed for the instrument to reach its operating temperature. Modern autoclaves include probes designed to assess the temperature within a load rather than that of the atmosphere. It is essential that the intended temperatures are reached, and most autoclaves produce a printout of the temperatures and pressures for each run. Sometimes, spores of *Geobacillus stearothermophilus* are introduced into a system along with the material to be sterilised; if subsequent testing shows that the spores have all been destroyed, it is reasonable to assume that the system has also destroyed any other biological entity present. Special tape that changes colour if the necessary temperature is reached can act as a more convenient but less reliable indicator.

Table 16.1   Temperature of steam at different pressures

Pressure		Temperature (°C)
Pounds/in^2 (psi)	Kilopascals (kPa)	
0	0	100
10	68.9	115
15	103.4	121
20	137.9	126
25	172.4	130

Figure 16.1 The autoclave. Steam enters the chamber, driving air out. As the pressure of the steam reaches 103 kPa (15 psi), the temperature reaches 121°C, sufficient to kill resistant endospores as well as vegetative cells.

Figure 16.2 Temperatures achieved in an autoclave at 103 kPa (15 psi) in different atmospheres. An autoclave only reaches maximum temperature in an atmosphere of pure steam, hence any air remaining in the system will reduce the final temperature achieved. Reproduced from Hardy, SP (2002) Human Microbiology, with permission from Taylor and Francis Group.

An effect similar to that achieved by autoclaving can be obtained by a method called *intermittent steaming*, or *tyndallisation* (after the Irish physicist, John Tyndall, who was one of the first to demonstrate the existence of heat-resistant microbial forms). This is used for those substances or materials that might be damaged by the high temperatures used in autoclaving. The material is heated to between 90 and 100°C for about 30 minutes on each of three successive days, and left at 37°C in the intervening periods. Vegetative cells are killed off during the heating period, and during the 37°C incubation any endospores that have survived will germinate. Once these have grown into more vegetative cells, they too are killed in the next round of steam treatment. Clearly this is quite a long-winded procedure, and it is therefore reserved for those materials that might be harmed by steam sterilisation.

High temperatures can cause damage to the taste, texture and nutritional value of many food substances, and in such instances it is sufficient to destroy vegetative cells by a process of *pasteurisation* (among his many other achievements, Pasteur demonstrated that the microbial spoilage of wines could be prevented by short periods of heating). Milk was traditionally pasteurised by heating large volumes at 63°C for 30 minutes, but the method employed nowadays is to pass it over a heat exchanger at 72°C for 15 seconds (HTST – high temperature, short time). This is not sterilisation as such, but it ensures the destruction of disease-causing organisms such as *Brucella abortus* and *Mycobacterium tuberculosis,* which at one time were frequently found in milk, as well as significantly reducing the organisms that cause food spoilage, thus prolonging the time the milk can be kept. Some protocols exceed these minimum values in order to reduce further the microbial content of the milk. One type of milk on sale in the shops is subjected to more extreme heating regimes; this is 'UHT' milk, which can be kept for several weeks without refrigeration, though many find that this is at some cost to its palatability! It is heated to 'ultra-high temperatures' (150°C) for a couple of seconds using superheated steam. The product is often referred to as being 'sterilised', but this is not true in the strictest sense. Milk is not the only foodstuff to be pasteurised; others include beer, fruit juices and ice cream, and each has its own time/temperature combination.

All the above methods employ a combination of heat and moisture to achieve their effect; the denaturation of proteins, upon which these methods depend, is enhanced in the presence of water. Heat is more readily transferred through water than through air, and the main reason that endospores are so resistant is because of their very low water content. In some situations, however, it is possible to employ dry heat, using an oven to sterilise metal instruments or glassware, for example. It is a more convenient procedure, but a higher temperature (160–170°C) and longer exposure time (2 hours) are required. Dry heat works by oxidising ('burning') the cell's components.

Microorganisms are quite literally burnt to destruction by the most extreme form of dry heat treatment, incineration. Soiled medical dressings and swabs, for example, are potentially hazardous, and are destroyed in this way at many hundreds of degrees celsius. As we saw in Box 4.1, sterilising the loops and needles used to manipulate microorganisms by means of flaming is a routine part of aseptic procedures.

Box 16.1 Metals as disinfectants

Several heavy metals, such as copper, mercury and silver, are potent antimicrobial agents. They act by interacting with proteins and inactivating them. In recent years, the value of using copper alloys for the manufacture of frequently touched surfaces such as door handles, hand rails and wash-basin taps in reducing infections has become appreciated. In a trial carried out in Birmingham, UK, replacement of a number of 'touch surfaces' with a copper alloy alternative reduced the microbial contamination levels by 90–100%. Similar results have been obtained in other trials elsewhere in the world.

16.1.1 Sterilisation by irradiation

Certain types of irradiation are used to control the growth of microorganisms. These include both ionising and non-ionising radiation.

The most widely used form of non-ionising radiation is *ultraviolet (UV) light*. Wavelengths around 260 nm are used because these are absorbed by the purine and pyrimidine components of nucleic acids, as well as certain aromatic amino acids in proteins. The absorbed energy causes a rupture of the chemical bonds, so that normal cellular function is impaired. You will recall from Chapter 11 that UV light causes the formation of *thymine dimers* (see Figure 11.21), where adjacent thymine nucleotides on the same strand are linked together, inhibiting DNA replication. Although many bacteria are capable of repairing this damage by enzyme-mediated photoreactivation, viruses are much more susceptible. UV lamps are commonly found in food preparation areas, operating theatres and specialist areas such as tissue culture facilities, where it is important to prevent contamination. Because they are also harmful to humans (particularly the skin and eyes), UV lamps can only be operated in such areas when people are not present. UV radiation has very poor penetrating powers – a thin layer of glass, paper or fabric can impede the passage of the rays. The chief application is therefore in the sterilisation of work surfaces and the surrounding air, although it is increasingly finding an application in the treatment of water supplies, where its ability to

Figure 16.3 The Radura. The logo is used on packaging of food that has been irradiated. It represents a plant (the foodstuff) surrounded by packaging, which has been penetrated in places by ionising radiation.

kill protozoan pathogens such as *Giardia* and *Cryptosporidium* gives it an advantage over conventional chlorine treatment. The use of UV irradiation in this context offers the additional advantage of degrading chemical pollutants such as pesticides.

Ionising radiations have a shorter wavelength and much higher energy, giving them greater penetrating powers. The effect of ionising radiations is due to the production of highly reactive free radicals, which disrupt the structure of macromolecules such as DNA and proteins. Surgical supplies such as syringes, catheters and rubber gloves are commonly sterilised employing gamma (γ) rays from the isotope cobalt-60 (^{60}Co).

Gamma radiation has been approved for use in over 50 countries for the preservation of food, which it does not only by killing pathogens and spoilage organisms but also by inhibiting processes that lead to sprouting and ripening. The practice has aroused a lot of controversy, largely due to concerns about health and safety, although the first patent applications for its use date back over a hundred years! Although the irradiated product does not become radioactive, there is a general suspicion on the part of the public about anything to do with radiation, which has led to its use on food being only very gradually accepted by consumers. Tight European Union regulations mean that the use of food irradiation here remains limited to a single product; however, in the United States, a positive attitude towards irradiation of food both by professional bodies and the media during the last 20 years has led to a more widespread acceptance of the technology. The *Radura symbol* (Figure 16.3) is used internationally to indicate that a food product has been irradiated. Gamma radiation is used in situations where heat sterilisation would be inappropriate, because of undesirable effects on the texture, taste or appearance of the product. This mainly relates to fresh produce such as meat, poultry, fruit and vegetables. Irradiation is not suitable for some foodstuffs, such as those with a high fat content, where unpleasant tastes and odours result. Ionising radiations have the great advantage over other methods of sterilisation that they can penetrate packaging.

Figure 16.4 Membrane filtration. Membrane filters are used to sterilise heat-labile substances. They are available in a variety of pore sizes, according to the specific application.

16.1.2 Filtration

Many liquids such as solutions of antibiotics or certain components of culture media become chemically altered at high temperatures, so the use of any of the heat regimes described above is not appropriate. Rather than killing the microorganisms, an alternative approach is simply to remove them. This can be done for liquids and gases by passing them through filters of an appropriate pore size. Filters used to be made from materials such as asbestos and sintered glass, but these have been largely replaced by membrane filters, commonly made of nitrocellulose or polycarbonate (Figure 16.4). These can be purchased ready-sterilised and the liquid passed through by means of pressure or suction. Supplies of air or other gases can also be filter-sterilised in this way. A pore size of 0.22 μm is commonly used; this will remove bacteria plus, of course, anything bigger, such as yeasts; however, *Mycoplasma* spp. and viruses are able to pass through pores of this size. With a pore size one-tenth this size, only the smallest of viruses can pass through, so it is important that an appropriate pore size is chosen for any given task. A drawback with all filters, but especially those of a small pore size, is that they can become clogged easily. Filters in general are relatively expensive, and are not the preferred choice if alternative methods are available.

High efficiency particulate air (HEPA) filters are used to create clean atmospheres in areas such as operating theatres and laboratory biological safety cabinets.

16.1.3 Sterilisation using ethylene oxide

Generally, chemical methods achieve only disinfection (see below); the use of the gas ethylene oxide, however, is effective against bacteria, bacterial spores and viruses. It is used for sterilising large items of medical equipment, and materials such as plastics that would be damaged by heat treatment.

$$H_2C \underset{O}{\overset{\diagup}{\text{———}}} CH_2 + R-SH \longrightarrow R-S-CH_2CH_2OH$$

Ethylene oxide

Figure 16.5 Ethylene oxide is an alkylating agent. Like other alkylating agents, ethylene oxide affects the structure of both proteins and nucleic acids. Labile hydrogen atoms such as those on sulphydryl groups are replaced with a hydroxyl ethyl radical.

Ethylene oxide is particularly effective in sterilising items such as dressings and mattresses, due to its great powers of penetration. In the food industry, it is used as an antifungal fumigant, for the treatment of dried fruit, nuts and spices. The materials to be treated are placed in a special chamber, which is sealed and filled with the gas in a humid atmosphere at 40–50°C for several hours. Ethylene oxide is highly explosive, so must be used with great caution; its use is rendered safer by administering it in admixture (10%) with a non-flammable gas such as carbon dioxide. It is also highly toxic, so all items must be thoroughly flushed with sterile air following treatment to remove any trace of it. Ethylene oxide is an alkylating agent; it denatures proteins by replacing labile hydrogens such as those on sulphydryl groups with a hydroxyl ethyl radical (Figure 16.5).

16.2   Disinfection

We saw at the beginning of this chapter how sterilisation is an absolute term, implying the total destruction of all microbial life. Disinfection, by comparison, allows the possibility that some organisms may survive, with the potential to resume growth when conditions become more favourable. A *disinfectant* is a chemical agent used to disinfect inanimate objects such as work surfaces and floors. In the food and catering industry, especially in the United States, the term *sanitisation* is used to describe a combination of cleaning and disinfection. Disinfectants are incapable of killing spores within a reasonable time period, and are generally effective against a narrower range of organisms than physical means. *Decontamination* is a term sometimes used interchangeably with disinfection, but its scope is wider, encompassing the removal or inactivation of microbial products such as toxins as well as the organisms themselves.

Disinfection is the elimination or inhibition of pathogenic microorganisms in or on an object so that they no longer pose a threat.

The lethal action of disinfectants is mainly due to their ability to react with microbial proteins, and therefore enzymes. Consequently, any chemical agent that can coagulate, or in any other way denature proteins will act as a disinfectant, and compounds belonging to a number of groups are able to do this.

16.2.1 Alcohols

The antimicrobial properties of ethanol have been known for over a century. It was soon realised that it worked more effectively as a disinfectant at less than 100% concentration, that is, when there was some water present. This is because denaturation of proteins proceeds much more effectively in the presence of water (recall that moist heat is more effective than dry heat for the same reason). It is important, however, not to overdo the dilution, as at low percentages some organisms can actually utilise ethanol as a nutrient! Ethanol and isopropanol are most commonly used at a concentration of 70%. As well as denaturing proteins, alcohols may act by dissolving lipids, and thus have a disruptive effect on membranes, and on the envelope of certain viruses. Both bacteria and fungi are killed by alcohol treatment, but spores are often resistant because of problems in rehydrating them; there are records of anthrax spores surviving in ethanol for 20 years! The use of alcohols is further limited to those materials that can withstand their solvent action.

Alcohols may also serve as solvents for certain other chemical disinfectants. The effectiveness of iodine, for example, can be enhanced by the iodine being dissolved in ethanol. Many commercially available hand sanitisers are ethanol-based.

16.2.2 Halogens

Chlorine is an effective disinfectant as a free gas, and as a component of chlorine-releasing compounds such as hypochlorite and chloramines. Chlorine gas, in compressed form, is used in the disinfection of municipal water supplies, swimming pools and the dairy industry. Sodium hypochlorite (household bleach) oxidises sulphydryl (–SH) and disulphide (S–S) bonds in proteins. Like chlorine, hypochlorite is inactivated by the presence of organic material. *Chloramines* are more stable than hypochlorite or free chlorine, and are less affected by organic matter. They are also less toxic and have the additional benefit of releasing their chlorine slowly over a period of time, giving them a prolonged bactericidal effect.

Iodine acts by combining with the tyrosine residues on proteins; its effect is enhanced by being dissolved in ethanol (1% I_2 in 70% ethanol) as tincture of iodine, an effective skin disinfectant. This is being superseded by *iodophores* (e.g. Betadine, Isodine), compounds in which iodine is combined with a solubilising agent, usually a detergent. Iodophores can combat bacteria, viruses and fungi, and are widely used in the food and drink industry.

Figure 16.6   Phenolics. The structure of (a) phenol and (b) hexachlorophene.

16.2.3   Phenolics

As we saw in Chapter 1, the germicidal properties of phenol (carbolic acid) were first demonstrated by Lister in the middle of the nineteenth century. Since phenol is highly toxic, its use in the disinfection of wounds has long since been discontinued, but derivatives such as cresols and xylenols continue to be used as disinfectants and antiseptics. These are both less toxic to humans and more effective against bacteria than the parent compound. Phenol is still used, however, as a benchmark against which the effectiveness of related disinfectants can be measured. The *phenol coefficient* compares the dilution at which the derivative is effective against a test organism with the dilution at which phenol achieves the same result. A phenol coefficient of more than one means that the new compound is more effective than phenol against the organism tested, whereas a value of less than one means that it is not as effective as phenol.

Phenol derivatives act by combining with and denaturing proteins, as well as disrupting cell membranes. Their advantages include the retention of activity in the presence of organic substances and detergents, and their ability to remain active for some time after application; hence their effect increases with repeated use. Familiar disinfectants such as Dettol, Lysol and Hibitane, are all phenol derivatives. Hexachlorophene (Figure 16.6) is very effective against Gram-positive bacteria such as staphylococci and streptococci, and used to be a component of certain soaps, surgical scrubs, shampoos and deodorants. Its use is now confined to specialist applications in hospitals since the finding that in some cases, prolonged application can lead to brain damage.

16.2.4   Surfactants

Surface active agents, or surfactants, such as *soaps* and *detergents*, have the property of orienting themselves between two interfaces to bring them into closer contact (Figure 16.7).

> A *surfactant* reduces the tension between two molecules at an interface.

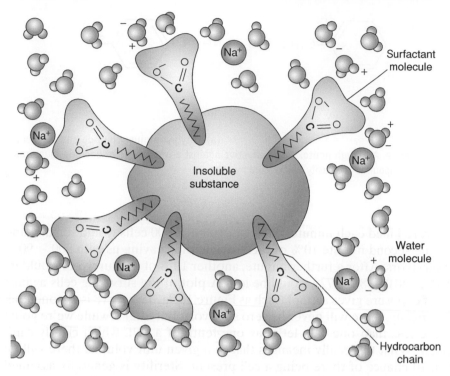

Figure 16.7 The action of surfactants. The long hydrophobic tails of the detergent are able to penetrate an insoluble grease particle. The negatively charged carboxyl group at the other end attracts the positive pole of water molecules, enhancing the water solubility of the grease. Reproduced from Black, JG (1999) Microbiology: Principles and Explorations, 4th edn, with permission from John Wiley & Sons.

The value of soap has less to do with its disinfectant properties than with its ability to facilitate the mechanical removal of dirt and microorganisms. It does this by emulsifying oil secretions, allowing the debris to be rinsed away. Detergents may be *anionic* (negatively charged), *cationic* (positively charged) or nonionic. Cationic detergents such as quaternary ammonium compounds (ammonium chloride with each chlorine replaced by an organic group, Figure 16.8) act by combining with phospholipids to disrupt cell membranes and affect cellular permeability.

16.3 The kinetics of cell death

When microorganisms are exposed to any of the treatments outlined in the preceding pages, they aren't all killed instantaneously. During a given time period, only a certain proportion of them will die. Suppose we had 1000 cells (an unrealistically small number, but it keeps the arithmetic simple) and that

(a)   (b)

Figure 16.8 Quaternary ammonium compounds. (a) Cetylpyridinium chloride and (b) benzalkonium chloride. Although nontoxic and effective against a wide range of bacteria, quaternary ammonium compounds are easily inactivated by soap.

10% were killed each minute. After one minute, 900 cells would remain, and after the second minute 10% of these would die, leaving us with $900 - 90 = 810$ survivors. After a further minute, another 10% of the survivors would be killed, so $810 - 81 = 729$ would be left. A plot of the surviving cells against time of exposure gives a graph such as Figure 16.9a. The curve is exponential; theoretically, there will never be zero survivors, but after a while we're going to have less than one cell, let's say one-tenth of a cell, which clearly can't happen. What this really means is that in a given unit volume, there will be a 1 in 10 chance of there being a cell present. Sterility is generally assumed when this figure falls as low as 1 in a million (see Figure 16.9b). Since only a proportion of the surviving population is killed per unit time, it follows that the more cells you have initially, the longer it will take to eliminate them (Figure 16.10).

The steepness of the slope in Figure 16.9b is an indication of the effectiveness of heat sterilisation. The *decimal reduction time*, or *D value*, is the time needed to reduce the population by a factor of 10 (i.e. to kill 90% of the population) using a particular heat treatment. The D value is a valuable parameter in settings such as the canning industry. It applies to a particular temperature; at higher temperatures, the rate of killing is enhanced, and so the D value is reduced (Figure 16.11). The increase in temperature required to reduce D by a factor of 10 is the *Z value*.

Since in real life, the microbial population is certain to be a mixed one, the critical factor is the death rate of the most resistant species, that is, the one with the highest percentage of survivors per minute. Sterilisation protocols should therefore be based on the rate of destruction of endospores.

It should be stressed that an effective treatment regime for one organism may be wholly inappropriate for another, since organisms differ in their susceptibility to different agents. Organism A may resist heat treatment better than organism B, but be more sensitive to a particular chemical treatment.

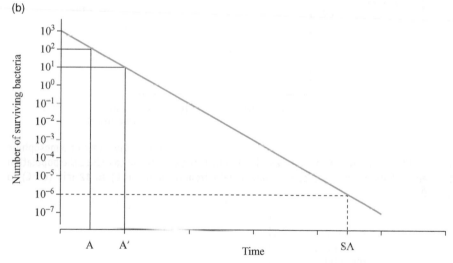

Figure 16.9 The kinetics of cell death. **(a)** During heat sterilisation, the number of living cells decreases by the same proportion per unit time, giving an exponential curve. **(b)** When plotted on a logarithmic scale, the decrease in numbers is seen as a straight line, whose slope is a reflection of the rate of killing. The time period between A and A′ is the decimal reduction time (D): the time taken to reduce the population to one-tenth of its size. The total period until the point SA is the sterility assurance value, when there is only a one in a million probability of any cells having survived. Reproduced from Hardy, SP (2002) Human Microbiology, with permission from Taylor & Francis Group.

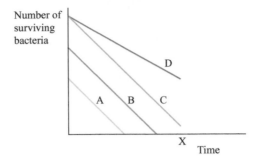

Figure 16.10   Sterilisation time is dependent on starting population. Populations A, B and C all have the same decimal reduction rate, but different starting populations, therefore after a given time they have different numbers of survivors. Population D has the same starting numbers as C, but because a smaller proportion are killed per minute, i.e. the slope is less steep, a greater number survive after time X.

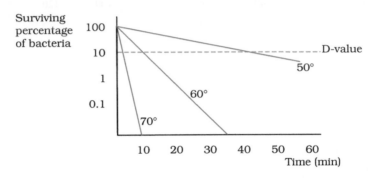

Figure 16.11   The D value is reduced at higher temperatures. The survival rate of a mesophilic bacterium at three different temperatures. The D value is the time taken to reduce the population to one-tenth of its starting value: note how it falls from 42 min (50°C) to 12 min (60°C) to 3 min (70°C).

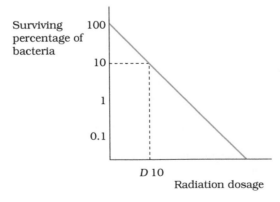

Figure 16.12   Killing by irradiation. The decimal reduction value (D10) is the dose of radiation necessary to reduce the population of microorganisms to one-tenth of its value. The value for D10 differs greatly between organisms.

16.3.1 Killing by irradiation

Microorganisms as a group are much more resistant to the effects of ionising radiation than are higher organisms, and some are more resistant than others. A plot of log surviving numbers against time of irradiation is shown in Figure 16.12. The *decimal reduction value* (D10) is analogous to the D value used for heat sterilisation (see earlier).

17

Antimicrobial Agents

By the end of the nineteenth century, the germ theory of infectious diseases had finally gained widespread acceptance. It is hardly surprising then, that the scientific community at that time should turn its collective attention towards ways of controlling infections and the organisms that cause them. In the following pages we shall describe the development during the twentieth century of chemical agents targeted against microbial pathogens, and the dramatic impact that they had on the treatment of infectious diseases.

In the 1880s, Paul Ehrlich observed how certain dyes would stain bacteria but not the surrounding tissues, leading him to formulate the idea of *selective toxicity*, whereby a substance could selectively target harmful microorganisms but leave human tissues undamaged. He tested hundreds of synthetic compounds in the search for his 'magic bullet' before finding, in 1910, an arsenic-containing drug (Salvarsan), which was effective against *Treponema pallidum*, the causative agent of syphilis. Ehrlich coined the term *chemotherapy* to describe any use of a drug or other chemical substance for the treatment of disease.

It was over 20 years before another significant antimicrobial drug was developed, when the German chemist Gerhard Domagk showed a synthetic red dye, Prontosil, to be active against a range of Gram-positive bacteria. Interestingly, this effect was not seen with cultured bacteria, but only when injected into infected animals. It was later shown that this was because it was necessary for Prontosil to be metabolised to *sulphanilamide* for it to become active. In the following decade, numerous derivatives of sulphanilamide were synthesised, many of which were more potent antimicrobial agents than the parent molecule. This class of compounds is known collectively as the *sulphonamides*, or sulfa drugs (Box 17.1). In the years leading up to World War II, sulphonamides dramatically improved the mortality rates associated with pneumonia and puerperal fever.

Essential Microbiology, Second Edition. Stuart Hogg.
© 2013 John Wiley & Sons, Ltd. Published 2013 by John Wiley & Sons, Ltd.

Box 17.1   Sulphonamides – the deadly mimics

Folic acid is required by cells as a coenzyme in the synthesis of nucleic acids. Whereas humans must ingest it preformed in their diet, bacteria possess the enzymes necessary to synthesise their own. Sulphonamides exploit this by fooling the bacterial cell into thinking they are molecules of *p*-aminobenzoic acid (PABA), a precursor of folic acid, because of their similar structures (see below).

$$H_2N-\underset{\text{OH}}{\overset{\displaystyle \bigcirc}{}}-\overset{}{\underset{|}{C}}=O$$

p- aminobenzoic acid (PABA)

$$H_2N-\bigcirc-\overset{O}{\underset{O}{\overset{\|}{\underset{\|}{S}}}}-NH_2$$

Sulphanilamide

    The sulphonamide acts as a competitive inhibitor of the enzyme dihydropteroate synthetase, and prevents the synthesis of folic acid, which in turn affects nucleic acid metabolism and leads to cell death. Because of their close structural resemblance to PABA, the sulphonamides are said to be *structural analogues*; another, equally descriptive name is *antimetabolites*. Some other antimicrobials employ a similar strategy.

    Nowadays, sulphonamides have largely been replaced by *antibiotics* because of (i) their side effects, and (ii) widespread bacterial resistance to sulphonamides because of their wholesale and indiscriminate use in the early years. Some other synthetic compounds are still useful as antimicrobial agents, however. *Isoniazid* is one of the principal agents used in the treatment of tuberculosis. It is nearly always given in association with another antimicrobial agent because of the high incidence of resistant forms of the mycobacteria that cause the disease. Like the sulphonamides, isoniazid is a structural analogue, and is thought to inhibit the production of mycolic acid in the mycobacterial cell wall (see Chapter 3). This specific mode of action means that isoniazid has a very narrow spectrum of use.

Another group of synthetic substances with antimicrobial properties are the *quinolones,* such as ciprofloxacin and norfloxacin. They interfere with nucleic acid synthesis by inhibiting DNA gyrase, an enzyme involved in DNA replication (see Chapter 11). Quinolones are used in the treatment of urinary infections, as are the *nitrofurans.* These are active against certain fungi and protozoans as well as a range of bacteria.

17.1 Antibiotics

The other major breakthrough in the treatment of infectious diseases was of course the discovery of naturally occurring antimicrobial agents, or *antibiotics* (Table 17.1). These are metabolites produced by certain microorganisms, which inhibit the growth of certain other microorganisms. As we shall see, the definition

> An *antibiotic* is a microbially produced substance (or a synthetic derivative) that has antimicrobial properties.

has been extended to include semi-synthetic derivatives of these naturally occurring molecules.

One of the best known of all stories of scientific discovery is that of how Sir Alexander Fleming discovered penicillin in 1928. Before we consider that, however, it is worth noting that a number of treatments for infectious

Table 17.1 Some antibiotics and their microbial source

Antibiotic	Microbial source
	Low GC Gram-positive bacteria
Bacitracin	*Bacillus subtilis*
Polymixin	*Bacillus polymixa*
	Actinomycetes
Rifampicin	*Amycolatopsis rifamycinica**
Gentamicin	*Micromonospora purpurea*
Actinomycin D	*Streptomyces parvulus*
Erythromycin	*Streptomyces erythreus*
Neomycin	*Streptomyces fradiae*
Nystatin	*Streptomyces noursei*
Streptomycin	*Streptomyces griseus*
Tetracycline	*Streptomyces rimosus*
Vancomycin	*Streptomyces orientalis*
	Fungi
Cephalosporins	*Cephalosporium acremonium*
Griseofulvin	*Penicillium griseofulvum*
Penicillin	*Penicillium chrysogenum*

*Formerly *Streptomyces mediterranei.*

diseases practised over the preceding centuries can, with the benefit of hind-sight, be regarded as antibiotic therapy. Many hundreds of years ago, for example, the Chinese used mouldy soybean curd in the treatment of boils, and South American Indians controlled foot infections by wearing sandals that had become furry with mould! In the decades leading up to Fleming's discovery, several other investigators had noticed the effects of *Penicillium* moulds on bacteria. In the late nineteenth century, Tyndall (see Chapter 16) made the observation that a culture medium cloudy with bacterial growth would clear when mould grew on the surface. Around the same time Pasteur and Joubert demonstrated that cultured anthrax bacilli could be inactivated in the presence of certain other microorganisms from the environment.

Most notably, in France, Ernest Duchesne carried out a more comprehensive investigation into the effects of *Penicillium*, including injecting it into animals infected with typhoid. Crucially, however, Fleming took the extra step of attempting to isolate the active component of the mould. Returning to St Mary's Hospital in | Fleming's lab at St Mary's has been turned into a museum, set out as it was when he made his famous discovery.

London after a holiday, Fleming noticed that a plate culture of *Staphylococcus aureus* had become contaminated by the growth of a mould; around it were clear areas, where the bacteria did not grow. The mould was subsequently identified as *Penicillium notatum,* a different species from that used by Duchesne. The substance that had diffused from it through the agar, preventing bacterial growth, became known as *penicillin.* Further investigation revealed that broth from a culture of the *Penicillium* mould was inhibitory towards the growth of a number of other Gram-positive pathogens, and remained so even when diluted several hundred times. Critically, when tested on mice, it caused few if any side effects.

When it came to purifying the active ingredient and using it *in vivo*, however, a number of problems were encountered. The penicillin proved to be impure, only produced in minute amounts, and unstable in the acid conditions of the stomach, thereby limiting its therapeutic potential. After publishing a few papers on the | Penicillin, like most naturally occurring antibiotics, is an example of a *secondary metabolite* (see Chapter 8).

subject, Fleming ceased work on penicillin and it was left to Howard Florey and Ernst Chain in 1939 to take up the challenge of producing it in sufficient quantities and in a pure enough form for therapeutic use. Early work in Oxford had to continue in the United States because of the German bombing campaign in Britain. The US entry into the World War II in late 1941 meant that the development of penicillin was awarded war project status, giving it greatly added impetus. As a result of their endeavours, the production of penicillin was greatly improved (see Box 17.2), and in 1945, Fleming, Chain and Florey shared the Nobel Prize for their work.

Box 17.2   How did Florey and Chain improve the yields of penicillin?

The work of Florey and Chain resulted in pure penicillin being produced on a large scale, suitable for therapeutic use. Among their achievements were:

- isolation of a better penicillin-producing species (*Penicillium chrysogenum,* famously isolated from a mouldy cantaloupe melon in Peoria, Illinois!) and selection of mutant strains induced by X-rays and UV irradiation;
- development of submerged culture technique, with sterile air being forced through the medium to supply essential oxygen;
- improvements in medium composition;
- addition of precursors to the medium;
- refinements to recovery methods.

In 1942 there was only enough penicillin in the world to treat a few hundred individuals, but by the end of the war production had grown to such an extent that seven million people a year could be treated. By the mid-1950s, such well-known antibiotics as tetracycline, chloramphenicol and neomycin had been isolated. The discovery of a few naturally occurring compounds had revolutionised the treatment of infectious diseases.

> Some penicillin tablets manufactured in 1945 were assayed some 60 years later and still found to retain appreciable antibacterial properties!

It should be mentioned that penicillin was not the first antibiotic to enter commercial production; that distinction went to *gramicidin*, discovered by René Dubos in 1939, but this is limited to topical (external) use only due to the fact that it causes haemolysis of red blood cells. Other antibiotics isolated during this period included *streptomycin*, isolated by Selman Waksman from *Streptomyces griseus*, which was the first antibiotic to prove effective against tuberculosis.

New antibiotics are still being sought today. Of the thousands isolated so far, only a small proportion have proved to be of any real therapeutic or commercial value. This is because a key prerequisite for any chemotherapeutic agent is *selective toxicity*, that is to say its effects

> A *side effect* is an undesirable and unintended effect of a therapeutic treatment on the recipient.

are limited to its microbial target and do not harm the human host. An obvious way of achieving this is for a compound to direct its effect against a metabolic or physiological function that is unique to microbial cells and not found in the host. One of the reasons why penicillin was, and continues to be, so successful, is that the target of its action is unique to bacteria, so its degree of selective toxicity is high. We shall look at further examples of this later in this chapter. If a chemotherapeutic agent inhibits the same process in the host as in the pathogen, or causes harm to the host in some other way, it is said to have *side effects*. These may include directly toxic effects, hypersensitivity (allergic) reactions or adverse effects on the host's normal resident microflora.

17.1.1   What other properties should an antibiotic have?

Selective toxicity is the most important single attribute of an antibiotic, but ideally it should also have as many of the following properties as possible:

- Antibiotics, like other chemotherapeutic agents, need to be *soluble in body fluids* in order to exert their effect by penetrating the body tissues. The compound must not be metabolised so quickly that it is excreted from the body before having a chance to act.
- If administered orally, it must *not be inactivated* by the acid environs of the stomach, and must be capable of being absorbed by the small intestine.
- An antibiotic should not have any significant effect on the *resident microflora* of the host.
- It should not be easy for the target pathogen to establish *resistance* against an antibiotic.
- It should not have a negative impact on the effect of other medication.
- Side effects such as *allergic reactions* should be minimal.
- It should be relatively inexpensive, and sufficiently stable to have a good *shelf life,* without special storage considerations.

> The *therapeutic index* provides a measure of the selective toxicity of a chemotherapeutic agent. It is the ratio between the concentration at which the substance causes harm to its host (toxic dose) and that at which it is required to be clinically effective (therapeutic dose). It is therefore desirable for an antibiotic to have a high therapeutic index.

> Antibiotics should not produce an allergic or *hypersensitivity* reaction in the host. This is caused by an extreme response by the host immune system, and is not the same as toxicity.

17.1.2   How do antibiotics work?

All antimicrobial agents have the common aim of interfering in some way with a normal, cellular function of the target bacterial cell, while leaving the

Table 17.2 Some commonly used antibiotic classes

Mode of inhibition	Representatives
Inhibitors of cell wall synthesis	Penicillins, cephalosporins
Disrupters of cell membranes	Polymixins, polyenes
Inhibitors of protein synthesis	Streptomycin, tetracyclines
Inhibitors of nucleic acid synthesis	Rifamycins

host human or animal cells unaffected (Table 17.2). The most commonly used antibiotics exert their effect by one of the following methods:

    I. Inhibition of cell wall synthesis
   II. Disruption of cell membranes
  III. Interference with protein synthesis
  IV. Interference with nucleic acid synthesis

Those antibiotics belonging to groups I and III are better able to discriminate between prokaryotic and eukaryotic cells, and consequently show more selective toxicity and a higher therapeutic index.

17.1.3   I: Inhibitors of cell wall synthesis

The main group that work in this way are the β-*lactam antibiotics*, so-called because they contain a β-lactam ring in their structure. Well-known members of this group are the *penicillins* and the *cephalosporins*. The β-lactams exert their effect by inhibiting cell wall synthesis in the target bacteria. You may recall from our discussion of bacterial cell wall structure in Chapter 3 that an important factor in the strengthening of the peptidoglycan component of the bacterial cell walls is the cross-linking of chains by transpeptidation. β-lactams disrupt this process by binding to the transpeptidase enzyme,[1] forming covalent bonds with a serine residue within the enzyme's active site. The

> Antibiotics may either kill their bacterial target, or merely prevent them from proliferating; they are said to be *bactericidal* or *bacteriostatic*, respectively. It is possible for an antibiotic to be bactericidal towards one organism and bacteriostatic towards another, or to be bactericidal at one dose and bacteriostatic at another.

cell wall continues to form, but becomes progressively weaker as more new, unlinked peptidoglycan is set down. Since bacteria are generally to be found in a hypotonic environment, as the wall weakens, water enters the cell, leading to swelling and then lysis. Because of the nature of their action, β-lactams

---

[1] This and other enzymes involved in the final step of peptidoglycan synthesis, such as carboxypeptidase and endopeptidase, are collectively known as penicillin-binding proteins (PBPs).

are most effective against rapidly growing cells. The β-lactams also act by preventing the natural regulation of enzymes called *autolysins* (Box 17.3).

Penicillins   The first β-lactam antibiotic to be discovered was benzylpenicillin, or penicillin G, whose action is restricted to Gram-positive bacteria, because it is unable to penetrate the Gram-negative cell wall. It is effective against Gram-positive bacteria when administered intramuscularly, but cannot be taken by mouth because it is broken down in the

> *Semi-synthetic penicillins are based on the core structure of the naturally occurring molecule, with the addition of chemically synthesised side chains.*

acid conditions of the stomach. Another naturally occurring penicillin, penicillin V, represented an advance inasmuch as it is less acid-labile and can therefore be taken orally. All the penicillins are based on a core structure or nucleus called 6-amino-penicillanic acid (Figure 17.1); most of those in use today are so-called *semi-synthetic penicillins,* which have a variety of novel side chains attached to this nucleus. Developed as a result of extensive research, these have overcome some of the problems inherent in naturally occurring penicillins such as instability and narrow specificity (Figure 17.2). *Ampicillin* is a semi-synthetic penicillin with a broader specificity than penicillin G; it is appreciably more effective against Gram-negative bacteria such as *Salmonella* and *E. coli*, its hydrophobic nature making it

Figure 17.1   Naturally occurring penicillins. (a) Penicillin G (benzylpenicillin) and (b) penicillin V. The dotted outline covers the 6-aminopenicillanic acid nucleus present in all variants of penicillin. The heavy outline denotes the β-lactam ring.

Figure 17.2 Some important semisynthetic penicillins. The shaded square represents the 6-aminopenicillanic acid nucleus common to all penicillins (see Figure 17.1).

better able to penetrate their outer membrane. It has the additional benefit of being acid-stable and can therefore be taken orally. *Amoxicillin* is another semi-synthetic compound that has the added benefit of being more readily absorbed when taken orally.

Box 17.3 β-Lactam antibiotics have a second mode of action

Autolysins are enzymes that function by breaking down peptidoglycan in a controlled fashion, causing breaks to allow for the addition of new peptidoglycan as the cell grows; they are normally regulated by naturally occurring inhibitors. The β-lactams neutralise the activity of these inhibitors, leading to further breakdown of the cell wall.

Box 17.4   Broad spectrum or narrow spectrum?

Certain antibiotics, due to the mechanism of their action, are only effective against a few different pathogens, while others can be used successfully against many different kinds. They are said to have, respectively, a *narrow spectrum* and a *broad spectrum* of activity.

On the face of it, all things being equal, you might expect your doctor to choose the antibiotic with the broadest possible spectrum of activity, but this isn't always the wisest option. When the cause of an infection isn't known, it makes sense to hedge one's bets and prescribe a broad-spectrum antibiotic ('whatever it is, this should sort it!'), but this policy is not without its dangers. The drug is likely to kill many members of the host's own resident microflora, which can lead to the occurrence of a *superinfection*, and the development of antibiotic-resistant strains is also made more likely. If the identity of the pathogen is suspected, an appropriate narrow-spectrum drug is to be preferred.

All of the penicillins described above are susceptible to naturally occurring bacterial β-*lactamases* (also called penicillinases), which break a bond in the β-lactam core of the penicillin molecule (Figure 17.3). This is one of the main mechanisms by which bacteria can display resistance towards β-lactam antibiotics. Sometimes, penicillins are taken in combination with a second β-lactam such as clavulanic acid that has little antibacterial activity of its own. These 'suicide β-lactams' bind to the β-lactamase with a high affinity, preventing it from acting on the penicillin. The commonest of these combinations is amoxiclav (Augmentin), made up of amoxicillin and potassium clavulanate. In

Box 17.5   Where did antibiotic resistance come from?

Have genes responsible for antibiotic resistance always existed in nature, or have they arisen since the development and widespread use of antibiotics? The answer, almost certainly, is the former. A sample of an *E. coli* strain, freeze-dried in 1946, was revived many years later, and found to have plasmid-encoded genes for resistance to streptomycin and tetracycline, neither of which was in clinical use until some years after the culture was preserved. It seems likely that bacteria possessed these genes to protect against naturally occurring antibiotics, an idea supported by the fact that R-plasmids have been found in non-pathogenic soil bacteria. Also, resistance to a number of antibiotics has been demonstrated in soil and water bacteria from sources sufficiently remote to be free from anthropogenic influence.

Figure 17.3 Action of β-lactamase on penicillin. A number of bacteria, especially staphylococci, possess the enzyme β-lactamase (penicillinase), which inactivates penicillin by cleavage of the β-lactam ring at the point marked by the arrow.

addition, some semi-synthetic penicillins such as meticillin (USA: methicillin) were found to be immune to attack by the β-lactamases and were therefore a valuable weapon against penicillin-resistant forms of bacteria. Meticillin is no longer used clinically, however, having been replaced by newer forms such as dicloxacillin and flucloxacillin, which are both more versatile and have fewer side effects.

Penicillin is not an appropriate treatment for the estimated 1–5% of adults who show an allergic reaction to it; in extreme cases, death from *anaphylactic shock* can result.

> An *anaphylactic shock* is an extreme form of hypersensitivity reaction.

Cephalosporins The cephalosporins, like the penicillins, have a structure based on a β-lactam ring (Figure 17.4). They also exert their effect on transpeptidases, but generally have a broader specificity and are more resistant to the action of β lactamases. Ceftriaxone, for example, is now used in the treatment of gonorrhoeal infections, caused by penicillin-resistant strains of *Neisseria gonorrhoeae*. In addition, cephalosporins are often used to treat patients who are allergic to penicillin. Cephalosporins were first

Figure 17.4 Cephalosporins are based on a nucleus of 7-amino-cephalosoranic acid, which, like the penicillins, features a β-lactam ring (shown as a square). Note that each molecule has two variable side chains.

isolated in the late 1940s from a marine fungus called *Cephalosporium acre-monium*, and came into general use in the 1960s. So-called second-, third- and fourth-generation cephalosporins have been developed to widen the spectrum of activity to include many Gram-negative organisms, and to keep one step ahead of pathogens developing resistance to earlier versions. Cefepime is a fourth-generation cephalosporin with a wide spectrum of activity and lower susceptibility to β-lactamases.

As well as treating existing infections, both penicillins and cephalosporins are also used prophylactically, that is, in the prevention of infections prior to surgery in particularly vulnerable patients.

Other antibiotics that affect the cell wall   *Carbapenems* are β-lactam antibiotics produced naturally by a species of *Streptomyces*. A semi-synthetic form, meropenem, is active against a wide range of Gram-positive and -negative bacteria, and is used when resistance to other β-lactams has developed.

Bacitracin and vancomycin are two other antibiotics that exert their effect on the cell wall, but by a different mechanism. Bacitracin is derived from species of *Bacillus* and acts on bactoprenol pyrophosphate, the lipid carrier molecule responsible for transporting units of peptidoglycan across the cell membrane to their site of incorporation into the cell wall. Its use is restricted to topical (surface) application, since if used internally it can cause kidney damage. Vancomycin is a highly toxic glycopeptide antibiotic that inhibits cell wall assembly in a narrow spectrum of Gram-positive organisms such as streptococci and staphylococci. It is particularly important in its use against infections caused by organisms resistant to meticillin and to the cephalosporins, such as meticillin resistant *Staphylococcus aureus* (MRSA: see Section 17.2). It is not absorbed from the gastrointestinal tract and is therefore most commonly administered intravenously.

17.1.4   II: Antibiotics that disrupt cell membranes

Polymyxins are a class of antibiotic that act by disrupting the phospholipids of the cytoplasmic membrane and causing leakage of cell contents. Produced naturally by a species of *Bacillus*, polymyxins are effective against pseudomonad infections of wounds and burns, often used in combination with bacitracin and neomycin (an inhibitor of protein synthesis). They are ineffective against Gram-positive bacteria due to their inability to penetrate the thick cell wall. The toxicity of polymyxins makes them unsuitable for internal use, so they are mainly used as a topical administration. Colistin (polymyxin E) is used as a drug of last resort against multiple drug-resistant forms of Gram-negative bacteria such as *Pseudomonas* and *Acinetobacter*. *Polyene antibiotics* such as amphotericin and nystatin are antifungal agents that act on the sterol components of membranes; they are discussed further in Section 17.5. *Daptomycin* belongs to a new class of naturally occurring cyclic lipopeptide antibiotics isolated from *Streptomyces roseosporus*. It is effective against a range of Gram-positive bacteria, including strains that have developed resistance to antibiotics such as meticillin and vancomycin. The lipophilic tail of the molecule is thought to insert itself into bacterial cell membranes, causing rapid depolarisation and a release of potassium ions. This leads to a cessation of synthesis of DNA and protein, and cell death.

17.1.5 III: Inhibitors of protein synthesis

Antibiotics that act by affecting protein synthesis generally have a relatively broad spectrum of action. They are able to exploit the difference between prokaryotic (70S) and eukaryotic (80S) ribosomes by specifically targeting the former. Such antibiotics consequently have a relatively high therapeutic index (although not as high as cell wall inhibitors). As we saw in our historical review earlier in this chapter, streptomycin was the first antibiotic shown to be effective against Gram-negative organisms. Its discovery in 1943 was particularly welcome since such organisms were unaffected by penicillin or sulphonamides. It proved to be particularly useful in the treatment of tuberculosis, the causative agent of which, *Mycobacterium tuberculosis*, is protected against the effects of penicillin by the waxy layer of mycolic acids in its cell wall.

Streptomycin belongs to a group of antibiotics called *aminoglycosides*, which act by binding to the 30S subunit of the bacterial ribosome, and disrupting formation of the initiation complex at the start of the translation process (Figure 17.5). Like some other 'wonder drugs', streptomycin has proved to have undesirable side effects, particularly involving the kidneys and auditory nerve, which have led to it being replaced in most applications by safer alternatives, although it may still be used against mycobacterial infections. Other members of the aminoglycosides include gentamicin, kanamycin and tobramycin. They are absorbed poorly by the gastrointestinal tract, and so are usually administered intravenously. Use of aminoglycosides is limited to the treatment of severe infections by aerobic Gram-negative bacteria, such as *Pseudomonas*, *Acinetobacter* and *Enterobacter*. Because of their toxicity, aminoglycosides are given for only a short period of time, and less toxic antibiotics are substituted once the cause of infection is positively identified. Aminoglycosides such as gentamycin have a synergistic effect when combined with β-lactam antibiotics such as penicillins ('pen and gent'), and have been used to treat streptococcal infections. Use of the aminoglycosides declined as a result of their toxic side effects, but with the development of resistance to many of the alternative antibiotics, their use is under consideration again.

Tetracyclines also work by binding to the 30S ribosomal subunit; they prevent the attachment of aminoacyl tRNA, and therefore extension of the peptide chain (Figure 17.5). They are yet another group of antibiotics produced

Figure 17.5   Inhibitors of protein synthesis. (a) By binding to the 30S subunit of the bacterial ribosome, aminoglycosides block the attachment of the 50S subunit. This prevents completion of the initiation complex, thus protein synthesis is inhibited. (b) Tetracyclines distort the shape of the 30S subunit, preventing the attachment of the appropriate aminoacyl-tRNA. (c) Chloramphenicol inhibits peptidyltransferase and prevents formation of new peptide bonds. (d) Macrolides such as erythromycin bind to the 50S subunit, preventing elongation of the growing peptide chain.

by *Streptomyces* spp., and the first group we have discussed that are entirely bacteriostatic in their action. Both natural and semi-synthetic tetracyclines are easily absorbed from the intestine, allowing them to be taken orally. Coupled with their broad specificity (the broadest of any antibiotic), this led to inappropriately widespread use in the years following their discovery, sometimes resulting in complications caused by the destruction of the normal resident microflora. Bacterial resistance to tetracyclines is common, because of their widespread use not only to treat infections but also as an additive to animal feed. Tetracyclines are still used for a number of applications, notably in the treatment of several sexually transmitted diseases.

Two important antibiotics that act on the larger, 50S, subunit of the prokaryotic ribosome are *erythromycin* and *chloramphenicol*. Both combine with the subunit, but in slightly different ways, to prevent the assembly of amino acids into a chain (Figure 17.5). Chloramphenicol was the first antibiotic to be discovered with a broad spectrum of activity; it also derives originally from *Streptomyces* spp., but is nowadays more easily produced synthetically. Its use has become severely restricted since it was shown to have some serious side effects, notably involving the bone marrow, and due to developing resistance it has been replaced as the agent of choice for the treatment of typhoid fever. It is, however, still extensively used in the treatment of conjunctivitis, as an ointment or eye drops.

Another antibiotic that exerts its effect on the 50S subunit is erythromycin, the best known of the *macrolide* group of antibiotics. Unlike chloramphenicol, it has a large hydrophobic molecule and is unable to gain access to most Gram-negative bacteria, thus restricting its spectrum of activity. Erythromycin can be taken orally and has a similar spectrum of activity to penicillin G. It is often used as an alternative to penicillin in the treatment of staphylococcal and streptococcal infections in children; it is particularly appropriate for this application as it is one of the least toxic of all commonly used antibiotics.

Lincosamides and streptogramins are further groups of antibiotics that act by binding to the 50S ribosomal subunit. They act in the same way as the macrolides, and because of this, resistance to one group will also confer resistance to all three ('MLS resistance').

17.1.6   IV: Inhibitors of nucleic acid synthesis

Rifampicin is the most important member of a group of agents called *rifamycins*. It acts by inhibiting the bacterial RNA polymerase, thereby preventing the production of mRNA. Its selective action is due to a far greater affinity for the bacterial version of the enzyme than the one found in the human host. Rifampicin is used against the mycobacteria that cause tuberculosis, an application for which its ability to penetrate tissues makes it well suited. Unlike most other antibiotics, rifampicin interacts with other drugs, often reducing or nullifying their effect. When used in high doses, it has the unusual side effect of turning secretions such as tears, sweat and saliva, as well as urine, an orange-red colour. Newer members of the rifamycins are rifabutin and rifapentine, which are excreted more slowly, and so don't have to be administered so often.

As mentioned earlier in this chapter, quinolones are a group of synthetic antimicrobials that also interfere with DNA replication.

17.2   Resistance to antibiotics

Fleming himself foresaw that the usefulness of penicillin might become limited should strains of pathogens arise with resistance to its effects. Indeed the β-lactamase enzymes responsible for resistance to penicillin had already been identified by the time the drug went into widespread clinical use in the early 1940s. By 1950 penicillin-resistant *S. aureus* was a common cause of infections in hospitals. A decade later, a semi-synthetic form of penicillin, meticillin was introduced; this was not affected by the β-lactamases that inactivated penicillin G, and was used to treat resistant forms. Within years, however, came the first reports of strains of *S. aureus* that did not respond to meticillin. The incidence of meticillin-resistant *S. aureus* (MRSA) has since increased greatly, and it represents a major source of *nosocomial infections*. In 1980, synthetic fluoroquinolones were introduced to counter the threat of MRSA, but within a year 80% of isolated strains had developed resistance to these too. Vancomycin has been regarded as a last resort treatment for MRSA, for a number of reasons; it has a number of serious side effects, its widespread use would encourage resistance against it, and it is very expensive. A case of vancomycin-resistant *S. aureus* (VRSA) emerged in Japan in 1996; a few months later it had reached the United States. This represents a serious threat; some of these strains respond to treatment with a cocktail of antibiotics, but

> *Nosocomial* infections are ones that are acquired in hospitals or similar locations. Some 5–10% of hospital patients acquire such an infection during their stay. This may prove to be fatal, especially among the elderly and immunocompromised. As well as the human cost, such infections extend the average time spent in hospital and therefore add greatly to the costs of treatment.

already people have died from untreatable VRSA infections. In 2002, a strain of VRSA was shown to have obtained its vancomycin resistance by cross-species transfer from a strain of *Enterococcus faecalis*. Since then, new VRSA isolates have been reported at a rate of one per year. *Clostridium difficile* is a major pathogen in nosocomial infections, causing diarrhoea in patients whose other gut bacteria have been eliminated by broad-spectrum antibiotic use. The incidence and severity of *C. difficile* infections are increasing and represent a major concern, as strains have been identified with resistance to the fluoroquinolones used to treat them.

Over 25 000 people are thought currently to die each year in Europe alone as a result of bacterial infections that do not respond to antibiotics. The development of resistance represents the greatest single challenge currently facing us in the fight against infectious diseases. In 2011, the World Health Organisation stated that the situation had reached a critical point, and without a concerted effort by governments and scientists, we could soon be dealing with a nightmare scenario of a worldwide spread of untreatable infections.

17.2.1   How does antibiotic resistance work?

We saw earlier in this chapter how antibiotics exert their effects in a variety of ways, so it should come as no surprise that there is no single mechanism of resistance. Resistance may be natural, that is, intrinsic to the microorganism in question, or it may be acquired. Some bacteria are able to resist antibiotic action simply by means of denying it entry to the cell; penicillin G, for example, is unable to penetrate the cell wall of Gram-negative bacteria. Others can pump the antibiotic back out of the cell before it has had a chance to act, by means of enzymes called *translocases*; this is fairly non-specific, so can lead to multiple drug resistance. Yet other bacteria are naturally resistant to a particular antibiotic because they lack the specific target for its action; for example, mycoplasmas do not possess peptidoglycan, the target for penicillin's action.

Bacteria may acquire resistance to an antibiotic by using or developing alternative biochemical pathways, so that its effect is cancelled out. Others can secrete enzymes that modify or degrade the antibiotic in question, causing it to lose its activity; we've already seen how penicillins can be inactivated by enzymatic cleavage of their β-lactam ring. Similarly, chloramphenicol can be acetylated, while members of the aminoglycoside family can be acetylated, adenylated or phosphorylated, all leading to loss of antimicrobial activity. Another way of acquiring resistance is by a mutation that modifies bacterial proteins in such a way that they are not affected by antimicrobial agents. You will recall that streptomycin normally acts by binding to part of the 30S subunit on the bacterial ribosome; the actual binding site is a protein called S12. Mutant forms of the S12 gene can lead to a product that still functions in protein synthesis, but loses its ability to bind to streptomycin. Similarly, mutations to genes in staphylococci means their penicillin-binding proteins (PBPs) no longer bind the antibiotic, so cross-linking of the cell wall is not inhibited.

A fifth-generation cephalosporin called ceftobiprole inhibits the modified PBP and is also resistant to β-lactamases; it represents one of the most promising defences against antibiotic-resistant bacteria such as MRSA and vancomycin-resistant enterococci (VRE).

17.2.2 How does resistance arise?

Occasionally, mutations occur spontaneously in bacteria, rendering them resistant to one antibiotic or another. Generally the mutation leads to a change in a receptor or binding site such as those just described, rendering the antibiotic ineffective. The changes are usually brought about by point mutations (see Chapter 11) occurring at very low frequency on chromosomal DNA. Under selective conditions, namely, in the presence of the antibiotic in question, such mutants are favoured, and outgrow their susceptible counterparts ('survival of the fittest'). Bacteria can, however, become resistant much more rapidly by acquiring the mutant resistance-causing gene from another bacterium. This is called *transmissible antibiotic resistance*; it occurs mainly as a result of bacterial conjugation, and is the cause of most of the resistance problems we presently face. Transmissible resistance was first reported in Japan in the late 1950s, when multi-drug resistance in *Shigella* was shown to have been acquired by conjugation with resistant *E. coli* in a patient's large intestine. *E. coli* is known to transfer R (resistance) plasmids to several other gut bacteria including *Klebsiella, Salmonella* and *Enterobacter,* as well as *Shigella*. Whereas chromosomal mutations usually result in a modification to the drug's binding site, genes carried on plasmids code for enzymes that inactivate a drug (e.g. β-lactamases) or lead to its exclusion from the cell

> **Box 17.6   What's in a name?**
>
> In 2010, the Indian government made a formal complaint about the name
> NDM-1, and the stigmatising effect it had on that country. In 2011, the
> editor of *The Lancet,* which had published the original findings, accepted
> that it had been an error to use a name that pointed to a country of
> origin. The name plasmid-encoding carbapenem-resistant metallo-beta-
> lactamase (PCM) has been proposed as an alternative.

(translocases). New Delhi metallo-β-lactamase (NDM-1, see Box 17.6) is
a plasmid-encoded β-lactamase that was first isolated in 2009, and confers
resistance to a range of β-lactams. Apparently originating in India, it has
now been found in other bacteria and in several countries around the world.
Transmissible resistance can also occur as a result of the transfer of *trans-
posons* (see Chapter 11). Chromosomal genes can pass to plasmids, facilitating transfer to other bacteria.

The incidence of bacterial strains resistant to an antibiotic in a particular area is related to use of the antibiotic in that area. The presence of an
antibiotic puts selective pressure on the microbial population and encour-
ages the proliferation of resistant forms. This is a particular problem in parts
of the world where the use of antibiotics is unregulated. Fortunately this resistance can, at least in part, be reversed, as several studies have shown, where
a more restricted use of certain antibiotics over several years was followed by
a reduction in the incidence of resistant bacterial forms.

17.3   Antibiotic susceptibility testing

In order to determine the most appropriate antimicrobial agent to use against
an infection, it is necessary to determine the susceptibility of the pathogen.

The *tube dilution assay* determines the *minimum inhibitory concentration*
(MIC) of the antibiotic, that is, the lowest concentration at which it prevents
growth of a given organism. A series of tubes containing increasingly dilute
preparations of the antibiotic are introduced into a broth with a standard
number of test organisms and incubated. The highest dilution (i.e. lowest
concentration) in the series to show no visible microbial growth is the MIC
(Figure 17.6). The MIC can be extended to show whether the antibiotic is just
inhibiting growth of the bacteria or actually killing them. This is done by plating out onto new, antibiotic-free agar and incubating again. The MBC (*minimum bactericidal concentration*) is the lowest concentration to have completely killed the bacteria, as shown by a lack of growth when plated out and
incubated. A value for MIC is also provided by the *E-test.* A commercially
produced plastic strip is impregnated with the antibiotic to give a calibrated

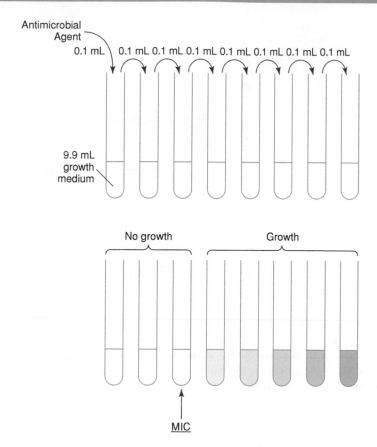

Figure 17.6 Minimal inhibitory concentration (MIC). The test organism is incubated with serially diluted antibiotic. The lowest concentration capable of preventing microbial growth is the MIC.

concentration gradient from one end to the other. It is laid on a lawn of bacteria on an agar plate and the antibiotic allowed time to diffuse out. The point where the elliptical area of clearing contacts the strip represents the MIC (Figure 17.7).

In the *disc diffusion method*, paper discs impregnated with the antibiotic are placed on the surface of an agar plate previously inoculated with the test organism (Figure 17.8). The antibiotic diffuses radially outwards, becoming less concentrated as it does so. A clear zone of inhibition appears where growth has been inhibited. The larger this is, the more susceptible the organism. Conditions may be standardised so that susceptibility (or otherwise) to the antibiotic can be determined by comparing the diameter of the zone of clearing with standard tables of values. From this, a suitable concentration for therapeutic use can be determined.

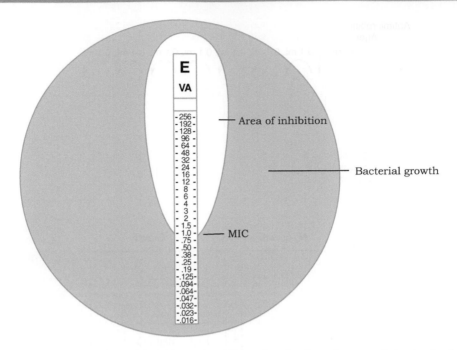

Figure 17.7 The E test. The point on the calibrated scale where the area of clearing begins represents the MIC of the antibiotic, the lowest concentration to prevent bacterial growth. In the example shown, the MIC for vancomycin is 1 μg/mL.

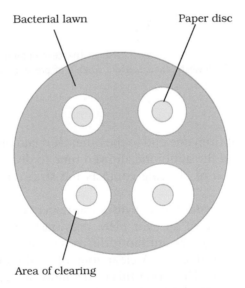

Figure 17.8 Antibiotic testing by disc diffusion. The bacterium to be tested is spread on an agar plate, and then paper discs impregnated with appropriate antibiotics are placed on the surface. Following incubation, susceptibility to an antibiotic is indicated by a clear ring surrounding the disc.

We have focused our attention up to now on antibacterial agents, but of course bacteria are not the only cause of infections. Now we review the relatively limited repertoire of compounds available for the treatment of infections caused by fungi, protozoa and viruses.

17.4   Antiviral agents

In spite of the looming threat of resistant strains, there is no doubt that antibiotics have been hugely successful in the control of bacterial diseases. We have, however, been a lot less successful when it comes to finding a treatment for diseases caused by viruses, in spite of the fact that roughly the same number of infectious diseases are caused by viruses as by bacteria. A quick revision of the *modus operandi* of viruses (see Chapter 10) should make it clear why effective antivirals are so elusive. Viruses survive by entering a host cell and hijacking its replicative machinery, thus a substance interfering with the virus is likely to harm the host as well. Antibacterials that target bacterial cell walls or protein synthesis are thus no use at all in this context. Antivirals are either directed against aspects of a virus's nucleic acid synthesis, or against virus-specific proteins. The first effective antivirals were not introduced until the 1960s, and their development was a largely hit-and-miss affair until the molecular biology revolution of the 1980s allowed a much better understanding of how viruses worked at the molecular level. Since then a number of compounds have been developed for clinical use that are able to target some specific property of the viruses, based on our detailed knowledge of their replication cycle (Table 17.3).

Most antiviral agents target nucleic acid synthesis, in many cases by acting as *nucleoside analogues*. These are molecules that are incorporated into viral DNA instead of the normal deoxynucleosides. They prevent extension of the chain because DNA polymerase is unable to act on them. The majority of viruses encode their own DNA polymerases, and the base analogues act by selectively inhibiting these, whilst having little effect on those of the host cell. They do this because they have a higher affinity for the viral polymerase than the host polymerase. An example is *aciclovir*, an analogue of guanosine that is used in the treatment of herpes simplex infections. It is converted to the nucleoside triphosphate by the action of thymidine kinase and then incorporated into the DNA chain instead of the 'correct' version (Figure 17.9). The aciclovir nucleotide has no $3'$-OH group for attachment to the next nucleotide, so further elongation of the chain is prevented. Vidarabine and azidothymidine (AZT) are other examples of base analogues, which are used in the treatment of retroviral infections. Here the enzyme affected is reverse transcriptase rather than DNA polymerase (recall from Chapter 10 that retroviruses use this enzyme to make a DNA copy from an RNA template). AZT (zidovudine; e.g. Retrovir) was one of the first substances shown to have an effect against HIV. Incorporation of the AZT triphosphate results

Table 17.3   Antiviral agents

Antiviral	Mode of action	Target virus
	Inhibition of DNA/RNA replication	
Aciclovir	Base analogue (G), inhibition of viral DNA polymerase	Herpes viruses
Ganciclovir	Base analogue (G), inhibition of viral DNA polymerase	Human cytomegalovirus
Vidarabine	Base analogue (A), inhibition of viral DNA polymerase	Herpes simplex, varicella-zoster
Azidothymidine (AZT)	Base analogue (T), inhibition of viral reverse transcriptase	HIV and other retroviruses
Lamivudine	Base analogue (C), inhibition of viral reverse transcriptase	HIV, hepatitis B
Nevirapine	Non-nucleotide inhibitor of reverse transcriptase	HIV
	Inhibition of viral proteins	
Atazanavir	Inhibitor of viral protease	HIV
Saquinavir	Inhibitor of viral protease	HIV
Zanamivir	Inhibitor of neuraminidase	Influenza A and B
Amantadine	Inhibitor of M_2 ion channel protein	
Raltegravir	Integrase inhibitor	HIV
	Inhibition of viral protein synthesis	
Fomivirsen	Antisense oligonucleotide: blocks translation of viral mRNA	Human cytomegalovirus
Enfuvirtide	Fusion inhibitor: mimics and replaces viral protein, preventing fusion to target cell	HIV

in chain termination. Reverse transcriptase can also be inhibited by other antivirals, known as non-nucleoside reverse transcriptase inhibitors (NNR-TIs; e.g. nevirapine).

A completely different approach to tackling viral infections is based on *antisense technology*, where a synthetic oligonucleotide is designed to bind to its complementary mRNA sequence and thus prevent it from being translated. Fomivirsen was the first of this class of antimicrobials to be introduced into clinical use; 21 nucleotides in length, it blocks the synthesis of a key protein of the human cytomegalovirus.

Other antiviral agents work by inhibiting the action or synthesis of viral proteins. One of the first antiviral agents to be approved for use was amantadine, which prevents uncoating of the influenza A virus. It does this by preventing the formation of acid conditions in the host cell's endocytotic vesicles by inhibiting the formation of an ion channel protein called M_2. Due to adverse side effects and the development of extensive viral resistance (see below), amantadine is no longer recommended for influenza therapy. Other anti-influenza drugs such as zanamivir and oseltamivir are analogues of sialic

Figure 17.9 Aciclovir inhibits viral DNA synthesis. (a) Aciclovir has a similar structure to the nucleoside deoxyguanosine, but lacks the −OH group (circled) necessary for chain extension. (b) Aciclovir needs to be phosphorylated to become active; virally encoded thymidine kinase is required for this. Aciclovir triphosphate (ACV-T) selectively inhibits viral, but not human, DNA polymerase. Any ACV-T that is incorporated into viral DNA acts as a chain terminator.

acid that block the active site of the enzyme neuraminidase, preventing the release and spread of new virus particles from infected cells. Zanamivir is inhaled as a fine powder directly into the lungs of patients, whilst oseltamivir (Tamiflu) is taken orally.

Protease inhibitors and *fusion inhibitors* are two further classes of antivirals used against HIV infections. Normally, multiple HIV proteins are translated together as a single polypeptide, and must be cleaved into the individual finished proteins by a viral protease. Protease inhibitors such as atazanavir and saquinavir act by preventing this. Enfuvirtide is a fusion inhibitor that mimics and binds to a viral envelope protein, gp41, preventing fusion of the virus to the target cell. The latest category of anti-HIV drugs are the *integrase inhibitors*. These inhibit the enzyme HIV-1 integrase, involved in the incorporation of viral DNA into the human genome. So far only raltegravir has been approved for use.

Despite the development of a range of antivirals, their effectiveness as therapeutic agents has been limited by the emergence of resistant strains. These result from mutations in the viral genome that alter the drugs' targets. This can be seen as even more serious than the emergence of antibiotic-resistant bacteria, because the choice of alternative antiviral agents is so restricted. The high rate of viral mutations means that treatment of HIV infections by a single drug is unlikely to be effective, so the simultaneous administration of a combination of drugs with different modes of action is preferred, a strategy known as *highly active anti-retroviral therapy* (HAART).

17.5 Antifungal agents

Fungi are eukaryotes, and therefore are unaffected by those agents that selectively target uniquely prokaryotic features such as peptidoglycans and 70S ribosomes. Therein lies the problem: anything that damages fungal cells is likely to damage human cells too. Because selective toxicity is harder to achieve, there is a very limited range of antifungal drugs available.

Several antifungals exert selective toxicity due to the fact that fungal membranes contain ergosterol rather than the cholesterol found in human and other animal cells. *Polyene* antibiotics such as *amphotericin* and *nystatin* (both produced by species of *Streptomyces*) bind to the ergosterol components of membranes, disrupting the membrane structure. Use of polyenes is limited, due to side effects including headaches, muscle pain and nausea; however, amphotericin B is still used against severe systemic infections of fungal origin, when medical staff have no effective alternative. Nystatin is used topically against *Candida* infections. Synthetic compounds such as the *imidazoles* and *triazoles* inhibit the synthesis of ergosterol, again resulting in a leaky membrane; they are effective against superficial mycoses (fungal infections of the skin, mouth and urinogenital tract). This group includes familiar drugs such as clotrimazole, ketoconazole and fluconazole. *Echinocandins* exploit another feature unique to fungi, their cell wall. They act as inhibitors to glucan synthase, an enzyme involved in the synthesis of glucan polymers, and not found in mammals.

Griseofulvin, a natural antibiotic produced by a species of *Penicillium,* is another antifungal agent whose use is restricted; it acts against skin infections by interfering with spindle formation during mitosis. Although used to treat superficial infections such as ringworm and athlete's foot, it is taken orally. Flucytosine is one of the very few antifungal agents currently in use that target nucleic acid synthesis, because of the similarity in molecular machinery in mammals and fungi. An analogue of cytosine, it is converted by fungi (but not host cells) into fluorouracil, which becomes integrated into RNA and inhibits fungal RNA and DNA synthesis. It is used against systemic infections, often in combination with other antifungals.

17.6 Antiprotozoal agents

When it comes to the development of therapeutic agents to counter infections caused by protozoans, we are again faced with the limitations imposed by the similarity between their cellular chemistry and that of their human hosts, with only a limited number of targets to exploit. Being such a diverse group, protozoans do not share a common feature that can be targeted by drugs, so antiprotozoals must use a range of strategies.

Protozoans, like bacteria but unlike humans, are able to synthesise their own folic acid, and are thus susceptible to drugs that inhibit its synthesis. *Pyrimethamine*, which inhibits the enzyme dihydrofolate reductase, is used to treat malaria, and also, in combination with the sulphonamide sulfadiazine, to treat cases of toxoplasmosis in immunocompromised individuals. Recall from Box 17.1 that sulphonamides also act to inhibit folic acid synthesis. Chloroquine also counters the malaria parasite *Plasmodium*, but utilises a completely different mechanism to do so. *Plasmodium* degrades haemoglobin within red blood cells, and modifies the haem that is so produced to a non-toxic form. Chloroquine prevents it doing this, and forms a complex with the haem that causes disruption of the parasite's cell membrane.

Metronidazole can be used to treat infections caused by a variety of protozoans including *Giardia, Trichomonas* and *Entamoeba*. It is unusual in also being effective against a number of bacterial pathogens including *Gardnerella* and *Clostridium difficile*. All the organisms in question are strict anaerobes, which have the ability to reduce metronidazole to an active form that inhibits nucleic acid synthesis. This ability is not shared by their aerobic hosts, and is the basis of the drug's selective toxicity.

Box 17.7 Bugs against bugs??

Scientists are always on the lookout for new weapons in the fight against infectious diseases. The emergence of resistant strains of pathogens means that new solutions continually need to be found. One novel line of research is hoping to utilise the bactericidal powers of a defence system used by certain insects. A sap-sucking species has been found that produces substances that interfere with bacterial protein synthesis. Most of these would harm protein synthesis in humans too, but certain peptides appear to be more selective in their action, protecting mice from *E. coli* and *Salmonella* infections. It may seem an unlikely source for a life saver, but then, so was *Penicillium*!

Box 17.8   Bacteriophages: our secret weapon against infections?

If bacteriophages are viruses that infect bacterial cells, why aren't they used in the fight against bacteria that cause infectious diseases? The short answer is: in some places they are! In the years following their discovery, there was considerable enthusiasm in some quarters for the notion that phages might be useful in the treatment of bacterial diseases. A particular attraction of phages as therapeutic agents is that they are extremely selective in their action, targeting one specific cell type. Early studies and trials had mixed results and before too long, antibiotics had revolutionised the treatment of infectious diseases in the West. This led to phage therapy being forgotten for several decades, but its use continued in the Soviet bloc countries, however, where even today phage preparations can be bought over the counter.

Western interest revived in the 1980s in the wake of the upsurge in antibiotic-resistant strains of bacteria, since when phages have attracted attention as possible allies in the control of infectious diseases, including some not responsive to antibiotic therapy. Several US companies have become involved in phage therapy, some in collaboration with the former Soviet state of Georgia, where research has been carried out over several decades.

Particular attention has been paid to veterinary applications, with a view to reducing the amount of antibiotic usage in animals, and it is hoped that one day phages may prove to be the weapon we need to fight antibiotic-resistant strains such as MRSA and VRE.

Box 17.9   The long road to a new antibiotic

Once a compound has been identified (by accident or design) as having antimicrobial properties, it undergoes exhaustive chemical modification to maximise its potency. Following tests in vitro to assess its antimicrobial efficacy, the selected compound must then be evaluated in vivo, not only for its antimicrobial properties, but also for any toxicity towards its host. Finally, clinical trials are essential to assess the effectiveness and safety in humans. Daptomycin illustrates this point: first described in 1986, it only received FDA approval for use against certain Gram-positive infections in 2003. It also serves as a reminder that not all new antibiotics are synthetic derivatives, and effective naturally occurring compounds are still being discovered.

17.7 The future

The development and widespread use of antibiotics must rank as the most remarkable of all medical advances made in the twentieth century. Over-confident assertions that infectious diseases would soon be a thing of the past, however, now have a hollow ring to them. Viruses have proved to be more difficult to deal with than bacteria, and many viral diseases continue to elude effective treatment. Most alarmingly, the threat of resistant strains casts a shadow over all the past achievements of antimicrobial therapy. The major aim of scientists now must be to develop new antimicrobials or other thera-peutic strategies at a pace greater than that at which pathogens are develop-ing resistance. Where once antibiotics were discovered largely by chance, we are now in a position where our understanding of the structure and function of pathogens can direct the design of an effective antimicrobial, for example by targeting a specific bacterial enzyme. However, bringing a new effective antibiotic to clinical usage is a notoriously long and drawn-out process; on average the process takes a full ten years from beginning to end (see Box 17.9). In 2000, the US Food and Drug Administration (FDA) approved a new synthetic agent shown to be effective against both MRSA and vancomycin-resistant *Enterococcus faecalis*. Linezolid (Zyvox), which works by blocking the initiation of protein synthesis, belongs to a new class of antibiotics called *oxazolidinones*. It was the first new anti-MRSA compound to be introduced for more than 40 years. In the decade since then, a handful of new antibi-otics have come to market, including two, tigecycline and telithromycin, that represent new chemical classes.

Another approach to countering resistant forms is to identify and target the mechanism by which they combat antibiotic therapy. A team at Rockefeller University in New York have identified two genes that enable resistant forms to rebuild their cell walls after antibiotic treatment. By targeting these genes, they hope to restore the potency of a cell wall inhibitor such as penicillin. Perhaps by the time you read this, other, less conventional approaches will have yielded promising results (see Boxes 17.6 and 17.7), but you can be just as sure that pathogens will have new tricks up their sleeves, and that the battle of humanity versus microorganisms will continue in the years to come.

VII
Microorganisms in Industry

18
Industrial and Food Microbiology

Many aspects of our everyday lives are influenced in some way by micro-organisms. In previous chapters we have noted how they can cause infectious diseases as well as providing a means of curing them, and the vital role they play in the environment. In addition, they are responsible for the production of much of what we eat and drink, synthesise industrially useful chemicals and can even extract precious metals from the earth (Table 18.1). In this chapter we shall look at some of the ways in which the activities of microorganisms have been harnessed for the benefit of humans, and developed on an industrial scale. The first applications of biotechnology many thousands of years ago were in the production of food and drink, so it is here that we shall begin our survey.

18.1 Microorganisms and food

To the general public, the association of microorganisms and food conjures up negative images of rotten fruit or food poisoning. On further reflection, some people may recall that yeast is involved in bread and beer production, but how many realise that microorganisms play a part in the manufacture of soy sauce, pepperoni and even chocolate? In the following we shall look at the contribution of microorganisms to the contents of our shopping baskets before considering one of the negative associations referred to above, the microbial spoilage of food.

The production of foodstuffs as a result of microbial fermentation reactions predates recorded history. The accidental discovery that such foods were less susceptible to spoilage than fresh foods must have made them an attractive proposition to people in those far-off days. Of course, until relatively recent

Essential Microbiology, Second Edition. Stuart Hogg.
© 2013 John Wiley & Sons, Ltd. Published 2013 by John Wiley & Sons, Ltd.

Table 18.1 Some applications of microorganisms

Food products
  Alcoholic drinks
  Dairy products
  Bread
  Vinegar
  Pickled foods
  Mushrooms
  Single-cell protein

Products from microorganisms
  Enzymes
  Amino acids
  Vitamins
  Antibiotics
  Vaccines
  Citric acid

Mining industries
  Metal extraction
  Desulphurisation of coal

Alternative fuels
  Ethanol
  Methane
  Hydrogen

Agriculture
  Microbial pesticides

Environment
  Bioremediation
  Sewage treatment/water purification
  Insect control
  Biosensors

times, nothing was known of the part played by microorganisms, so the production of beer, cheese and vinegar would not have entailed the carefully controlled processes in use today. Indeed, it was only with the development of isolation techniques towards the end of the nineteenth century (recall Chapter 1), that it became possible to use pure cultures of microorganisms in food production for the first time. The fermentation of foodstuffs, hitherto an art, became a science.

18.1.1 Alcoholic fermentations

There is evidence that alcoholic drinks including beer and wine were being produced thousands of years before the Christian era, making them among the earliest known examples of the exploitation of microorganisms by humans. Ethanol results from the fermentation process because the conversion of sugar to carbon dioxide and water is incomplete:

$$C_6H_{12}O_6 \rightarrow 2CH_3CH_2OH + 2CO_2$$

Wine can be made from almost any fruit juice with a high sugar content. The vast majority of commercially produced wines, however, derive from the fermentation of the sugar present in grapes (Figure 18.1). Such fermentation reactions may be initiated by yeasts found naturally on the grape skin; however, the results of such fermentations are erratic and may be unpalatable. In commercial winemaking the *must* (juice) resulting from the crushed grapes is treated with sulphur dioxide to kill the natural microflora, and then inoculated with the yeast *Saccharomyces cerevisiae*, variety *ellipsoideus*. Specially

Figure 18.1 Wine production. The figure outlines the essential steps involved in the production of red wine. White wine production differs in certain details, but shares the main steps of crushing the grapes, fermenting their sugar content into alcohol and aging to allow flavour development.

developed strains are used, which produce a higher percentage of alcohol (ethanol) than naturally occurring yeasts. Fermentation proceeds for a few days at a temperature of 22–27°C for red wines (lower for whites), after which the wine is separated from the skins by pressing. This is followed by *ageing* in oak barrels, a process that may last several months, and during which the flavour develops. *Malolactic* fermentation is a bacterial secondary fermentation carried out on certain types of wine. Malic acid, which has a sharp taste, is converted to the milder lactic acid, imparting smoothness to the wine. A secondary product of malolactic fermentation is diacetyl, which imparts a 'buttery' flavour to the wine.

Only black grapes (including the skins) are used in the production of red wines; white wines may use white grapes or black ones with the skins removed. When all or most of the sugar has been converted to alcohol, dry wines result; when some sugar remains, we get a sweeter wine. Most wines have an alcohol content of around 10–12%. For sparkling wines, an additional fermentation is carried out in the bottle to generate the bubbles of CO_2 characteristic of such wines.

Spirits such as brandy and rum result from the products of a fermentation process being concentrated by *distillation*. This gives a much higher alcohol content than that of wines.

Beer is produced by the fermentation of barley grain. The procedure varies according to the type of beer, but follows a series of clearly defined steps (Figure 18.2). Grain, unlike grapes, contains no sugar to serve as a substrate for the yeast, so before fermentation can begin it is soaked in water and allowed to germinate. This stimulates the production of the enzymes necessary for the conversion of starch to maltose (*'malting'*). An additional source of starch may be introduced during the next stage, *mashing*, in which the grains are ground up in warm water and further digestion takes place. The liquid phase, or *wort*, is drained off and boiled; this has the effect of inactivating the enzymes, precipitating proteins and killing off any microorganisms. It is at this point that *hops* are added. They impart flavour and colour to the finished product and also possess antimicrobial properties, thereby helping to prevent contamination. In the next stage, the wort is filtered and transferred to the fermentation vessel where yeast is introduced.

Two species of yeast are commonly used in the brewing process, both belonging to the genus *Saccharomyces*. *S. cerevisiae* is mainly used in the production of darker beers such as traditional English ales and stouts, whereas *S. carlsbergensis* (no prizes for guessing where this one was developed!) gives lighter coloured, less cloudy, lager-type beers. Cells of *S. cerevisiae* are carried to the surface of the fermentation by carbon dioxide bubbles (top fermenters), while *S. carlsbergensis* cells form a sediment at the bottom (bottom fermenters).

Fermentation takes about a week to complete, at a temperature appropriate for each type of yeast (*S. carlsbergensis* prefers somewhat lower

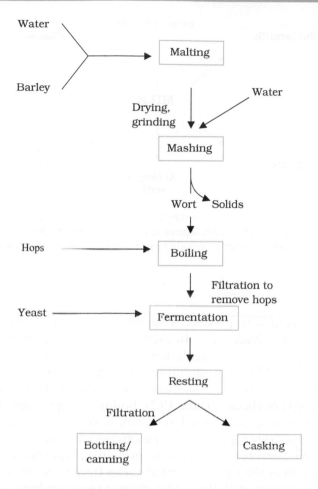

Figure 18.2 Beer production. The early stages serve to convert the carbohydrate present in the grain into a form that can be fermented by the yeast.

temperatures than *S. cerevisiae*). Following fermentation, the beer is allowed to age or 'rest' for some months in the cold. Beers destined for canning or bottling are filtered to remove remaining microorganisms. 'Spent' yeast may be dried and used as an animal food supplement.

Beers typically have an alcohol content of around 4%. Small amounts of other secondary products such as amyl alcohol and acetic acid are also produced, and contribute to the beer's flavour. 'Light', or low-carbohydrate beers are produced by reducing the levels of complex carbohydrates. The yeasts do not possess the enzymes necessary to cope with these branched molecules, so a supplement of debranching enzymes may be added to aid their breakdown.

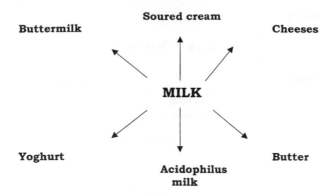

Figure 18.3   Fermented dairy products. Fermentation is initiated by the inoculation of a starter culture of lactic acid bacteria to convert lactose to lactic acid. Heterolactic fermenters such as *Leuconostoc* are added when aromatic flavouring compounds such as diacetyl are required.

18.1.2   Dairy products

Milk can be fermented to produce a variety of products, including butter, yoghurt and cheese (Figure 18.3). In each case, acid produced by the action of lactic acid bacteria causes coagulation or *curdling* of the milk proteins; in cheesemaking, the addition of rennin results in the separation of the semi-solid *curd* from the liquid *whey*. The subsequent steps in the process depend on the specific type of cheese (Table 18.2). Following separation, the curd of most cheeses is pressed and shaped, removing excess liquid and firming the texture. During the *ripening* process, salt is often added, and flavour develops due to continuing microbial action on the protein and fat components of the cheese. The length of the ripening period varies from a month to more than a year according to type, with the harder cheeses requiring longer periods. In some cases a fresh inoculation of microorganisms is made at this point, such as the addition of *Penicillium* spores to Camembert and Brie. For Emmental,

Table 18.2   Types of cheese. Cheeses are classified according to their texture. Unripened cheeses are those that have not undergone the aging or ripening process, during which additional flavours develop

Soft	Semi-soft	Semi-hard	Hard
Unripened: Mozzarella Cottage	Roquefort Stilton	Cheddar Emmental	Parmesan Pecorino
Ripened: Camembert Brie			

Propionibacterium freudenreichii is added, leading to the generation of carbon dioxide bubbles that form the holes characteristic of this type of cheese.

Yoghurt is another milk derivative. Thickened milk is exposed to the action of two types of bacteria, *Streptococcus salivarius* ssp. *thermophilus* and *Lactobacillus delbrueckii* ssp. *bulgaricus*, both of which ferment lactose present in milk into lactic acid. In addition, the latter contributes aromatics responsible for imparting flavour to the yoghurt.

Other dairy products such as soured cream and buttermilk are also produced by means of the fermentative properties of species of streptococci and lactobacilli.

18.1.3 Bread

The biological agent responsible for bread production is yeast. In fact, baker's yeast and brewer's yeast are just different strains of the same species, *Saccharomyces cerevisiae*. In breadmaking, aerobic rather than anaerobic conditions are favoured, so sugar present in the dough is converted all the way to carbon dioxide rather than to alcohol. It is CO_2 production that causes the bread to rise. Any small amount of ethanol that may be produced is evaporated during the baking process.

Many other popular foodstuffs are the result of microbial fermentation processes (see Table 18.3). These include vinegar, soy sauce and sauerkraut. *Silage* is animal fodder made from the fermentation of grass and other plant material by the action of lactic acid bacteria.

Table 18.3 Fermented food products

Product	Source	Fermentative microorganisms
Sauerkraut	Cabbage	*Leuconostoc brevis, Lactobacillus plantarum*
Olives	Olives	*Leuconostoc brevis, Lactobacillus plantarum, Lactobacillus brevis*
Pepperoni	Ground beef, pork	*Lactobacillus plantarum, Pediococcus pentosaceus*
Pickles	Cucumber	*Leuconostoc brevis, Lactobacillus plantarum, Pediococcus* spp.
Soy sauce	Soybean curd	*Aspergillus oryzae, Saccharomyces rouxii, Pediococcus soyae*
Tempeh*	Soybean	*Rhizopus oligosporus*
Kombucha	Tea	*Gluconobacter, Saccharomyces,* etc.
Sake	Rice	*Aspergillus oryzae, Saccharomyces cerevisiae*
Vinegar	Wine, cider	*Acetobacter, Gluconobacter*
Cocoa, chocolate	Cacao beans	*Saccharomyces cerevisiae, Candida rugosa, Acetobacter, Geotrichum*

*Tempe (or tempeh) is a solid fermented soya bean 'cake' that is widely consumed as a meat substitute in Indonesia. It forms an important part of the diet of many Indonesians.

18.1.4   Microorganisms as food

As we have seen in the previous sections, a number of microorganisms are involved in the production of food products. Others, however, *are* foodstuffs! Perhaps the most obvious of these are mushrooms, the stalked fruiting bodies of certain species of Basidiomycota (see Chapter 8), notably *Agaricus bisporus*. These are grown in the dark at favourable temperatures in order to stimulate the production of fruiting bodies. Another fungus, *Fusarium venenatum*, forms the basis of Quorn™, a processed mycoprotein that has been used as a meat substitute for some years in the UK. Whereas mushrooms are grown as agricultural products, Quorn™ must be produced under highly regulated sterile conditions. Other microbial food sources include certain algae (seaweed), which form an important part of the diet in some parts of the world, and bacteria and yeast grown in bulk as *single-cell protein* (SCP) for use as a protein-rich animal food supplement. The cyanobacterium *Spirulina* has been collected from dried-up ponds in parts of central Africa for use as a food supplement since time immemorial and is now available at health stores in the West.

18.1.5   The microbial spoilage of food

We have described in previous chapters the nutritional versatility of microorganisms and their role in the global recycling of carbon. Unfortunately for us, fresh foods such as meats, fruit and vegetables provide a rich source of nutrients, which a wide range of heterotrophic microorganisms find just as attractive as we do. Certain microbial types are associated with particular foodstuffs, depending on their chemical composition and physical factors such as pH and water content. Acidic foods such as fruits, for example, tend to favour the growth of fungi rather than bacteria.

Often, spoilage organisms come from the same source as the food – for example, soil on vegetables, or meat exposed to the contents of the animal's intestine following slaughter. Others are introduced as contaminants during transport, storage or preparation. Among the most commonly found spoilage organisms are a number of human pathogens, including *Pseudomonas, Salmonella, Campylobacter* and *Listeria*. Thus, although microbial spoilage may merely lead to foodstuffs being rendered unpalatable, in other cases it can also result in serious and even fatal illness ('food poisoning'). Whilst observable changes to foodstuffs are only likely after the microbial population has reached a considerable size, food poisoning can result from the presence of much smaller numbers of contaminants.

Some foodstuffs are more susceptible to spoilage than others: fresh items such as meat, fish, dairy produce, and fruit and vegetables are all highly perishable. Foods such as rice and flour, on the other hand, are much more resistant, because having no water content they do not provide suitable conditions for microbial growth. Drying is one of a number of methods of food preservation, all designed to prevent growth of microorganisms by making conditions unfavourable. Other methods include heating/canning, drying, pickling, smoking and, in many countries, irradiation. Such methods all have to strike a balance between inhibiting microbial growth and not having an adverse effect on the appearance, texture and palatability of the foodstuffs.

18.2 Microorganisms in the production of biochemicals

Many products of microbial metabolism find an application in the food and other industries. These include amino acids, enzymes, steroids and antibiotics (Table 18.4). Microbial growth conditions are adjusted so that production of the metabolite in question takes place at an optimal rate. Often an unnaturally high rate of production is achieved by the use of a mutated or genetically engineered strain of microorganism, or by manipulating culture conditions to favour excess metabolite production.

The development of a microbial means of producing *acetone* was vital to the allied effort in World War I. Acetone was a crucial precursor in explosives manufacture, and the demands of war soon outstripped supply by traditional methods of production. The problem was solved when Chaim Weizmann isolated a strain of *Clostridium acetobutylicum* that could ferment molasses to acetone and butanol (another industrially useful product). Nowadays, acetone is made more cheaply from petrochemicals.

Microbially produced *amino acids* are used in the food industry, in medicine and as raw materials in the chemical industry. The one produced in the greatest quantities by far is *glutamic acid* (almost two million tons per year), with most of it ending up as the flavour enhancer monosodium glutamate. The amino acids *aspartic acid* and *phenylalanine* are components of the artificial sweetener aspartame and are also synthesised on a large scale.

Table 18.4 Commercially useful products of microbial metabolism

Product	Use
Amino acids:	
Glutamic acid	Flavour enhancer
Lysine	Animal feed additive
Aspartic acid + phenylalanine	Artificial sweetener (aspartame)
Citric acid	Antioxidant, flavour enhancer, emulsifier
Enzymes	Numerous – see Table 18.5
Antibiotics	Treatment of infectious diseases
Vitamins	Dietary supplements
Steroids	Anti-inflammatory drugs, oral contraceptives

CH₂OH		CH₂OH		CH₂OH		O=C⌐
HO–CH		HO–CH		C=O		C–OH
HO–CH		HO–CH		HO–CH		C–OH ⎤ O
HC–OH	**Chemical** →	HC–OH	**Biological** →	HC–OH	**Chemical** →	HC ⌐
HO–CH		HO–CH		HO–CH		HO–CH
HC=O		CH₂OH		CH₂OH		CH₂OH
D-Glucose		D-Sorbitol		L-Sorbose		L-Ascorbic Acid

Figure 18.4   Ascorbic acid is produced by a combination of chemical and microbial reactions. Most steps in the synthesis of ascorbic acid are purely chemical, but the conversion of sorbitol to sorbose is carried out by the sorbitol dehydrogenase enzyme of *Acetobacter suboxydans*.

A number of organic acids are produced industrially by microbial means, most notably *citric acid*, which has a wide range of applications in the food and pharmaceutical industries. This is mostly produced as a secondary metabolite by the large-scale culture of the mould *Aspergillus niger*.

Certain microorganisms serve as a ready source of *vitamins*. In many cases these can be synthesised less expensively by chemical means; however, *riboflavin* (by the mould *Ashbya gossypii*) and *vitamin B₁₂* (by the bacteria *Propionibacterium shermanii* and *Pseudomonas denitrificans*) are produced by large-scale microbial fermentation. Microorganisms play a partial role in the production of *ascorbic acid* (*vitamin C*). Initially, glucose is reduced chemically to sorbitol, which is then oxidised by a strain of *Acetobacter suboxydans* to the hexose sorbose. Chemical modifications convert this to ascorbic acid (Figure 18.4).

Enzymes of fungal and bacterial origin have been utilised for many centuries in a variety of processes. It is now possible to isolate and purify the enzymes needed for a specific process, and the worldwide market is worth over a billion pounds. The most useful industrial enzymes include proteases, amylases, lipases and pectinases. Thanks to advances in protein engineering it is now becoming possible to 'design' completely new enzymes with specific properties for industrial applications. Some applications of enzymes are listed in Table 18.5, and two examples are briefly described below. Syrups and modified starches are used in a wide range of foodstuffs, including soft drinks, confectionery and ice cream, as well as having a wealth of other applications. Different enzymes or combinations of enzymes are used to produce the desired consistencies and physical properties. *High fructose corn syrup* (HFCS) is a sweetener used as a component of a multitude of processed foods since the 1970s. It is some 75% sweeter than sucrose and has several other advantages. HFCS is a mixture of fructose, dextrose (a form of glucose) and disaccharides, and is produced by the action of a series of three enzymes on

Table 18.5 Some industrial applications of microbially produced enzymes

Industry	Enzyme	Application
Food and drink	Rennet	Cheese manufacture
	Lipase	
	Pectinase	Fruit juice production
		Coffee bean extraction
	Amylase	Improved bread dough quality
		Haze removal in beer
	Amylase	Fructose syrup production
	Glucoamylase	
	Glucose isomerase	
Animal feed	Amylase	Improved digestibility
	Cellulase	
	Protease	
Detergent	Protease	Stain and grease removal
	Lipase	
	Amylase	
	Cellulase	Fabric softener
Paper	Cellulase	Pulp production
Textiles	Cellulase	'Stone-washed' jeans
Leather	Protease	Dehairing, softening, fat removal
	Lipase	
Molecular biology	*Taq* polymerase	Polymerase chain reaction

the starch (amylose and amylopectin) of corn (maize). *Alpha amylase* hydrolyses the internal α-1,4 glycosidic bonds of starch, but is not able to degrade ends of the chain. The resulting di- and oligosaccharides are broken down to the monomer glucose by the action of *glucoamylase*, then finally *glucose isomerase* converts some of the glucose to its isomer fructose.

Enzymes have been added to *cleaning products* such as washing powders, carpet shampoos and stain removers since the 1960s, and this remains one of the principal industrial applications of enzymes. *Proteases* are the most widely used enzymes in this context; working in combination with a surfactant, they hydrolyse protein-based stains such as blood, sweat and various foods. Greasy and oily stains present a different challenge, made all the more difficult by the move towards lower washing temperatures. The inclusion of *lipases* aids the removal of stains such as butter, salad dressing and lipstick, while *amylases* deal with starch-based stains such as cereal or custard. The food and detergent industries between them account for around 80% of all enzyme usage.

We have already seen in Chapter 17 that *antibiotics* are now produced on a huge scale worldwide. Figure 18.5 outlines the stages in the isolation, development and production of an antibiotic.

Isolating an antibiotic from a natural source is not all that difficult, but finding a new one that is therapeutically useful is another matter. Initially, the

Figure 18.5   Stages in the isolation and development of an antibiotic. See the text for details.

antimicrobial properties of a new isolate are assessed by streaking it across an agar plate, then inoculating a range of bacteria at right angles (Figure 18.6). As the antibiotic diffuses through the agar, it will inhibit growth of any susceptible species. Isolates that are still deemed worth persisting with are then grown up in a laboratory scale fermenter; it is essential for commercial culture that the antibiotic-producing organism can be cultured in this way.

Before committing to large-scale production, exhaustive further tests must be carried out on two fronts: to ascertain the potency of the preparation and the breadth of its antimicrobial spectrum, and to carry out toxicity testing on animals to determine its *therapeutic index* (see Chapter 17). The final stages of development involve pilot-scale production, followed by clinical trials on human volunteers.

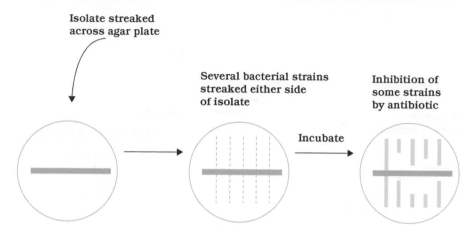

Figure 18.6   Assessing the antimicrobial properties of an antibiotic. Candidate antibiotic is streaked onto an agar plate along with several bacterial isolates. Following incubation, areas of clearing indicate inhibition of growth and thus susceptibility to the antibiotic.

When an antibiotic or any other fermentation product finally goes into production, it is cultured in huge stirred fermenters or *bioreactors*, which may be as large as 200 000 litres. A typical stirred fermenter has impellers for mixing the culture, an air line for aeration and microprocessor-controlled probes for the continuous monitoring and regulation of temperature, pH and oxygen content (Figure 18.7). Cultures with a high protein content may also have an antifoaming agent added. The process of *scale-up* is a complex operation, and not simply a matter of growing the microorganism in question in ever larger vessels. Factors such as temperature, pH and aeration must all be considered at the level of the individual cell if scale-up is to be successful. Fermenters are usually made from stainless steel, which can withstand heat sterilisation; the economic consequences of microbial contamination when working on such a large scale can be immense.

18.3   Products derived from genetically engineered microorganisms

In Chapter 12 we saw how microorganisms can be genetically modified so that they produce commercially important proteins. This is done by incorporating the gene for the desired protein into an appropriate vector and inserting it into a host cell such as *E. coli* or *Saccharomyces cerevisiae*. The initial application of this technology was in the microbial production of medically important proteins such as insulin and epidermal growth factor (Table 18.6); however, bacteria and other microorganisms can be genetically modified to produce a range of other products such as pharmaceuticals, vaccines and modified enzymes. These include enzymes used in diagnostic

Figure 18.7 A continuous flow stirred tank reactor. Parameters such as pH and concentrations of specific metabolites are closely monitored to ensure the maintenance of optimum conditions. Outlets allow for the collection of samples during fermentation as well as the collection of cells and medium at the conclusion of the reaction. Additions and collections are carried out under aseptic conditions.

Table 18.6 Medically important proteins made by recombinant microorganisms

Protein	Application	Produced in
Insulin	Treatment of type 1 diabetes	*E. coli*
Human growth hormone	Treatment of pituitary dwarfism	*E. coli*
Hepatitis B vaccine	Vaccination of susceptible personnel, e.g. healthcare workers, drug users	*Saccharomyces cerevisiae*
Epidermal growth factor	Treatment of wounds, burns	*E. coli*
Acyltransferase	Used in synthesis of ovarian cancer drug taxol	*E. coli*
Endostatin	Antitumour agent	*Pichia pastoris* (yeast)
Tissue plasminogen activator (tPA)	'Clot-busting' drug	*E. coli*

and analytical applications, where a higher purity of preparation is required than, for example, the enzymes used in detergents. These are often derived originally from other microorganisms; for example, the thermostable DNA polymerase from *Thermus aquaticus* used in the polymerase chain reaction (PCR) is now commonly made by recombinant *E. coli* cells transformed with the *T. aquaticus* gene. Many of the more recent recombinant human proteins to be developed for therapeutic use have been too complex for expression in a microbial system (e.g. factor VIII), so it has been necessary to employ cultured mammalian cells.

18.4 Microorganisms in wastewater treatment and bioremediation

These applications of microbial processes in an environmental context are discussed in Chapter 14.

18.5 Microorganisms in the mining industry

An unexpected application for microorganisms is to be found in the mining industry. Acidophilic bacteria, including *Acidithiobacillus ferrooxidans*, are increasingly being used to extract valuable metals, notably copper, from low-grade ores that would not be worth working by conventional technologies. You may recall from Chapter 14 that *A. ferrooxidans* is the organism largely responsible for the phenomenon of acid mine drainage; by carrying out the same reactions in a different context, however, it can be put to a beneficial use. Tailings, that is, mineral waste with a low metal content, are gathered in huge tips and acidified water is sprinkled over them (Figure 18.8), stimulating the growth of indigenous bacteria. Bacterial oxidation results in soluble copper sulphate leaching from the tip and being collected for copper extraction. This bacterial action is known as direct bioleaching, but if you follow the

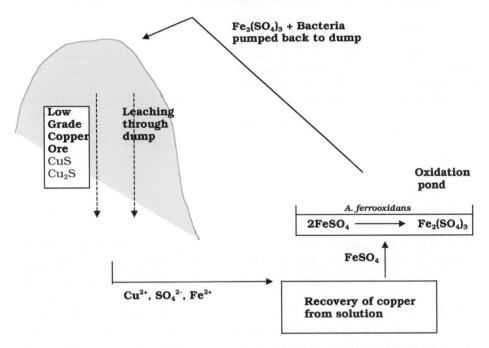

Figure 18.8 The bacterial extraction of copper. Solubilisation of copper sulphides occurs by a combination of direct (biological) and indirect (chemical) leaching. The ferric iron necessary for chemical oxidation is produced by bacterial oxidation of ferrous iron in the oxidation pond. Reproduced from Prescott, LM, et al. (2002) Microbiology, 5th edn, with permission from McGraw-Hill.

process in Figure 18.8, you will see that *A. ferrooxidans* has not finished yet! This remarkable organism can also oxidise iron from its ferrous to ferric form; the resulting ferric sulphate is a potent oxidising agent, which, when recycled to the tip, carries out indirect (chemical) bioleaching, and so the cycle continues. *A. ferrooxidans* has a number of other unusual features that enable it to survive in this hostile environment; it thrives in acidic conditions (pH <2.0), and has an unusually high tolerance of metal ions such as copper. Operations such as this must be carefully controlled to avoid serious adverse effects on the environment.

Bacteria are also involved in the extraction of other metals such as uranium and gold; the methodologies differ slightly, but still involve the conversion of an insoluble compound to a soluble one. It is only in the last couple of decades or so that the economic possibilities of *biohydrometallurgy* have been fully appreciated, and now a significant proportion of the world's copper and other metals is produced in this way. The method is inexpensive but rather slow; it may take years to extract the copper from a large tip. However, as high-grade copper-bearing ores become increasingly scarce, it seems likely to play an increasingly important role.

Sulphur-oxidising bacteria also have a role to play in the coalmining industry. Increased environmental awareness in many countries means that it is no longer acceptable to burn off the sulphur content of coal as sulphur dioxide; so an alternative must be found. One possibility is the *biodesulphurisation* of coal, using sulphur-oxidising bacteria to remove the sulphur before combustion. Whilst technically feasible, economic considerations mean that this has not yet been widely adopted.

Glossary

Acid-fast stain: a procedure for assessing the ability of an organism to retain hot carbol fuchsin stain when rinsed with acidic alcohol.

Acidophilic: 'acid-loving'; a term applied to organisms that show optimal growth in acid conditions (pH <5.5).

Activated sludge treatment: a method of wastewater treatment involving aeration in tanks that have been seeded with a mixed microbial sludge.

Activation energy: the energy necessary to initiate a chemical reaction.

Active site: the part of an enzyme involved in binding its substrate; the site of catalytic action.

Active transport: an energy-requiring process by which a substance is transported against an electrochemical gradient.

Adenosine triphosphate (ATP): the principal compound used in the storage and transfer of energy in cellular systems.

Adhesin: a protein or glycoprotein found on surface structures of bacteria, such as fimbriae, pili or capsules. The term is sometimes extended to describe the structures themselves.

Aerobe: an organism that grows in the presence of molecular oxygen, which it uses as a terminal electron acceptor in aerobic respiration.

Aerotolerant anaerobe: an anaerobe that is able to tolerate the presence of oxygen, even though it does not use it.

Aetiology: the cause or origin of a disease.

Aldose: a sugar molecule that contains an aldehyde group.

Alga (pl. algae): a photosynthetic, eukaryotic plant-like organism. It may be unicellular or multicellular.

Essential Microbiology, Second Edition. Stuart Hogg.
© 2013 John Wiley & Sons, Ltd. Published 2013 by John Wiley & Sons, Ltd.

Alternation of generations: a pattern of sexual reproduction that includes both haploid and diploid mature individuals.

Ames test: a test to assess the mutagenicity of a substance.

Amino acid: the building block of proteins: an organic acid bearing both amino and carboxyl groups, and a variable side group.

Amphipathic: a word used to describe a compound having a polar region at one end and a nonpolar region at the other.

Anabolism: the reactions involved in the synthesis of macromolecules, usually requiring an input of energy.

Anaerobe: an organism that grows in the absence of molecular oxygen.

Anammox: the formation of nitrogen gas by the anaerobic oxidation of ammonia and nitrite.

Angstrom unit (symbol: Å): one ten-billionth (10^{-10}) of a metre.

Anisogamy: the fusion of unequally sized gametes.

Anoxygenic photosynthesis: a form of photosynthesis in which oxygen is not generated; found in the purple and green photosynthetic bacteria.

Antibiotic: a microbially produced substance (or a semi-synthetic derivative) that has antimicrobial properties.

Antibody: a protein of high binding specificity produced by the immune systems of higher animals in response to infection by a foreign organism.

Anticodon: the three-nucleotide sequence carried by a tRNA, that base-pairs with its complementary mRNA codon.

Antigenic shift: a process by which major variations in viral antigens occur.

Antiseptic: a chemical disinfection agent that is mild enough to be used on human skin or tissues.

Archaea: a group of prokaryotes that diverged from all others (see Bacteria) at an early stage in evolution, and that show a number of significant differences from them. One of the three domains of life.

Ascospore: a haploid spore produced by members of the Ascomycota.

Ascus (pl. asci): a sac-like structure containing the ascospores in members of the Ascomycota.

Aseptic technique: a set of practical measures designed to prevent the growth of unwanted contaminants from the environment.

Atomic mass: the average of the mass numbers of an element's different isotopes, taking into account the proportions in which they occur.

Atomic number: the number of protons in the nucleus of an element.

Autoclave: an appliance that uses steam under pressure to achieve sterilisation.

Autotroph: an organism that can derive its carbon from carbon dioxide.

Auxotroph: a mutant that lacks the ability to synthesise an important nutrient such as an amino acid or vitamin, and must therefore have it provided (cf. prototroph).

Axenic culture: a pure culture containing one type of organism only, and completely free from contaminants.

Bacillus (pl. bacilli):   a rod-shaped bacterium.

Bacteria: prokaryotes excluding the Archaea (q.v.). Less formally, the term is frequently used to describe all prokaryotes.

Bactericidal: causing the death of bacteria.

Bacteriophage: a virus whose host is a bacterial cell.

Bacteriostatic: inhibiting the growth of bacteria, but not necessarily killing them.

Basidiocarp: the fruiting body of members of the Basidiomycota.

Basidiospore: a haploid spore produced by members of the Basidiomycota.

Basidium: a club-shaped structure carrying the basidiospores in members of the Basidiomycota.

Batch culture: a microbial culture grown in a closed vessel with no addition of nutrients or removal of waste products.

Beta-lactamase: an enzyme that breaks a bond in the β-lactam core of penicillins and cephalosporins.

Bioaugmentation: the deliberate introduction of specific microorganisms into an environment in order to assist with bioremediation.

Biochemical oxygen demand (BOD):   a measure of the amount of oxygen needed by microorganisms to oxidise the organic content of a water sample.

Biodegradation: the natural processes of breakdown of organic matter by microorganisms.

Biodeterioration: the damage caused to materials of economic importance due to biological (mainly microbial) processes.

Biofilm: a complex system of microorganisms and their surrounding polysaccharide matrix.

Biomass: the total amount of cellular material in a system.

Bioreactor: a fermentation vessel for the controlled growth of microorganisms.

Bioremediation: the use of biological processes to improve a specific environment, such as by the removal of a pollutant.

Calvin cycle: a pathway for the fixation of carbon dioxide, used by photosynthetic organisms and some chemolithotrophs.

Capsid: the protein coat of a virus particle.

Capsomer: a protein subunit of a viral capsid.

Capsule: a clearly defined polysaccharide layer surrounding the cell of certain prokaryotes.

Carcinogen: an agent capable of causing cancer.

Catabolism: the reactions that break down large molecules, usually coupled to a release of energy.

Catabolite repression: the mechanism by which the presence of a preferred nutrient (generally glucose) has the effect of preventing the synthesis of enzymes that act on other nutrients.

Central dogma of biology: the proposal that information (= genetic) flow in organisms is in one direction only, from DNA to RNA to protein.

Chemotroph: an organism that obtains its energy from chemical compounds.

Chloroplast: a chlorophyll-containing organelle found in photosynthetic eukaryotes.

Chromosome: a nuclear structure on which most of a eukaryotic cell's genetic information (DNA) is carried, in association with specialised proteins called histones. The nucleoid (q.v.) of prokaryotes is also often referred to as a chromosome.

Cilium (pl. cilia):   a short motile hair-like structure found on the surface of some eukaryotic cells.

Citric acid cycle: see Tricarboxylic acid cycle.

Clamp connection: a mechanism, unique to members of the Basidiomycota, for ensuring the maintenance of the dikaryotic state.

Cloning: 1.   the production of multiple copies of a specific DNA molecule.

2.   the production of genetically identical cells or whole organisms.

Coccus (pl. cocci):   a spherically-shaped bacterium.

Codon: a sequence of three nucleotide bases that corresponds to a specific amino acid.

Coenocytic: containing many nuclei within a single plasma membrane.

Coenzyme: a loosely bound organic cofactor that influences the activity of an enzyme.

Cofactor: a non-protein component of an enzyme (often a metal ion) essential for its normal functioning.

Commensal: an organism that lives in or on another organism, deriving some benefit from the association but not harming the other party.

Competence (of a bacterial cell): the state of being able to take up naked DNA from outside the cell.

Compound: a substance comprising the atoms of two or more elements.

Conidiophore: an aerial hypha that bears conidia.

Conidium (conidiospore; pl. conidia): an asexual spore, found in members of the Ascomycota and Actinomycetes. Often forms chains.

Conjugation: a process of genetic transfer in bacteria involving intimate contact between cells and direct transfer of DNA across a sex pilus.

Consumer: a heterotroph that derives energy from the consumption of other organisms.

Continuous culture: a microbial culture in which nutrient concentrations and other conditions are kept constant by the addition of fresh medium and the removal of old medium.

Contractile vacuole: a fluid-filled vacuole involved in the osmoregulation of certain protists.

Cosmid: a hybrid cloning vector capable of accommodating large inserts of up to 50 kb.

Covalent bond: a bond formed by the sharing of a pair of electrons between atoms.

Cyanobacteria: a group of mostly unicellular prokaryotes that carry out oxygenic photosynthesis. Commonly known as the blue-greens.

Cytokine: a class of protein signal molecules secreted by cells of the immune system, and involved in cellular communication.

Decimal reduction time (D value): the time needed to reduce a cell population by a factor of ten (i.e. to kill 90% of the population) using a particular heat treatment.

Decomposer: an organism that breaks down the remains and waste products of producers and consumers.

Defined medium: a growth medium whose precise chemical composition is known.

Definitive host: the host species in which a parasite undergoes sexual reproduction.

Denitrification: the reduction of nitrite and nitrate to nitrogen gas under anaerobic conditions.

Diauxic growth: a form of growth that has two distinct phases due to one carbon source being used preferentially over a second.

Differential medium: a medium that allows colonies of a particular organism to be differentiated from others growing in the same culture.

Differential stain: a stain that employs two or more dyes to distinguish between different cellular structures or cell types.

Dikaryon: a structure formed by two cells whose contents, but not nuclei, have fused.

Dimorphic: existing in two distinct forms.

Dioecious: having male and female reproductive structures on separate individuals.

Diploid: having two sets of chromosomes.

Disaccharide: a carbohydrate formed by the joining of two monosaccharides.

Disc diffusion method: a method for assessing the antimicrobial properties of a substance.

Disinfection: the elimination or inhibition of pathogenic organisms in or on an object so that they no longer pose a threat.

Dissimilatory sulphate reduction: the reduction of sulphate to hydrogen sulphide by obligate anaerobes, using either an organic compound or hydrogen gas as electron donor.

DNA library: a collection of cloned DNA fragments.

Domain: the highest level of taxonomic grouping.

Eclipse period: the part of a viral replication cycle when no complete viral particles are present in the infected host cell.

Ecosystem: the organisms of a particular habitat, together with their inanimate surroundings.

Ectoparasite: a parasite that attaches to the outside of its host.

Electron: a subatomic particle carrying a negative charge.

Electron transport chain: a series of donor/acceptor molecules that transfer electrons from donors (e.g. NADH) to a terminal electron acceptor (e.g. O_2).

Embden–Meyerhof pathway: see Glycolysis.

Endoparasite: a parasite that fully enters its host and lives inside it.

Endoplasmic reticulum: a tubular network found in the cytoplasm of eukaryotic cells.

Endospore: a highly resistant spore found within certain bacteria.

Endotoxin: the lipid A component of the Gram-negative bacterial cell wall, released when the cell dies.

Enrichment culture: a culture that uses a selective medium to encourage the growth of an organism present in low numbers.

Entner–Doudoroff pathway: an alternative pathway for the oxidation of glucose, producing a mixture of pyruvate and glyceraldehyde-3-phosphate.

Enzyme: a cellular catalyst (usually protein), specific to a particular reaction or group of reactions.

Epiphyte: an organism that grows on the surface of a plant.

Eukaryote: an organism whose cells contain a true nucleus and membrane-bound organelles such as mitochondria and endoplasmic reticulum.

Excision repair: a repair mechanism in which damaged sections of DNA are cut out and replaced.

Exon: a coding region of a gene. cf. Intron.

Exotoxin: a toxin secreted into their surroundings by (mostly) Gram-positive bacteria.

Expression vector: a vector that allows the transcription and translation of a foreign gene inserted into it.

Facilitated diffusion: the transport of molecules across membranes with the help of carrier proteins. Transport takes place down a concentration gradient, and energy is not required.

Facultative anaerobe: an organism that can grow in the absence of oxygen, but utilises it when available.

Fastidious (of an organism): unable to synthesise a range of nutrients and therefore having complex requirements in culture.

Fatty acid: a long-chain hydrocarbon with a carboxyl group at one end.

Feedback inhibition: a control mechanism whereby the final product of a metabolic pathway acts as an inhibitor to an enzyme that catalyses an early step (usually the first) in the pathway.

Fermentation: a microbial process by which an organic substrate (usually a carbohydrate) is broken down without the involvement of oxygen or an electron transport chain, generating energy by substrate-level phosphorylation

Flagellum (pl. flagella): a long hair-like extracellular structure associated with locomotion. Found in both prokaryotes and eukaryotes, although each has its own distinctive ultrastructure.

F plasmid: a plasmid containing genes that code for the construction of the sex pilus, across which it is transferred to a recipient cell.

Frameshift mutation: a mutation that results in a change to the reading frame, and thus an altered sequence of amino acids downstream of the point where it occurs.

Fungi: a kingdom of non-photosynthetic eukaryotes characterised by absorptive heterotrophic nutrition.

Gamete: a haploid reproductive cell arising from meiosis. Fusion with another gamete forms a diploid zygote.

Gametophyte: the haploid, gamete-forming stage in a life cycle with alternation of generations.

Gene: a sequence of DNA that usually encodes a polypeptide.

Generalised transduction: the transfer of a gene from one bacterial cell to another as a result of being inadvertently packaged into a phage particle.

Generation (doubling) time: the time taken for a population of cells to double in number under specific conditions.

Genetic code: the 64 triplet codons and the corresponding amino acids or termination sequences.

Genome: the complete genetic material of an organism.

Genotype: the genetic make-up of an organism.

Gluconeogenesis: a series of reactions by which glucose is synthesised from compounds such as amino acids and lactate.

Glycolysis (or Embden–Meyerhof pathway): a series of reactions by which glucose is oxidised to two molecules of pyruvate with the synthesis of two molecules of ATP.

Glycosidic linkage: a covalent linkage formed between sugar molecules.

Golgi apparatus: an organelle of eukaryotes comprising flattened membranous sacs. It functions in the packaging of various molecules for transport elsewhere in the cell.

Gram stain: a differential stain that divides bacteria into Gram-positive (purple) or Gram-negative (pink). The reaction depends on the constitution of the organism's cell wall.

Group translocation: a form of active transport in which a solute is modified to prevent its escape from the cell.

Haemolysin: a substance that causes the lysis of red blood cells.

Haloduric: able to tolerate high salt concentrations.

Halophilic: showing a requirement for moderate to high salt concentrations.

Haploid: having only one set of chromosomes.

Heterotroph: an organism that must use one or more organic molecules as its source of carbon.

Hexose: a six-carbon sugar.

Hexose monophosphate shunt: see Pentose phosphate pathway.

Hfr cell: a bacterial cell that has a transferred F plasmid integrated into its chromosome.

Histone: a basic protein found associated with DNA in eukaryotic chromosomes.

Horizontal gene transfer: the transfer of genetic material as a result of mechanisms independent of reproduction (cf. Vertical gene transfer).

Human Genome Project: an international effort to map and sequence all the DNA in the human genome. The project has also involved sequencing the genomes of several other organisms.

Hydrogen bond: a relatively weak bond that forms between covalently bonded hydrogen and any electronegative atom, most commonly oxygen or nitrogen.

Hydrophilic: ('water-loving'): having an affinity for water.

Hydrophobic: ('water-fearing'): repelled by water.

Hypha (pl. hyphae): a thread-like filament of cells characteristic of fungi and actinomycetes.

Immersion oil: a viscous oil used to improve the resolution of a light microscope at high power. It has the same refractive index as glass.

In vitro: ('in glass'): outside of the living organism, in test tubes, etc. (cf. In vivo).

In vivo: ('in life'): within the living organism.

Inoculum: the cells used to 'seed' a new culture.

Intermediate host: a host species in which a parasite completes the asexual stages of its life cycle.

Interrupted mating: a technique in which the order of gene transfer during conjugation is determined by terminating transfer at different time intervals.

Intron: a non-coding sequence within a gene. cf. Exon.

Ion: an atom or group of atoms that carries a charge due to losing or gaining one or more electrons.

Isogamy: fusion of identical motile gametes.

Isotopes: variant forms of an element, having the same number of protons and electrons but differing in the number of neutrons.

Karyogamy: the fusion of nuclei from two different cells.

Ketose: a sugar molecule that contains a ketone group.

Kinetoplast: a specialised structure within the mitochondria of certain flagellated protozoans.

Koch's postulates: a set of criteria, proposed by Robert Koch, which must be satisfied in order to link a specific organism to a specific disease.

Krebs cycle: see Tricarboxylic acid cycle.

Latent period: the period in a viral replication cycle between infection of the host and release of newly synthesised viral particles.

Latent virus: a virus that remains inactive in the host for long periods before being reactivated.

LD_{50} (lethal dose 50): the dose of an antimicrobial agent that is required to kill 50% of a group of experimental animals within a certain time period.

Leukocidin: a type of exotoxin that destroys white blood cells.

Lichen: a symbiotic association of a fungus (usually an ascomycete) and an alga or cyanophyte.

Lipase: an enzyme that digests lipids.

Lithotroph: an organism that uses inorganic molecules as a source of electrons.

Lysogeny: a form of bacteriophage replication in which the viral genome is integrated into that of the host, and is replicated along with it. See Prophage.

Lytic cycle: a process of viral replication involving the bursting of the host cell and release of new viral particles.

Magnetosome: a particle of magnetite (iron oxide) found in certain bacteria, allowing them to orient themselves in a magnetic field.

Malting: the stage in beer-making in which grain is soaked in water to initiate germination, and activate starch-digesting enzymes.

Mashing: the stage in beer-making in which soluble material is released from the grain in preparation for fermentation.

Mass number: the combined total of protons and neutrons in the nucleus of an element.

Meiosis: a form of nuclear division in diploid eukaryotic cells that results in haploid daughter nuclei.

Merodiploid: a genome that is partly haploid and partly diploid.

Mesophile: an organism that grows optimally at moderate temperatures (20–45°C).

Messenger RNA (mRNA):   a form of RNA that is synthesised as a complementary copy of the template strand of DNA.

Methanogen: a member of the Archaea that is capable of generating methane under anaerobic conditions.

Methanotroph: a bacterium capable of using methane as a carbon and energy source.

Michaelis constant (K_m):   the substrate concentration in an enzyme reaction that results in a rate of reaction equal to one-half of the maximum.

Minimum inhibitory concentration (MIC):   the lowest concentration of an antimicrobial substance that prevents growth of a particular organism.

Mismatch repair: a repair system that replaces incorrectly inserted nucleotides.

Missense mutation: a mutation that alters the sense of the message encoded in the DNA, and results in an incorrect amino acid being produced at the point where it occurs.

Mitochondrion (pl. mitochondria):   a spherical to ovoid organelle found in eukaryotes. The site of ATP generation via the tricarboxylic acid cycle and oxidative phosphorylation.

Mitosis: a form of nuclear division in eukaryotic cells, resulting in daughter nuclei each with the same chromosome complement as the parent.

Monoecious: having reproductive structures of both sexes on the same individual.

Monosaccharide: the simplest form of carbohydrate molecule.

Most probable number (MPN): a statistical method of estimating microbial numbers in a liquid sample based upon the highest dilution able to support growth.

mRNA: see Messenger RNA.

Mutagen: a chemical or physical agent capable of inducing mutations.

Mutation: any heritable alteration in the DNA sequence

Mutualism: an association between two species from which both participants derive benefit.

Mycelium (pl. mycelia): a tangled mass of branching hyphae.

Mycorrhiza (pl. mycorrhizae): a mutualistic relationship between a fungus and a plant root.

Mycosis (pl. mycoses): a disease caused by a fungus.

Mycotoxin: a toxin produced as a secondary metabolite by many fungi that can cause disease in humans and other animals.

Neutron: a subatomic particle carrying neither positive nor negative charge.

Nitrification: the two-step process by which ammonia is oxidised to nitrite and then to nitrate

Nonsense codon: see Terminator codon.

Nonsense mutation: a mutation that results in a 'stop' codon being inserted into the mRNA at the point where it occurs, and the premature termination of translation.

Nosocomial infection: an infection that is acquired in a hospital or other healthcare setting.

Nuclear membrane: the double membrane surrounding the nucleus of a eukaryotic cell.

Nucleocapsid: the genome of a virus and its surrounding protein coat.

Nucleoid: another name for the bacterial chromosome. The site of most of a prokaryotic cell's DNA.

Nucleolus: a discrete region of the eukaryotic nucleus, where ribosomes are assembled.

Nucleotide: the building block of nucleic acids, comprising a pentose sugar, a nitrogenous base and one or more phosphate groups.

Nucleus: 1. the central, membrane-bound structure in eukaryotic cells that contains the genetic material. **2.** the region of an atom that contains the protons and neutrons.

Obligate anaerobe: an organism that is incapable of growth in the presence of oxygen.

Okazaki fragment: a discontinuous fragment of single-stranded DNA synthesised complementary to the lagging strand during DNA replication.

Oncogene: a gene associated with the conversion of a cell to a cancerous form.

Oogamy: the fusion of a small, motile sperm cell and a larger, immobile egg cell.

Operon: a group of genes under the control of a single operator sequence.

Organotroph: an organism that uses organic molecules as a source of electrons.

Outer membrane: the outermost part of the Gram-negative cell wall, comprising phospholipids and lipopolysaccharide.

Oxidation: a chemical reaction in which an electron is lost.

Oxygenic photosynthesis: a form of photosynthesis in which oxygen is produced; found in algae, cyanobacteria (blue greens) and also green plants.

Parasitism: an association between two species from which one partner derives some or all of its nutritional requirements by living either in or on the other (the host), which usually suffers harm as a result.

Pasteurisation: a mild heating regime used to destroy pathogens and spoilage organisms present in food and drink, especially milk.

Pathogen: an organism with the potential to cause disease.

Pellicle: a semi-rigid structure composed of protein strips found surrounding the cell of many unicellular protozoans and algae.

Pentose: a five-carbon sugar.

Pentose phosphate pathway (or hexose monophosphate shunt): a secondary pathway for the oxidation of glucose, resulting in the production of pentoses that serve as precursors for nucleotides.

Peptide bond: the bond formed between the amino group of one amino acid and the carboxyl group of another.

Peptidoglycan: a polymer comprising alternative units of N-acetylmuramic acid and N-acetylglucosamine that forms the major constituent of bacterial cell walls.

Phagemid: a hybrid cloning vector, comprising elements of plasmid and phage.

Phagocytosis: the ingestion and digestion of particulate matter by a cell, a process unique to eukaryotes.

Phenol coefficient: a measure of the efficacy of a disinfectant against a given organism, compared to that of phenol.

Phenotype: the observable characteristics of an organism.

Phospholipid: an important constituent of all membranes, comprising a triacylglycerol in which one fatty acid is replaced by a phosphate group.

Phosphorylation: the addition of a phosphate group.

Photophosphorylation: the synthesis of ATP using light energy.

Photoreactivation: a DNA repair mechanism involving the light-dependent enzyme DNA photolyase.

Photosynthesis: a process by which light energy is trapped by chlorophyll and converted to ATP, which is used to drive the synthesis of carbohydrate by reducing CO_2.

Phototroph: an organism that is able to use light as its source of energy.

Phylogenetic: relating to the evolutionary relationship between organisms.

Pilus (pl. pili): a short, hairlike appendage found on the surface of prokaryotes and assisting with attachment. Specialised sex pili are involved in bacterial conjugation.

Plankton: the floating microscopic organisms of aquatic ecosystems.

Plasma membrane: the membrane that surrounds a cell.

Plasmid: a small, self-replicating loop of extrachromosomal DNA, found in bacteria and some yeasts.

Plasmodium (pl. plasmodia): a mass of protoplasm containing several nuclei and bounded by a cytoplasmic membrane.

Plasmogamy: the fusion of the cytoplasmic content of two cells.

Plasmolysis: the shrinkage of the plasma membrane away from the cell wall, due to osmotic loss of water from the cell.

Point mutation: a mutation that involves the substitution of one nucleotide by another.

Polar: having unequal charge distribution, caused by unequal sharing of atoms.

Polymerase chain reaction (PCR): a technique that selectively replicates a specific DNA sequence by means of *in vitro* enzymatic amplification.

Polypeptide: a chain of many amino acids joined together by peptide bonds.

Polyribosome (polysome): a chain of ribosomes attached to the same molecule of mRNA.

Polysaccharide: a carbohydrate polymer of monosaccharide units.

Portal of entry: the route by which a pathogen gains entry to the body.

Primary producer: an autotroph that obtains energy from sunlight or chemical sources.

Primer: a short sequence of single-stranded DNA or RNA required by DNA polymerase as a starting point for chain extension.

Prion: a self-replicating protein responsible for a range of neurodegenerative disorders in humans and other mammals.

Prokaryote: an organism lacking a true nucleus and membrane-bound organelles.

Promoter: a sequence upstream of a gene, to which RNA polymerase binds to initiate transcription.

Prophage: the DNA of a temperate phage that has integrated into the host genome. It remains inactive whilst in this form.

Prostheca (pl. prosthecae): a stalked structure formed by an extension of the cell wall and plasma membrane of certain bacteria.

Prosthetic group: a non-polypeptide component of a protein, such as a metal ion or a carbohydrate

Protease: an enzyme that digests proteins.

Protista: a eukaryotic kingdom, comprising mostly unicellular organisms.

Protocooperation: a form of mutualistic relationship that is not obligatory for either partner.

Proton: a positively charged subatomic particle.

Protoplast: a cell that has had its cell wall removed.

Prototroph: a microorganism that is able to synthesise organic cellular constituents such as amino acids and nucleic acids from inorganic nutrients.

Protozoa: a group of single-celled eukaryotes with certain animal-like characteristics.

Pseudomurein: a modified form of peptidoglycan found in some archaean cell walls.

Pseudopodium (pl. pseudopodia): a projection of the plasma membrane of the amoebas that causes the cell to change shape and allows movement.

Psychrophile: an organism that grows optimally at low temperatures ($<15°C$).

Psychrotroph: an organism that is able to tolerate low temperatures, but grows better at more moderate values.

Reading frame: the way in which a sequence of nucleotides is read in triplets, depending on the starting point.

Real image: an image that can be projected onto a flat surface such as a screen.

Recombinant DNA: DNA that comprises material from more than one source, joined together.

Recombination: any process that results in new combinations of genes.

Redox potential (E_0): the tendency of a compound to lose or gain electrons.

Reduction: a chemical reaction in which an electron is gained.

Refractive index: the ratio between the velocity of light as it passes through a substance and its velocity in a vacuum.

Regulatory gene: a gene whose protein product has an effect in controlling the expression of other genes.

Replication fork: a Y-shaped structure formed by the separating strands of DNA during replication.

Repressor protein: a protein that prevents transcription of a gene by binding to its operator.

Resolution: the capacity of an optical instrument to discern detail.

Restriction endonuclease: an enzyme of microbial origin that cleaves DNA at a specific nucleotide sequence.

Reverse transcriptase: an enzyme found in retroviruses that can synthesise DNA from an RNA template.

Rhizosphere: the region around the surface of a plant's root system.

Ribosomal RNA (rRNA):   a type of RNA that forms part of the structure of ribosomes.

Ribosome: an organelle made up of protein and RNA, found in both prokaryotes and eukaryotes. The site of protein synthesis.

rRNA: see Ribosomal RNA.

Saprobe: an organism that feeds on dead and decaying organic materials. Previously termed saprophyte.

Saturated fatty acid: a fatty acid with only single covalent bonds between adjacent carbon atoms. cf. Unsaturated fatty acid.

Secondary metabolite: a substance produced by a microorganism after the phase of active growth has ceased.

Selectable marker: a gene that allows cells containing it to be identified by the expression of a recognisable characteristic.

Selective medium: a medium that favours the growth of a particular organism or group of organisms, often by suppressing the growth of others.

Semi-conservative replication: the process of DNA replication by which each strand acts as a template for the synthesis of a new complementary strand. Each resultant double-stranded molecule thus comprises one original strand and one new one.

Septate: separated by septa or cross-walls.

Sex pilus: a narrow extension of the bacterial cell, through which genetic material is transferred during conjugation.

Shuttle vector: a cloning vector that can replicate in both bacterial and yeast host cells.

Siderophore: a class of low molecular weight compounds with a high affinity for binding ferric iron. They enable bacteria and fungi to scavenge iron from environments where it is present in very low concentrations.

Silent mutation: a mutation that has no effect on the amino acid encoded by the triplet.

Single-cell protein: bacteria or yeast grown in bulk for use as a protein-rich food supplement.

Slime layer: a diffuse and loosely attached polysaccharide layer surrounding the cell of certain prokaryotes.

Specialised transduction: the transfer of a limited selection of genes due to imprecise excision of a prophage in a lysogenic infection cycle.

Sporangiophore: a specialised aerial hypha that bears the sporangia.

Sporangium (pl. sporangia):   a structure inside which spores develop.

Spore: a resistant, non-motile reproductive cell.

Sporophyte: the diploid, spore-forming stage in a life cycle with alternation of generations.

Sporozoite: a motile infective stage of members of the Sporozoa that gives rise to an asexual stage within the new host.

Sterilisation: the process by which all microorganisms present on or in an object are destroyed or removed.

Steroid: a member of a group of lipids based on a four-ring structure.

Stop codon: see Terminator codon.

Substrate-level phosphorylation: the synthesis of ATP by the direct transfer of a phosphate group from a phosphorylated organic compound to ADP.

Superantigen: a bacterial protein that elicits a massive inflammatory reaction in a host by causing T-lymphocytes to overproduce cytokines.

Tautomerism: the ability of a molecule such as a nucleotide base to exist in two alternative forms.

Taxonomy: the science of classifying living (and once-living) organisms.

Temperate phage: a bacteriophage with a lysogenic replication cycle.

Terminator: a sequence of DNA that indicates transcription should stop.

Terminator codon: one of the three triplet sequences (also called **stop codons** or **nonsense codons**) that indicate translation should stop.

Thallus: a simple vegetative plant body showing no differentiation into root, stem and leaf.

Therapeutic index: a measure of the selective toxicity of a chemotherapeutic agent.

Thylakoid: a photosynthetic membrane found in chloroplasts or free in the cytoplasm (in cyanobacteria). It contains photosynthetic pigments and components of the electron transport chain.

Total cell count: an enumeration method that counts all cells, living or dead.

Toxoid: a bacterial toxin that has lost its toxicity but retained its immunogenicity.

Trace element: an element required in minute amounts for growth.

Transamination: a reaction involving the transfer of an amino group from one molecule to another.

Transcription: the process by which single-stranded mRNA is synthesised from a complementary DNA template.

Transduction: the bacteriophage-mediated transfer of genetic material between bacteria.

Transfer RNA (tRNA): a form of RNA that carries specific amino acids to the site of protein synthesis.

Transformation: the uptake of naked DNA from the environment and its integration into the host genome.

Transition: a mutation in which a purine replaces a purine or a pyrimidine replaces a pyrimidine.

Translation: the process by which the message encoded in mRNA is converted into a sequence of amino acids.

Transposable element: a sequence of DNA that is able to relocate to another position on the genome.

Transversion: a mutation in which a purine replaces a pyrimidine, or a pyrimidine replaces a purine.

Triacylglycerol (triglyceride): a lipid formed by the joining of three fatty acids to a molecule of glycerol.

Tricarboxylic acid cycle (TCA cycle): a series of reactions that oxidise acetate to CO_2, generating reducing power in the form of NADH and $FADH_2$ for use in the electron transport chain. Also known as **citric acid cycle** or **Krebs cycle**.

Triose: a three-carbon sugar.

tRNA: see Transfer RNA.

Undefined or complex medium: a growth medium whose precise chemical composition is not known.

Unsaturated fatty acid: a fatty acid that contains one or more double bonds between adjacent carbon atoms.

Vaccination: inoculation with a vaccine to provide protective immunity.

Vaccine: a preparation of dead or inactivated living pathogens or their products used to provide protective immunity.

Vector: **1.** a DNA molecule such as a plasmid that is able to carry a piece of foreign DNA into a recipient cell. **2.** an organism, usually an arthropod, that transmits a parasite from one host to another.

Viable cell count: an enumeration method that counts only those cells capable of reproducing to form a visible colony.

Virion: a complete, intact viral particle.

Viroid: a plant pathogen that comprises only single-stranded RNA (ssRNA) and does not code for protein product.

Virtual image: an image that has no physical existence in space, and cannot be projected onto a screen.

Virulence factor: a structural or physiological feature that helps a pathogen to enter and colonise a host.

Virulent phage: a bacteriophage with a lytic replication cycle.

Virus: a submicroscopic, non-cellular parasite, comprising protein and RNA or DNA.

Wildtype: the naturally occurring, non-mutant form of an organism or gene.

Wobble: the degree of flexibility allowed in the third base in a codon when pairing with tRNA. The wobble hypothesis explains how a single tRNA can pair with more than one codon.

Yeast artificial chromosome (YAC): a cloning vector able to accommodate inserts of several hundred kilobases in size.

Zoonosis (pl. zoonoses): a disease normally found in animals, but transmissible to humans under certain circumstances.

Zoospore: a flagellated, asexual spore.

Zygospore: a thick-walled resistant diploid structure formed by certain fungi.

Further Reading

The declared intention of *Essential Microbiology* is to serve as an introduction to microbiology for students who do not intend to specialise in the subject. It would please the author greatly, however, if this book stimulated readers to deepen their knowledge and understanding by referring to other, more specialised texts, and to broaden it by reading more general titles, in which microbiology is discussed in a wider context. Below are listed a number of suggested sources for further reading; the list is by no means comprehensive, and is just a selection of the huge number of available texts. The reader may well find others to suit his or her individual needs. Although titles are listed under the heading of a particular chapter, many of course will also be equally useful in the context of one or more other chapters. Both academic texts and more popular science titles have been included.

General microbiology

For a general overview of the world of microorganisms and their applications, John Postgate's classic *Microbes and Man*, 4th edition (Cambridge University Press, 2000) provides an excellent starting point. Those wishing to study microbiology in greater depth may want to consider one of the many excellent general microbiology texts available. The following are well-established examples that embrace the whole field of microbiology, and cover many topics in greater depth than is possible or indeed desirable in the present book.

Black JG (2012) *Microbiology: Principles and Explorations*, 8th edn. John Wiley & Sons, Ltd.
Madigan MT, Martinko JM, Stahl DA and Clark DP (2011) *Brock Biology of Microorganisms*, 13th edn. Pearson Education.
Sharma JL (2009) *Dictionary of Microbiology*. CBS Publishers and Distributors.

Essential Microbiology, Second Edition. Stuart Hogg.
© 2013 John Wiley & Sons, Ltd. Published 2013 by John Wiley & Sons, Ltd.

Tortora GJ, Funke BR and Case CL (2012) *Microbiology: an Introduction*, 11th edn. Pearson Education.

Willey JM, Sherwood LM and Woolverton CJ (2011) *Prescott's Microbiology*, 8th edn. McGraw-Hill.

Chapter 1

Krasner RI (2007) *20th Century Microbe Hunters*. Jones & Bartlett.

Lagerkvist U (2003) *Pioneers of Microbiology and the Nobel Prize*. World Scientific Publishing.

Murphy DB and Davidson DW (2012) *Fundamentals of Light Microscopy and Electronic Imaging*, 2nd edn. Wiley-Blackwell.

Chapter 2

Crowe J and Bradshaw A (2010) *Chemistry for the Biosciences: The Essential Concepts*, 2nd edn. Cambridge University Press.

Hames DB and Hooper NM (2011) *BIOS Instant Notes in Biochemistry*, 4th edn. Garland Science.

Lewis R and Evans W (2011) *Chemistry*, 4th edn. Palgrave Macmillan.

Chapter 3

Alberts B, Bray D, Hopkin K, Johnson A, Lewis J, Raff M, Roberts K and Walker P (2009) *Essential Cell Biology*, 3rd edn. Garland Science.

Bolsover SR, Shephard EA, White HA and Hyams JS (2011) *Cell Biology: A Short Course*, 3rd edn. Wiley-Blackwell.

Singleton P (2004) *Bacteria in Biology, Biotechnology and Medicine*, 6th edn. John Wiley & Sons, Ltd.

Chapter 4

Atlas RM (2010) *Handbook of Microbiological Media*, 4th edn. CRC Press.

Barrow GI and Feltham RKA (eds) (2004) *Cowan and Steel's Manual for the Identification of Medical Bacteria*, 3rd edn. Cambridge University Press.

Cappuccino JG and Sherman N (2010) *Microbiology: A Laboratory Manual*, 10th edn. Pearson.

Chapter 5

Gerday C and Glansdorff N (eds) (2007) *Physiology and Biochemistry of Extremophiles*. American Society of Microbiology.

Isaac S and Jennings D (1995) *Microbial Culture*. Garland Science.

Chapter 6

Kim BH and Gadd GM (2008) *Bacterial Physiology and Metabolism*. Cambridge University Press.

Tymoczko JL, Berg JM and Stryer L (2012) *Biochemistry: A Short Course*, 2nd edn. Freeman.

Chapter 7

Various editors: *Bergey's Manual of Systematic Bacteriology*, 2nd edn. Springer-Verlag:
  Vol. 1 (2001) *The Archaea and the Deeply Branching and Phototrophic Bacteria*.
  Vol. 2 (2005) *The Proteobacteria*.
  Vol. 3 (2009) *The Firmicutes*.
  Vol. 4 (2010) *The Bacteroidetes, Spirochaetes, Tenericutes (Mollicutes), Acidobacteria, Fibrobacteres, Fusobacteria...* [etc.]
  Vol. 5 (2012) *The Actinobacteria*.

Dahl C and Friedrich CG (2007) *Microbial Sulfur Metabolism*. Springer.

Maczulak A (2010) *Allies and Enemies: How the World Depends on Bacteria*. Pearson Education.

Wassenaar TM (2011) *Bacteria: The Benign, the Bad and the Beautiful*. Wiley-Blackwell.

Chapter 8

Deacon JW (2005) *Fungal Biology*, 4th edn. Wiley-Blackwell.

Money NP (2006) *The Triumph of the Fungi: A Rotten History*. Oxford University Press.

Stephenson SL (2010) *The Kingdom Fungi: The Biology of Mushrooms, Molds and Lichens*. Timber Press.

Chapter 9

Bonner JT (2008) *The Social Amoebae: The Biology of Cellular Slime Molds*. Princeton University Press.

Castillo V and Harris R (2013) *Protozoa: Biology, Classification & Role in Disease*. Nova Science Publishers.

Graham JW, Wilcox LW and Graham LE (2008) *Algae*, 2nd edn. Benjamin Cummings.

Chapter 10

Cann AJ (2011) *Principles of Molecular Virology*, 5th edn. Academic Press.

Carter C and Saunders V (2012) *Virology: Principles and Applications*, 2nd edn. John Wiley & Sons.

Dimmock NJ, Easton AJ and Leppard KN (2006) *Introduction to Modern Virology*, 6th edn. John Wiley & Sons.

Wagner EK, Hewlett MJ, Bloom DC and Camerini D (2007) *Basic Virology*, 3rd edn. Wiley-Blackwell.

Chapter 11

Brown TE (2011) *Introduction to Genetics: A Molecular Approach*. Garland Science.

Dale JW and Park SF (2010) *Molecular Genetics of Bacteria*, 5th edn. Wiley-Blackwell.

McLennan A, Bates A, Turner P and White M (2012) *BIOS Instant Notes in Molecular Biology*, 4th edn. Garland Science.

Chapter 12

Brown TE (2010) *Gene Cloning and DNA Analysis: An Introduction*, 6th edn. Wiley-Blackwell.

Dale JW, von Schantz M and Plant N (2011) *From Genes to Genomes: Concepts and Applications of DNA Technology*, 3rd edn. Wiley-Blackwell.

Kreuzer H and Massey A (2008) *Recombinant DNA and Biotechnology: A Guide for Students*, 3rd edn. American Society for Microbiology.

Lodge J, Lund P and Minchin S (2006) *Gene Cloning: Principles and Applications*. Taylor & Francis.

Chapter 13

Camisao AF and Pedroso CC (eds) (2012) *Symbiosis: Evolution, Biology & Ecological Effects*. Nova Science Publishers.

Nash TH (ed.) (2008) *Lichen Biology*, 2nd edn. Cambridge University Press.

Paracer S and Ahmadjian V (2000) *Symbiosis: An Introduction to Biological Associations*, 2nd edn. Oxford University Press.

Purvis W (2007) *Lichens*. Natural History Museum Press.

Smith SE and Read DJ (2007) *Mycorrhizal Symbiosis*. Academic Press.

Chapter 14

Alsop D, Seal KJ and Gaylarde CC (2004) *Introduction to Biodeterioration*, 2nd edn. Cambridge University Press.

Atlas RM and Philp J (eds) (2005) *Bioremediation: Applied Microbial Solutions for Real-World Environmental Cleanup*. American Society for Microbiology.

Bitton G (2011) *Wastewater Microbiology*, 4th edn. Wiley-Blackwell.

Maier RM, Pepper IL and Gerba CP (2008) *Environmental Microbiology*, 2nd edn. Academic Press.

Mitchell R and Gu JD (eds) (2010) *Environmental Microbiology*, 2nd edn. Wiley-Blackwell.

Chapter 15

Crawford DA (2009) *Deadly Companions: How Microbes Shaped Our History.* Oxford University Press.

Ford M (2010) *Medical Microbiology (Fundamentals of Biomedical Science).* Oxford University Press.

Greenwood D, Slack RCB, Peutherer JF and Barer MR (2007) *Medical Microbiology: A Guide to Microbial Infections: Pathogenesis, Immunity, Laboratory Diagnosis and Control*, 17th edn. Churchill Livingstone.

Kasper DL and Fauci AS (2010) *Harrison's Infectious Diseases.* McGraw-Hill Medical.

Richardson MD and Warnock DW (2012) *Fungal Disease: Diagnosis and Management*, 4th edn. Wiley-Blackwell.

Salyers AA, Wilson BA, Whitt DD and Winkler M (2011) *Bacterial Pathogenesis: A Molecular Approach*, 3rd edn. American Society for Microbiology.

Wiser MF (2010) *Protozoa and Human Disease.* Garland Science.

Wolfe ND (2011) *Viral Storm: The Dawn of a New Pandemic Age.* Allen Lane.

Chapter 16

Fraise A, Lambert PA and Maillard J-Y (2003) *Russell, Hugo and Ayliffe's Principles and Practice of Disinfection, Preservation and Sterilization*, 4th edn. Wiley-Blackwell.

Chapter 17

Brown K (2005) *Penicillin Man: Alexander Fleming and the Antibiotic Revolution.* The History Press.

Drlica KS and Perlin DS (2011) *Antibiotic Resistance: Understanding and Responding to an Emerging Crisis.* Financial Times/Prentice Hall.

Finch F, Davey P, Wilcox M and Irving W (2012) *Antimicrobial Chemotherapy*, 6th edn. Oxford University Press.

Franklin TJ (2005) *Biochemistry and Molecular Biology of Antimicrobial Drug Action*, 6th edn. Springer-Verlag.

Gallagher JC and MacDougall C (2011) *Antibiotics Simplified*, 2nd edn. Jones & Bartlett.

Chapter 18

Adams MR and Moss MO (2008) *Food Microbiology*, 3rd edn. Royal Society of Chemistry.

Hornsey IS (1999) *Brewing.* Royal Society of Chemistry.

Jay JM, Locssner MJ and Golden DA (2006) *Modern Food Microbiology*, 7th edn. Springer.

Ratledge C and Kristianssen B (eds) (2006) *Basic Biotechnology*, 3rd edn. Cambridge University Press.

Russell J and Cohn R (2012) *Bioleaching.* Bookvika Publishing.

Smith JE (2009) *Biotechnology*, 5th edn. Cambridge University Press.

Index

Note: 'f' following an entry indicates a footnote.

Essential Microbiology, Second Edition. Stuart Hogg.
© 2013 John Wiley & Sons, Ltd. Published 2013 by John Wiley & Sons, Ltd.